Ecology Basics

Ecology Basics

Volume 1

Acid deposition—Lichens

edited by
The Editors of Salem Press

SALEM PRESS, INC.
Pasadena, California   Hackensack, New Jersey

Library of Congress Cataloging-in-Publication Data
Ecology basics / edited by the editors of Salem Press.
    p. cm. — (Magill's choice)
  Includes bibliographical references.
  ISBN 1-58765-174-2 (set : alk. paper) — ISBN 1-58765-175-0 (v. 1 : alk. paper) — ISBN 1-58765-176-9 (v. 2 : alk. paper)
    1. Ecology—Encyclopedias. I. Salem Press. II. Series.

QH540.4.E39 2003
577′.03—dc21

                                                                    2003011370

Second Printing

PRINTED IN THE UNITED STATES OF AMERICA

Contents

Publisher's Note ix
Contributors xiii
Complete List of Contents xvii

Acid deposition 1
Adaptations and their mechanisms 7
Adaptive radiation 12
Allelopathy 15
Altruism 18
Animal-plant interactions 24

Balance of nature 28
Biodiversity 32
Biogeography 37
Biological invasions 40
Bioluminescence 43
Biomagnification 47
Biomass related to energy 50
Biomes: determinants 55
Biomes: types 59
Biopesticides 65
Biosphere concept 69

Camouflage 72
Chaparral 76
Clines, hybrid zones, and introgression 80
Coevolution 86
Colonization of the land 90
Communication 95
Communities: ecosystem interactions 100
Communities: structure 104
Competition 111
Conservation biology 119
Convergence and divergence 120

Deep ecology 123
Defense mechanisms 125
Deforestation 131
Demographics 137
Dendrochronology 145
Desertification 149
Deserts 154
Development and ecological strategies 161
Displays 167

Ecology: definition 171
Ecology: history 179
Ecosystems: definition and history 184
Ecosystems: studies 191
Endangered animal species 196
Endangered plant species 205
Erosion and erosion control 211
Ethology 215
Eutrophication 222
Evolution: definition and theories 227
Evolution: history 236
Evolution of plants and climates 241
Extinctions and evolutionary explosions 246

Food chains and webs 255
Forest fires 258
Forest management 263
Forests 269

Gene flow 274
Genetic diversity 278
Genetic drift 281
Genetically modified foods 284
Geochemical cycles 288
Global warming 292
Grasslands and prairies 298
Grazing and overgrazing 304
Greenhouse effect 308

Habitats and biomes 313
Habituation and sensitization 319

Contents

Herbivores 326
Hierarchies 329
Human population growth 333
Hydrologic cycle 338

Insect societies 343
Integrated pest management 351
Invasive plants 354
Isolating mechanisms 358

Lakes and limnology 364
Landscape ecology 374
Lichens 381

Publisher's Note

Magill's Choice: Ecology Basics offers 132 essays, each of which covers a fundamental ecological concept taught in biology, environmental science, and introductory ecology courses. Alphabetically arranged, these topics range from the level of individual *organisms* and their interactions with the environment through *populations* of organisms and *communities* of more than one species, to the level of *ecosystems* and *global* ecology. These two volumes provide coverage of ecology in its broadest scientific sense: not only according to Ernst Haeckel's original definition—the interaction of organisms with their environment—but also in the now-conventional sense first expounded in 1954 by Herbert George Andrewartha and Louis Charles Birch as the processes that influence the "abundance and distribution" of organisms, and, further, in the sense of ecosystem ecology, introduced in 1971 by Eugene Odum. All levels taught in most general and introductory courses, as well as key environmental issues and ecological impacts of pollution, are considered here.

Previously appearing in three Salem publications—*Magill's Encyclopedia of Science: Animal Life* (2002), *Magill's Encyclopedia of Science: Plant Life* (2003), and *Earth Science* (2001)—the essays begin by listing the subdiscipline of ecology into which the topic is generally categorized; where the topic is central to more than one subdiscipline, all are listed. There follows a brief synopsis defining the topic and its significance. The text of each essay, ranging from two to six pages, is subheaded to flag the core concepts addressed. Each essay ends with the signature of the academician who wrote it, a full set of cross-references to other essays in this publication that treat related concepts, and a list of sources for further study to assist students and general readers searching for fuller information.

The subdisciplines into which the essays are classified are as follows:

- **Agricultural ecology:** Also called "agroecology," the study of agricultural ecosystems, their components (such as crop species), functions, interactions, and impact on natural ecosystems and abiotic factors such as atmospheric and water systems—often with an emphasis on the development of sustainable systems.
- **Aquatic and marine ecology:** The study of the ecology of freshwater systems (rivers, lakes), estuaries, and marine environments (both coastal and open ocean), including the physical, chemical, and biological processes associated with them.

- **Behavioral ecology:** The study of how individual organisms interact through behavior with other organisms and their environment to survive and reproduce, which has an impact on population.
- **Biomes:** The primary, large-scale ecosystems of the world, largely identified with geographical regions and classified on the basis of precipitation, temperature, climate, soil types, flora, and fauna.
- **Chemical ecology:** Concerns the biochemicals (or semiochemicals) that organisms produce and release, which have physiological and behavioral effects on other organisms.
- **Community ecology:** The study of the impacts that populations of different species have on populations of other species with which they interact, be those interactions of plants with other plants, animals with other animals, or plants with animals. The emphasis is on how these populations of different species change, enhance, or delimit one another. Population ecology is related but is focused on growth and change within populations of a single species.
- **Ecoenergetics:** The flow of energy through ecological systems at all levels, from individual organisms, populations, and communities to ecosystems and the global environment. Includes abiotic factors (such as geochemical cycles) as well as biotic factors.
- **Ecosystem ecology:** The study of the flow of energy into, through, and out of large-scale systems, and how that flow influences all abiotic factors and living organisms in the ecosystem.
- **Ecotoxicology:** The study of natural and human-made pollutants and their toxic effects on organisms, populations, communities, and ecosystems, as well as the ways these pollutants impact ecological processes to change ecosystems and their components.
- **Evolutionary ecology:** The study of how evolutionary processes such as selection and adaptation influence the interactions of organisms with their environments and shape species and ecosystems.
- **Global ecology:** The study of the impacts of such factors as global warming, pollution, and disease on organisms and ecosystems worldwide. Much of global ecology considers the ecological impacts of human-driven influences such as international travel, trade, the built environment, and the use of petrochemicals.
- **History of ecology:** The development of the discipline of ecology.
- **Landscape ecology:** The science of managing the habitat components of modified landscapes—a burgeoning field concerned with preserving the naturalness of modified landscapes while minimizing the negative impact of human intrusion in natural habitats within these landscapes.

- **Paleoecology:** The study of past ecosystems and environments.
- **Physiological ecology:** Sometimes called "autoecology," "ecophysiology," or "comparative physiology," a type of ecology that focuses on individual organisms, examining how they function mechanically and physiologically in their environments and how such factors as temperature, seasons, soil, and nutrients affect survival and reproduction of those organisms. Unlike morphology or physiology, the emphasis is on linking individual organisms, via their performance attributes, to populations and communities.
- **Population ecology:** The study of the growth and decline of groups of individuals of the same species, and how these fluctuations function in relation to other populations in the same ecosystem. Examines such factors as the availability of food and hence predation, herbivory, and mutualisms. Community ecology is closely related to population ecology but focuses on the interactions between populations of different species.
- **Restoration and conservation ecology:** Restoration ecology is the study and implementation of ways to return degraded or deteriorating communities and ecosystems to their original condition. Restoration ecologists work to restore habitat and return endangered species to viable numbers; they do not seek to restore extinct species or recreate ancient habitats. Conservation ecology is the use of biological science to design and implement methods to ensure the survival of species, ecosystems, and ecological processes. Conservation biologists develop strategies to preserve biodiversity before it becomes degraded.
- **Soil ecology:** The study of soil as an ecosystem, including the interactions of both abiotic and biotic components of soil: water, minerals, bacteria, fungi, plant matter, microbial organisms, and small animals such as insects and worms. Soil ecology extends beyond the physical borders of soil to include the impact of soil on aboveground lifeforms such as larger plants and animals, as well as processes (geochemical cycles, erosion, human agricultural practices) that impact soil.
- **Speciation:** The study of the processes whereby new species arise.
- **Theoretical ecology:** The study of the fundamental theories, concepts, and models of ecological relationships, from the simplest predator-prey, host-parasite models to population, community, ecosystem, global, and evolutionary models.

For convenience, both volumes of *Ecology Basics* contain a full list of the contents. At the end of volume 2, several research tools are offered: a Glossary, a list of Web Sites, a Categorized Index (by type of ecology), and a Subject Index.

All essays were prepared by qualified academicians and experts, without whose invaluable contributions these volumes would not be possible. Their names and affiliations follow.

Contributors

Richard Adler
University of Michigan, Dearborn

Steve K. Alexander
University of Mary Hardin-Baylor

Richard W. Arnseth
Science Applications International

George K. Attwood
Maharishi International University

David Landis Barnhill
Guilford College

Erika L. Barthelmess
St. Lawrence University

Margaret F. Boorstein
C. W. Post College of Long Island University

P. E. Bostick
Kennesaw State College

Catherine M. Bristow
Michigan State University

William R. Bromer
Pepperdine University

Steven D. Carey
University of Mobile

Richard W. Cheney, Jr.
Christopher Newport University

David L. Chesemore
California State University, Fresno

Sneed B. Collard
University of West Florida

J. A. Cooper
Independent Scholar

Alan D. Copsey
Central University of Iowa

Mark S. Coyne
University of Kentucky

Greg Cronin
University of Colorado at Denver

James F. Crow
University of Wisconsin

Gordon Neal Diem
ADVANCE Education and Development Institute

John P. DiVincenzo
Middle Tennessee State University

Allan P. Drew
SUNY, College of Environmental Science and Forestry

Frank N. Egerton
University of Wisconsin, Parkside

David K. Elliott
Northern Arizona University

Jessica O. Ellison
Clarkson University

Danilo D. Fernando
*SUNY, College of Environmental
    Science and Forestry*

James F. Fowler
State Fair Community College

Roberto Garza
San Antonio College

Ray P. Gerber
Saint Joseph's College

D. R. Gossett
*Louisiana State University,
    Shreveport*

Jerry E. Green
Miami University

Linda Hart
University of Wisconsin, Madison

Thomas E. Hemmerly
Middle Tennessee State University

John S. Heywood
Southwest Missouri State University

Joseph W. Hinton
Independent Scholar

Carl W. Hoagstrom
Ohio Northern University

Virginia L. Hodges
*Northeast State Technical
    Community College*

David Wason Hollar, Jr.
Rockingham Community College

Robert Hordon
Rutgers University

Richard D. Howard
Independent Scholar

Jason A. Hubbart
California State University, Fresno

Samuel F. Huffman
University of Wisconsin, River Falls

Lawrence E. Hurd
Washington and Lee University

Diane White Husic
East Stroudsburg University

Jeffrey A. Joens
Florida International University

Christopher Keating
Angelo State University

Kenneth M. Klemow
Wilkes University

P. R. Lannert
Independent Scholar

David M. Lawrence
John Tyler Community College

Walter Lener
Nassau Community College

W. David Liddell
Utah State University

Contributors

Robert Lovely
University of Wisconsin, Madison

Yiqi Luo
University of Oklahoma

Michael L. McKinney
University of Tennessee, Knoxville

Kristie Macrakis
Harvard University

Paul Madden
Hardin-Simmons University

Nancy Farm Männikkö
Independent Scholar

Linda Mealey
College of St. Benedict

John S. Mecham
Texas Tech University

Randall L. Milstein
Oregon State University

Eli C. Minkoff
Bates College

Richard F. Modlin
University of Alabama, Huntsville

Thomas C. Moon
California University of Pennsylvania

Randy Moore
Wright State University

Christina J. Moose
Independent Scholar

Edward N. Nelson
Oral Roberts University

Bryan Ness
Pacific Union College

John G. New
Loyola University of Chicago

Edward B. Nuhfer
University of Wisconsin, Platteville

Oghenekome U. Onokpise
Florida A&M University

Robert W. Paul
St. Mary's College of Maryland

Rex D. Pieper
New Mexico State University

Noreen D. Poor
University of South Florida

Robert Powell
Avila College

Donald R. Prothero
Occidental College

Carol S. Radford
Maryville University, St. Louis

P. S. Ramsey
Independent Scholar

C. Mervyn Rasmussen
Independent Scholar

Ronald J. Raven
State University of New York at Buffalo

Darrell L. Ray
University of Tennessee, Martin

David D. Reed
Michigan Technological University

Gregory J. Retallack
University of Oregon

Mariana Louise Rhoades
St. John Fisher College

James L. Robinson
University of Illinois at Urbana-Champaign

David W. Rudge
Western Michigan University

James L. Sadd
Occidental College

Lisa M. Sardinia
Pacific University

Samuel M. Scheiner
Northern Illinois University

John Richard Schrock
Emporia State University

Donna Janet Schroeder
College of St. Scholastica

Jon P. Shoemaker
University of Kentucky

Sanford S. Singer
University of Dayton

Elizabeth Slocum
Independent Scholar

Dwight G. Smith
Southern Connecticut State University

Roger Smith
Independent Scholar

Valerie M. Sponsel
University of Texas, San Antonio

Joan C. Stevenson
Western Washington University

Dion Stewart
Adams State College

Toby R. Stewart
Independent Scholar

Marshall D. Sundberg
Emporia State University

Frederick M. Surowiec
Independent Scholar

Leslie V. Tischauser
Prairie State College

Yujia Weng
Northwest Plant Breeding Company

Samuel I. Zeveloff
Weber State College

Ming Y. Zheng
Gordon College

Complete List of Contents

Volume 1

Acid deposition, 1

Adaptations and their mechanisms, 7

Adaptive radiation, 12

Allelopathy, 15

Altruism, 18

Animal-plant interactions, 24

Balance of nature, 28

Biodiversity, 32

Biogeography, 37

Biological invasions, 40

Bioluminescence, 43

Biomagnification, 47

Biomass related to energy, 50

Biomes: determinants, 55

Biomes: types, 59

Biopesticides, 65

Biosphere concept, 69

Camouflage, 72

Chaparral, 76

Clines, hybrid zones, and introgression, 80

Coevolution, 86

Colonization of the land, 90

Communication, 95

Communities: ecosystem interactions, 100

Communities: structure, 104

Competition, 111

Conservation biology, 119

Convergence and divergence, 120

Deep ecology, 123

Defense mechanisms, 125

Deforestation, 131

Demographics, 137

Dendrochronology, 145

Desertification, 149

Deserts, 154

Development and ecological strategies, 161

Displays, 167

Ecology: definition, 171

Ecology: history, 179

Ecosystems: definition and history, 184

Ecosystems: studies, 191

Endangered animal species, 196

Endangered plant species, 205

Erosion and erosion control, 211

Ethology, 215

Eutrophication, 222

Evolution: definition and theories, 227

Evolution: history, 236

Evolution of plants and climates, 241

Extinctions and evolutionary explosions, 246

Food chains and webs, 255

Forest fires, 258

Forest management, 263

Forests, 269

Gene flow, 274

Genetic diversity, 278

Genetic drift, 281

Genetically modified foods, 284

Geochemical cycles, 288
Global warming, 292
Grasslands and prairies, 298
Grazing and overgrazing, 304
Greenhouse effect, 308
Habitats and biomes, 313
Habituation and sensitization, 319
Herbivores, 326
Hierarchies, 329

Human population growth, 333
Hydrologic cycle, 338
Insect societies, 343
Integrated pest management, 351
Invasive plants, 354
Isolating mechanisms, 358
Lakes and limnology, 364
Landscape ecology, 374
Lichens, 381

Volume 2

Mammalian social systems, 385
Marine biomes, 391
Mediterranean scrub, 399
Metabolites, 402
Migration, 407
Mimicry, 415
Mountain ecosystems, 419
Multiple-use approach, 422
Mycorrhizae, 425
Natural selection, 428
Nonrandom mating, genetic drift,
    and mutation, 435
Nutrient cycles, 440
Ocean pollution and
    oil spills, 444
Old-growth forests, 452
Omnivores, 455
Ozone depletion and
    ozone holes, 457
Paleoecology, 464
Pesticides, 470
Pheromones, 476
Phytoplankton, 482
Poisonous animals, 486
Poisonous plants, 490
Pollination, 495
Pollution effects, 500

Population analysis, 507
Population fluctuations, 513
Population genetics, 520
Population growth, 528
Predation, 536
Punctuated equilibrium
    vs. gradualism, 543
Rain forests, 549
Rain forests and the atmosphere,
    554
Rangeland, 560
Reefs, 564
Reforestation, 572
Reproductive strategies, 576
Restoration ecology, 583
Savannas and deciduous tropical
    forests, 586
Slash-and-burn agriculture, 590
Soil, 594
Soil contamination, 601
Speciation, 604
Species loss, 608
Succession, 612
Sustainable development, 618
Symbiosis, 621
Taiga, 629
Territoriality and aggression, 633

Complete List of Contents

Trophic levels and ecological niches, 641

Tropisms, 650

Tundra and high-altitude biomes, 655

Urban and suburban wildlife, 659

Waste management, 667

Wetlands, 672

Wildlife management, 677

Zoos, 681

Glossary, 687

Web Sites, 729

Categorized Index, 735

Subject Index, 741

Ecology Basics

ACID DEPOSITION

Types of ecology: Aquatic and marine ecology; Ecotoxicology

Electric utilities, industries, and automobiles emit sulfur dioxide and nitrogen oxides that are readily oxidized into sulfuric and nitric acids in the atmosphere. Long-range transport and dispersion of these air pollutants produce regional acid deposition. Acid deposition alters aquatic—and possibly forest—ecosystems and accelerates corrosion of buildings, monuments, and statuary.

In 1872 Robert Angus Smith used the term "acid rain" in his book *Air and Rain: The Beginnings of a General Climatology* to describe precipitation affected by coal-burning industries. Today, "acid rain" refers to the deposition of acidic gases, particles, and precipitation (rain, fog, dew, snow, or sleet) on the surface of the earth. The normal acidity of rain is pH 5.6, which is caused by the formation of carbonic acid from water-dissolved carbon dioxide. The acidity of precipitation collected at monitoring stations around the world varies from pH 3.8 to 6.3 (pH 3.8 is three hundred times as acidic as pH 6.3). The acidity is created when sulfur dioxide and nitrogen oxides react with water and oxidants in the atmosphere to form water-soluble sulfuric and nitric acids. Ammonia, as well as soil constituents such as calcium and magnesium that are often present in suspended dust, neutralizes atmospheric acids, which helps explain the geographical variation of precipitation acidity.

Increasing Acidity

Between the mid-nineteenth century and World War II, the Industrial Revolution led to a tremendous increase in coal burning and metal ore processing in both Europe and North America. The combustion of coal, which contains an average of 1.5 percent sulfur by weight, and the smelting of metal sulfides released opaque plumes of smoke and sulfur dioxide from short chimneys into the atmosphere.

Copper, nickel, and zinc smelters fumigated nearby landscapes with sulfur dioxide and heavy metals. One of the world's largest nickel smelters, located in Sudbury, Ontario, Canada, began operation in 1890 and by 1960 was pouring 2.6 million tons of sulfur dioxide per year into the atmosphere. By 1970 the environmental damage extended to 72,000 hectares of injured vegetation, lakes, and soils surrounding the site; within this area 17,000 hectares were barren. The land was devastated not only by acid de-

1

Activists blame coal-burning power plants and factory emissions for acid rain problems. After being emitted by large, stationary sources, especially those that have very high smokestacks, pollutants can travel thousands of kilometers in the atmosphere. Those that are transformed into sulfuric and nitric acid aerosols are incorporated into precipitation, which eventually makes contact with the earth's surface. (PhotoDisc)

position but also by the accumulation of toxic metals in the soil, the clear-cutting of forested areas for fuel, and soil erosion caused by wind, water, and frost heave.

In urban areas, high concentrations of sulfur corroded metal and accelerated the erosion of stone structures. During the winter, the added emissions from home heating and stagnant weather conditions caused severe air pollution episodes characterized by sulfuric acid fogs and thick, black soot. In 1952 a four-day air pollution episode in London, England, killed an estimated four thousand people.

After World War II, large coal-burning utilities in Western Europe and the United States built their plants with particulate control devices and tall stacks (higher than 100 meters) to improve the local air quality. Huge industrial facilities throughout Eastern Europe and the Soviet Union operated without air pollution controls for most of the twentieth century. The tall stacks increased the dispersion and transport of air pollutants from tens to hundreds of kilometers. Worldwide emissions of sulfur dioxide increased; in the United States emissions climbed from 18 million tons in 1940 to a peak of 28 million tons in 1970. Acid deposition evolved into an interstate and even an international problem.

In major cities, exhaust from automobiles combined with power plant and industrial emissions to create a choking, acrid smog of ozone, and nitric and organic acids formed by photochemical processes. The rapid deterioration of air quality in cities, with the attendant health and environmental consequences, spurred the passage of the U.S. Clean Air Act (CAA) of 1963, which was amended and expanded in 1970, 1977, and 1990. Each amendment to the CAA brought new requirements for air pollution controls.

Effects on Aquatic Ecosystems
The nature and extent of the environmental impact of acid deposition are in dispute. Landscapes or surface waters impoverished by limestone or acid-buffering soils are more sensitive to acid deposition. Regions that are both sensitive and exposed to acid deposition include the eastern United States, southeastern Canada, southern Sweden and Norway, central and Eastern Europe, the United Kingdom, southeastern China, and the northern tip of South America. Scientists hypothesize that within these regions acid rain disrupts aquatic ecosystems and contributes to forest decline.

In southern Norway, for example, fish have been virtually extinct since the late 1970's in four-fifths of the lakes and streams in an area of 2 mil-

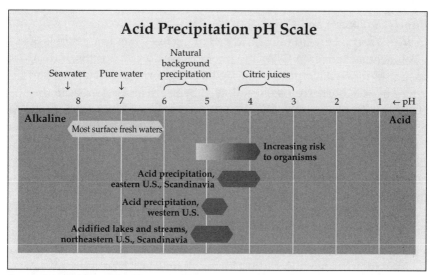

Source: Adapted from John Harte, "Acid Rain," in *The Energy-Environment Connection,* edited by Jack M. Hollander, 1992.

Note: The acid precipitation pH ranges given correspond to volume-weighted annual averages of weekly samples.

lion hectares. Records and long-term monitoring showed that the decline of fish populations began in the early twentieth century with dramatic losses in the 1950's. A strong correlation has been found between fish extinction and lake acidity. Researchers have also found that the diversity of not only fish but also phytoplankton, zooplankton, invertebrates, and amphibian species diminishes by more than 50 percent as lake water pH drops from 6.0 to 5.0. Below pH 5.6, aluminum released from lake sediments or leached from the surrounding soils interferes with gas and ion exchange in fish gills and can be toxic to aquatic life. Below pH 4.0, no fish survive.

In the United States, 11 percent of the lakes in the Adirondack Mountains in New York are too acidic to sustain fish life. Much controversy surrounds the claim that these lakes were acidified by plumes of air pollutants carried by prevailing winds from the Ohio River Valley. Like the lakes in Norway, the Adirondack lakes have low acid-neutralizing capacity. Fish declines that began in the early twentieth century and continued through the 1980's corresponded to reductions in pH. Fish kills often followed spring snowmelt, which filled the waterways with acid accumulated in winter precipitation. Historical records, field observations, and laboratory experiments contradict arguments that overfishing, disease, or water pollution killed the fish.

Effects on Forests and Cities
In areas exposed to acid rain, dead and dying trees stand as symbols of environmental change. In Germany the term *Waldsterben*, or forest death, is used to describe the rapid declines of Norway spruce, Scotch pine, and silver fir trees in the early 1980's, followed by beech and oak trees in the late 1980's, especially at high elevations in the Black and Bavarian Forests. At higher altitudes, clouds frequently shroud mountain peaks, bathing the forest canopy in a mist of heavy metals, and sulfuric and nitric acids. Under drought conditions, invisible plumes of ozone from sources hundreds of kilometers distant intercept the mountain slopes. Several forests within the United States are likewise affected, including the forests of ponderosa pine in the San Bernardino Mountains of California, balsam fir in the Smokey Mountains of North Carolina and Tennessee, and red spruce in the Green and White Mountains of New England.

After more than one decade of intensive field and laboratory investigations of forest decline in North America and Europe, the link between dead trees and acid deposition remained little more than circumstantial. Laboratory experiments often showed that acid rain had no effect, or even a fertilizing effect, on trees. Changes in foliage color, size, and shape; destruction of fine roots and associated fungi; and stunted growth are symptoms of tree stress. Many researchers attribute these symptoms and forest decline to the interactions of acid precipitation, ozone, excessive nitrogen deposition, land management practices, climate change, drought, and pestilence.

Ambient air concentrations of sulfur dioxide and nitrogen oxides are typically higher in major cities, a result of the high density of emission sources. The acids they form accelerate the weathering of exposed stone, brick, concrete, glass, metal, and paint. For example, the calcite in limestone and marble reacts with water and sulfuric acid to form gypsum (calcium sulfate). The gypsum washes off stone with rain or, if eaves protect the stone, accumulates as a soot-darkened crust. The acid-induced weathering obscures the details of elaborate carvings on medieval cathedrals, ancient Greek columns, and Mayan ruins at alarming rates. Graffiti, pigeon excrement, and the growth of bacteria and fungi on rock surfaces may compound the damage.

Prevention Efforts

In the United States, the Acidic Deposition Control Program, Title IV of the Clean Air Act amendments of 1990, directed the Environmental Protection Agency (EPA) to reduce the adverse effects of acid rain. Public law mandated that the United States achieve 40 percent and 10 percent annual reductions in sulfur dioxide and nitrogen dioxide emissions, respectively, by the year 2000 from a 1980 base. The National Acid Precipitation Assessment Program coordinates interagency acid deposition monitoring and research, and assesses the cost, benefits, and effectiveness of acid deposition control strategies. This echoes the 1985 30 Percent Protocol of the Convention on Long-Range Transboundary Air Pollution. Twenty-one nations signed the protocol, thereby agreeing to reduce sulfur dioxide emissions 30 percent from 1980 levels by 1993.

Strategies to reduce acid deposition in the United States target large electric utilities responsible for 70 percent of the sulfur dioxide and 30 percent of the nitrogen oxide emissions. Utilities participate in a novel market-based emission allowance trading and banking system that permits great flexibility in controlling sulfur dioxide emissions. For example, a utility may choose to remove sulfur from coal by cleaning it, burn a cleaner fuel such as natural gas, or install a gas desulfurization system to reduce emissions. The $6 billion international Clean Coal Technology Demonstration Program, funded by governments and private industries, continues to develop technologies—such as catalytic conversion of nitrogen oxides to inert nitrogen—to radically decrease emissions of acid gases from coal-fired power plants.

Acid deposition

Computer models of acid deposition in the northeastern United States predicted that a 50 percent reduction in sulfur dioxide emissions would decrease sulfur deposition by 44 to 48 percent. Between 1980 and 1996, U.S. electric utilities lowered annual emissions of sulfur dioxide by 30 percent from 17.5 million tons and nitrogen oxides by 14 percent from 7 million tons. Average ambient air concentrations of sulfur dioxide decreased 37 percent and nitrogen oxide concentrations 10 percent between 1987 and 1996. However, a similar trend in sulfate and nitrate deposition has not been observed.

Noreen D. Poor

See also: Lakes and limnology; Marine biomes; Ocean pollution and oil spills; Reefs.

Sources for Further Study

Adriano, D. C., and A. H. Johnson, eds. *Biological and Ecological Effects.* Vol. 2 in *Acidic Precipitation.* New York: Springer-Verlag, 1989.
Ahrens, C. Donald. *Meteorology Today.* 6th ed. Pacific Grove, Calif.: Brooks/Cole, 2000.
Canter, Larry W. *Acid Rain and Dry Deposition.* Chelsea, Mich.: Lewis, 1989.
Ellerman, A. Denny, et al. *Markets for Clean Air: The U.S. Acid Rain Program.* New York, N.Y.: Cambridge University Press, 2000. Illustrated.
Gunn, John M., ed. *Restoration and Recovery of an Industrial Region: Progress in Restoring the Smelter-Damaged Landscape near Sudbury, Canada.* New York : Springer-Verlag, 1995.
Lutgens, Frederick K., and Edward J. Tarbuck. *The Atmosphere.* 8th ed. Upper Saddle River, N.J.: Prentice Hall, 2001.
U.S. Geological Survey. *Acid Rain and Our Nation's Capital.* Author, 1997.

ADAPTATIONS AND THEIR MECHANISMS

Types of ecology: Evolutionary ecology; Physiological ecology

Adaptations are structures, physiological mechanisms, or behaviors that are shaped by the environment and enable organisms to cope with specific environmental conditions. Studying adaptations helps scientists understand how organisms live with environmental constraints and allows them to examine the mechanisms of evolution.

Why so many species exist is one of the most intriguing questions of ecology. The study of adaptations offers an explanation. Because there are many ways to cope with the environment, and because natural selection has guided the course of evolutionary change for billions of years, the vast variety of species existing on the earth today is simply an extremely complicated variation on the theme of survival. Many of the features that are most interesting and beautiful in biology are adaptations.

Adaptations are the result of long evolutionary processes in which succeeding generations of organisms become better able to live in their environments. Specialized structures, physiological processes, and behaviors are all adaptations when they allow organisms to cope successfully with the special features of their environments. Adaptations ensure that individuals in populations will reproduce and leave well-adapted offspring, thus ensuring the survival of the species.

Mutation and Natural Selection
Adaptations arise through mutations—heritable changes in an organism's genetic material. These rare events are usually harmful, but occasionally they give specific survival advantages to the mutated organism and its offspring. When certain individuals in a population possess advantageous mutations, they are better able to cope with their specific environmental conditions and, as a result, will contribute more offspring to future generations compared with those individuals in the population that lack the mutation. Over time, the number of individuals that have the advantageous mutation will increase in the population at the expense of those that do not have it. Individuals with an advantageous mutation are said to have a higher "fitness" than those without it, because they tend to have comparatively higher survival and reproductive rates. This is natural selection.

Over very long periods of time, evolution by natural selection results in increasingly better adaptations to environmental circumstances. Natural selection is the primary mechanism of evolutionary change, and it is the force that either favors or selects against mutations. Although natural selection acts on individuals, a population gradually changes as those with adaptations become better represented in the total population. Predaceous fish, for example, which rely on speed to pursue and overtake prey, would benefit from specific adaptations that would increase their swimming speed. Therefore, mutations causing a sleeker and more hydrodynamically efficient form would be beneficial to the fish predator. Such changes would be adaptations if they resulted in improved predation success, diet, and reproductive success, compared with slower members of the population. Natural selection would favor the mutations because they confer specific survival advantages to those that carry the mutations and impose limitations on those lacking these advantages. Thus, those individuals with special adaptations for speed would have a competitive advantage over slower-swimming individuals. These attributes would be passed to their more numerous offspring and, in evolutionary time, speed and hydrodynamic efficiency would increase in the population.

General vs. Specific Adapations
Adaptations can be general or highly specific. General adaptations define broad groups of organisms whose general lifestyle is similar. For example, mammals are homeothermic, provide care for their young, and have many other adaptations in common. At the species level, however, adaptations are more specific and give narrow definition to those organisms that are more closely related to one another. Slight variations in a single characteristic, such as bill size in the seed-eating Galápagos finches, are adaptive in that they enhance the survival of several closely related species. An understanding of how adaptations function to make species distinct also furthers the knowledge of how species are related to one another.

Although natural selection serves as the instrument of change in shaping organisms to very specific environmental features, highly specific adaptations may ultimately be a disadvantage. Adaptations that are specialized may not allow sufficient flexibility (generalization) for survival in changing environmental conditions. The degree of adaptative specialization is ultimately controlled by the nature of the environment. Environments, such as the tropics, that have predictable, uniform climates and have had long, uninterrupted periods of climatic stability are biologically complex and have high species diversity. Scientists generally believe that this diversity results, in part, from complex competition for resources and from intense predator-prey interactions. Because of these factors, many narrowly specialized adaptations have evolved when environmental stability and predictability prevail. By contrast, harsh physical environments with unpredictable or erratic climates seem to favor organisms with general adaptations, or adaptations that allow flexibility. Regardless of the environment type, organisms with both general and specific adaptations exist because both types of adaptation enhance survivorship under different environmental circumstances.

Structural Adaptations

Structural adaptations are parts of organisms that enhance their survival ability. Camouflage, enabling organisms to hide from predators or their prey; specialized mouth parts that allow organisms to feed on specific food sources; forms of appendages, such as legs, fins, or webbed toes, that allow efficient movement; protective spines that make it difficult for the organism to be eaten—these are all structural adaptations. These adaptations enhance survival because they assist individuals in dealing with the rigors of the physical environment, obtaining nourishment, competing with others, or hiding from or confusing predators.

Metabolic and Physiological Adaptations

Metabolism is the sum of all chemical reactions taking place in an organism, whereas physiology consists of the processes involved in an organism carrying out its function. Physiological adaptations are changes in the metabolism or physiology of organisms, giving them specific advantages for a given set of environmental circumstances. Because organisms must cope with the rigors of their physical environments, physiological adaptations for temperature regulation, water conservation, varying metabolic rate, and dormancy or hibernation allow organisms to adjust to the physical environment or respond to changing environmental conditions.

Desert environments, for example, pose a special set of problems for organisms. Hot, dry environments require physiological mechanisms that enable organisms to conserve water and resist prolonged periods of high temperature. Highly efficient kidneys and other excretory organs that assist organisms in retaining water are physiological adaptations related to the metabolisms of desert organisms. The kangaroo rat is a desert rodent extremely well adapted to its habitat. Kangaroo rats do not drink, but rather can obtain all of their water from the seeds they eat. They produce highly concentrated urine and feces with very low water content.

Adaptation to a specific temperature range is also an important physiological adaptation. Organisms cannot live in environments with temperatures beyond their range of thermal tolerance, but some organisms are adapted to warmer and others to colder environments. Metabolic response to temperature is quite variable among animals, but most animals are either homeothermic (warm-blooded) or poikilothermic (cold-blooded). Homeotherms maintain constant body temperatures at specific temperature ranges. Although a homeotherm's metabolic heat production is constant when the organism is at rest and when environmental temperature is constant, strenuous exercise produces excess heat that must be dissipated into the environment, or overheating and death will result. Physiological adaptations that enable homeotherms to rid their bodies of heat are the ability to increase blood flow to the skin's surface, sweating, and panting, all of which promote heat loss to the atmosphere.

Behavioral Adaptations
Behavioral adaptations allow organisms to respond appropriately to various environmental stimuli. Actions taken in response to various stimuli are adaptive if they enhance survivorship. Migrations are behavioral adaptations because they ensure adequate food supplies or the avoidance of adverse environmental conditions. Courtship rituals that help in species recognition prior to mating, reflex and startle reactions allowing for quick retreats from danger, and social behavior that fosters specialization and cooperation for group survival are behavioral adaptations.

Coevolution
Because organisms must also respond and adapt to an environment filled with other organisms—including potential predators and competitors—adaptations that minimize the negative effects of biological interactions are favored by natural selection. Many times the interaction between species is so close that each species strongly influences the others in the interaction and serves as the selective force causing change. Under these circumstances, species evolve together in a process called coevolution. The adaptations resulting from coevolution have a common survival value to all the species involved in the interaction. The coevolution of flowers and their pollinators is a classic example of these tight associations and their resulting adaptations.

The Peppered Moth
A classic example of recent evolutionary change and adaptation comes from England. The peppered moth with a mottled gray color, is well adapted to resting quietly on pale tree bark, with which it blends nicely. This adaptive coloration (camouflage) enhanced the moth's survival be-

cause the moths could remain largely undetected by predators during daylight hours. Between 1850 and 1950, however, industrialization near urban centers blackened tree trunks with soot, making the gray form disadvantageous, as it stood out on the contrasting background. During this period, the gray moths began to disappear from industrial areas, but a black-colored variant, previously rare, became increasingly common in the population. These circumstances made it possible for scientists to test whether the peppered moth's camouflage was adaptive.

In a simple experiment, moths were raised in the laboratory, and equal numbers of gray and black moths were released in both industrial and unpolluted rural areas. Sometime later, only half of the gray-colored moths could be recovered from the industrial sites, while only half of the black forms could be recovered from the rural sites, compared with the total number released. These results enabled the scientists to conclude that increased predation on the gray moths in industrial areas led to a greater fitness of the black moths, so the frequency of black moths increased in the population. The reverse was true at the rural sites. This is the first well-documented case of natural selection causing evolutionary change, and it illustrates the adaptive significance of camouflage.

Robert W. Paul

See also: Adaptive radiation; Biodiversity; Biogeography; Camouflage; Clines, hybrid zones, and introgression; Coevolution; Defense mechanisms; Evolution: history; Isolating mechanisms; Natural selection; Nonrandom mating, genetic drift, and mutation; Population genetics; Punctuated equilibrium vs. gradualism; Speciation; Species loss; Trophic levels and ecological niches.

Sources for Further Study

Birkhead, Mike, and Tim Birkhead. *The Survival Factor.* New York: Facts on File, 1990.

Brandon, Robert N. *Adaptation and Environment.* Princeton, N.J.: Princeton University Press, 1990.

Gould, Stephen J. *Ever Since Darwin.* New York: W. W. Norton, 1977.

Ricklefs, Robert E. *Ecology.* 4th ed. New York: W. H. Freeman, 1999.

Rose, Michael R., and George V. Lauder, eds. *Adaptation.* San Diego, Calif.: Academic Press, 1996.

Weibel, Ewald R. *Symmorphosis: On Form and Function in Shaping Life.* Cambridge, Mass.: Harvard University Press, 2000.

Whitfield, Philip. *From So Simple a Beginning: The Book of Evolution.* New York: Macmillan, 1995.

ADAPTIVE RADIATION

Types of ecology: Evolutionary ecology; Population ecology; Speciation

*In adaptive radiation, numerous species evolve from a common ancestor intro-
duced into an environment with diverse ecological niches. The progeny evolve genetically into customized variations of themselves, each adapting to survive in a
particular niche.*

In 1898 Henry F. Osborn identified and developed the concept of adaptive radiation, whereby different forms of a species evolve, quickly in
evolutionary terms, from a common ancestor. According to the principles
of natural selection, organisms that are the best adapted (most fit) to compete will live to reproduce and pass their successful traits on to their offspring. The process of adaptive radiation illustrates one way in which natural selection can operate when members of one population of a species
are cut off from another or migrate to a different environment that is isolated from the first. Such isolation can occur from one patch of plantings to
another, from one mountaintop or hillside to another, from pond to pond,
or from island to island. Faced with different environments, the group will
diverge from the original population and in time become different enough
to form a new species.

Divergent Populations and Speciation
In a divergent population, the relative numbers of one form of allele (characteristic) decrease, while the relative numbers of a different allele increase. New environmental pressures will select for favorable alleles that
may not have been favored in the old environment. Over successive generations, therefore, a new gene created by random mutation may replace the
original form of the gene if, for example, the trait encoded by that gene allows the divergent group to cope better with environmental factors, such
as food sources, predators, or temperature. The result in the long term is
that molecular material that forms genes, deoxyribonucleic acid (DNA),
changes sufficiently through the growth of divergent populations to allow
new generations to become significantly different from the original population. In time, the new population is unable to reproduce with members of
the original population and becomes a new species.

Adaptive Radiation of Animals
Adaptive radiation occurs dramatically when a species migrates from one

landmass to another. This may occur between islands or between continents and islands. A classic example of adaptive radiation is the evolution of finches noted by Charles Darwin during his trips to the Galápagos Islands off the west coast of South America. Several species of plants and animals had migrated to these islands from the South American mainland by means of flight, wind, ocean debris, or other means of transport. Finches from the mainland—perhaps aided by winds—settled on fifteen of the islands in the Galápagos group and began to adapt to the various unoccupied ecological niches on those islands, which differed. Over several generations, natural selection favored a variety of finch species with beaks adapted for the different types of foods available on the different islands. As a result, several species of different finches evolved, roughly simultaneously, on these islands.

A more recent example of adaptive radiation in its early stages has taken place in an original population of brown bears. The brown bear can be found throughout the Northern Hemisphere, ranging from the deciduous forests up into the tundra. During one of the glacier periods, a small population of the brown bear was separated from the main group; according to fossil evidence, this small population, under selection pressure from the Arctic environment, evolved into the polar bear. Although brown bears are classified as carnivores, their diets are mostly vegetarian, with occasional fish and small animals as supplements. On the other hand, the polar bear is mostly carnivorous. Besides its white coat, the polar bear is different from the brown bear in many ways, including its streamlined head and shoulders and the stiff bristles that cover the soles of its feet, which provide traction and insulation, enabling it to walk on ice.

Adaptive Radiation of Plants

Although plants seem unable to "migrate" as birds and other animals do, adaptive radiation occurs in the plant world as well. In the Hawaiian Islands, for example, twenty-eight species of the *Asteraceae* family are known together as the Hawaiian silversword alliance. The entire group appears to be traceable to one ancestor, thought to have arrived on the island of Kauai from western North America. The silverswords—which compose three genera, *Argyroxiphium*, *Dubautia*, and *Wilkesia*—have since evolved into twenty-eight species, and this speciation came about due to major ecological shifts. These plants are therefore prime examples of adaptive radiation.

Within the silversword alliance, different species have adapted to widely varying ecosystems found throughout the islands. *Argyroxiphium sandwicense*, for example, is endemic to the island of Maui and grows at high elevations from 6,890 to 9,843 feet (2,100-3,000 meters) on the dry, alpine slopes of the volcano Haleakala. This species has succulent leaves covered with silver hairs. It is thought that the hairs lessen the pace of evaporative moisture loss and protect the leaves from the sun. In contrast, species of the genus *Dubautia* that grow in wet, shady forests have large leaves that lack hairs.

Despite their "customized" physiologies, the silverswords that have evolved in Hawaii are all closely related to one another, so much so that any two can hybridize. Studies of the silverswords have provided what geneticist Michael Purugganan called a "genetic snapshot of plant evolution."

Jon P. Shoemaker, updated by Bryan Ness

See also: Adaptations and their mechanisms; Biodiversity; Biogeography; Clines, hybrid zones, and introgression; Coevolution; Competition; Convergence and divergence; Evolution: definition and theories; Evolution: history; Evolution of plants and climates; Extinctions and evolutionary explosions; Gene flow; Genetic diversity; Genetic drift; Isolating mechanisms; Natural selection; Nonrandom mating, genetic drift, and mutation; Population genetics; Punctuated equilibrium vs. gradualism; Speciation; Trophic levels and ecological niches.

Sources for Further Study

Givnish, Thomas J., and Kenneth J. Sytsma, eds. *Molecular Evolution and Adaptive Radiation*. New York: Cambridge University Press, 2000.

Robichaux, Robert, et al. "'Radiating' Plants." *Endangered Species Bulletin Update*, March/April, 1999, S4-S5.

Schluter, Dolph. *The Ecology of Adaptive Radiation*. Oxford, England: Oxford University Press, 2000.

ALLELOPATHY

Types of ecology: Chemical ecology; Community ecology

Allelopathy refers to all the biochemical interactions between species, including microorganisms.

For an allelopathic interaction to occur, chemicals must be released into the environment by one organism that will affect the growth of another. In this way allelopathy differs from competition, which involves removal of some factor from the environment that is shared with other organisms.

Allelopathy was recognized as early as Theophrastus (300 B.C.E.), who pointed out that chick pea plants destroy weeds growing around them.

Methods of Action

A variety of different allelochemicals are produced by plants, usually as secondary metabolites that do not have a specific function in the growth and development of the host plant but that do affect the growth of other plants. Originally plant physiologists thought these secondary products were simply metabolic wastes that plants had to store because they do not have an excretory system as animals do. Their various functions are now beginning to be understood.

One class of allelochemicals, coumarins, block or slow cell division in the affected plant, particularly in root cells. In this way growth of competing plants is inhibited, and seed germination can be prevented. Several kinds of allelochemicals, including flavonoids, phenolics, and tannins, suppress or alter hormone production or activity in competing plants. Other chemicals, including terpenes and certain antibiotics, alter membrane permeability of host cells, making them either leaky or impermeable. In some cases, membrane uptake can be enhanced, particularly for micronutrients in low concentration in the soil. Finally, a variety of allelochemicals have both positive and negative effects on metabolic activity of the affected plant.

Allelopathy in Agriculture

Most of the negative effects of weeds on crop plants have been attributed to competition; however, experiments using weed extracts have demonstrated that many weeds produce allelochemicals. Similarly, some crop plants are allelopathic to others and themselves, including wheat, corn,

Some plant species, including peach trees, release chemicals into the soil that inhibit the growth of other plants that might otherwise compete with them. (PhotoDisc)

and rice. In these cases the residues of one year's crop can interfere with crop growth in subsequent years. This is increasingly important for farmers to consider who are incorporating low-tillage methods to reduce soil erosion. To minimize these effects, some of the traditional techniques of cover cropping, companion cropping, and crop rotation must be employed. Known allelopaths are also beginning to be used as biological control agents to manage invasive and weedy plant species.

Allelopathy in Nature
Several tree species, including black walnut, black locust, and various pines, are known to produce allelochemicals that inhibit the growth of understory species. In some cases this is a result of drip from the foliage or leachate from fallen leaves and fruit. In other cases, roots secrete allelochemicals that kill seedlings of other plants. Bracken fern (*Pteridium aquilinum*) is known to affect the growth of many other plants.

Marshall D. Sundberg

See also: Animal-plant interactions; Biological invasions; Coevolution; Communities: ecosystem interactions; Competition; Defense mechanisms;

Invasive plants; Metabolites; Poisonous plants; Trophic levels and ecological niches.

Sources for Further Study

Moore, Randy, W. Dennis Clark, and Darrell S. Vodopich. *Botany*. 2d ed. New York: McGraw-Hill, 1998.

Rice, Elroy L. *Allelopathy*. 2d ed. Orlando, Fla.: Academic Press, 1984.

ALTRUISM

Type of ecology: Behavioral ecology

Altruistic behavior involves an individual's sacrifice of self in order to help others. In some animals, altruism appears to be genetically determined.

Those who study animal behavior (ethology) have observed that on occasion individuals act altruistically. In other words, they appear voluntarily to put the needs of their group or of another individual ahead of their own needs. According to some scientists, there are examples in nature where a particular species might not have survived had there not been sacrifice by some on behalf of the many. One important question is whether this so-called altruism has been a matter of voluntary choice or whether it has occurred as a part of the selection process, making it, therefore, an involuntary response.

Group Formation

Of interest to a wide group, including psychologists, sociologists, philosophers, and political scientists, are the questions of whether altruism is desirable behavior—perhaps even to the exclusion of egoism—and whether altruism may be necessary for human survival. Some wonder whether such behavior is necessary, whether it can be learned, and whether humans will voluntarily choose to learn it. Biologists and geneticists have been left the problem of determining, if possible, whether the tendency for altruism is inherited or learned behavior. Unfortunately for scientists, the study of human beings in social groups in the wild is virtually impossible. However, the study of animal behavior, primarily in native habitats, has provided some insight, although it must be recognized that different species have solved problems of survival in different ways.

Animals of the same species are bound to consort, if only for mating purposes. Most species are, in fact, found to live in groups, not only for purposes of reproduction but also because sources of food attract individuals to the same places and because congregation provides better protection from predators. It is common in nature for groups to form because their individual members have the same physical needs, and such groups may stay together as long as the needs of those individuals can be met. This does not necessarily mean that there exists in the group any loyalty or even any recognition of individuals as members of the group. In more highly developed societies, however, groups such as families or tribes develop.

Offspring and Reproduction

In animal life, two or more adults and their offspring often form close bonds and tend to exclude those who are not related. Each recognizes the others as being members, and membership is restricted to those who are among the founders or who are born into the smaller group and who conform in recognizable ways to the norms of the group. Hierarchy or rank is recognized, and often there is a division of labor within the group.

It has been demonstrated that species that spend a large amount of time providing for their young tend to have developed higher social orders. Humans, for example, must care for their young much longer, before they are able to become independent, than must many of the lower forms of animal life. Humans are aware of a bond that almost always exists between parent and child and of the spirit of mutual support and a cooperation that may exist even in the extended family. Cooperative behavior within such a familial group may be considered to benefit all members. Because such behavior is not consistent, however—there are times when such bonds do not exist and when families are not cooperative—such behavior cannot necessarily be attributed to predisposition. Some have argued that in primitive animal societies, so-called altruism may have evolved of necessity in order to achieve reproductive success, but that in human society there may be no evolutionary explanation for the phenomenon. Indeed, it could be argued that pure altruism, for humans, might be self-defeating and therefore unlikely to have developed as an inherited trait.

Evidence has been gathered in the study of some *Hymenoptera* (the order of insects that includes bees, wasps, ants, sawflies, and other colony-forming insects) that certain members of the population forage for the group while others lay eggs and remain at the nest to guard them. Where such behavior has evolved, through the necessity of feeding and protecting those that will propagate their kind, the foragers may be labeled altruistic: They have sacrificed their own reproductive possibilities for survival of the group. Some have questioned whether this phenomenon can truly be labeled altruism, however, because the donor appears to be "programmed" to perform such behaviors rather than having a choice not to perform the behaviors (conscious purpose is very difficult to assess in animals). Moreover, some researchers wonder how the traits that favor altruistic behavior can survive and become dominant in a group if those having the traits deemed desirable are not allowed to reproduce. With the use of mathematical models, it has been demonstrated that such traits can be preserved only within the family unit.

Among close relatives, the traits appear with enough strength that they will be reproduced in a greater concentration, thereby compensating for the loss suffered by the sacrifice of the donors. This phenomenon has been referred to as kin selection, because it occurs in groups that have strong recognition of membership—to the extent that there exists aggressive defense against intruders, even of the same species. Discrimination against outsiders is an important facet of altruism of this type. The willingness of an individual to provide for others at the expense of its own interests diminishes as the degree of relatedness decreases.

Reciprocal Sacrifice

Most parental behavior would not be labeled altruistic, since it is in the interest of the parent to care for the offspring in order to ensure the survival of the parent's genes. Of perhaps more interest than what happens among closely related members of a group and even between parent and offspring is the question of what motivates sacrifice on the part of an individual when no close relationship with the recipient exists—for example, a male animal coming to the rescue of an unrelated male animal who is being attacked by a third male of the same species.

One theory maintains that these acts of personal sacrifice are performed on the chance that reciprocal sacrifice may occur at some future time. Whether this type of altruism can occur through natural selection, which acts through individuals, is an interesting question. Models have shown that in a population where individuals are likely to encounter and recognize one another on a frequent basis, it is possible that reciprocal exchanges can take place. Individual A might be the donor on the first encounter, individual B on the second. This theory requires that the two must have a high probability of subsequent encounters and that the tendency for altruism must already have been established through kin selection. Because animals are usually suspicious of strangers on first encounter, it is necessary to speculate that in its beginning, altruism was a selected-for trait in very small groups where strangers were not only nonhostile but also likely to be relatives and likely to be met again. This type of behavior, in which individuals act in a manner not to their own advantage and not in order that their own genes or the genes of relatives will survive, is performed, in theory, with some expectation of imagined reciprocal gain. How this type of behavior has come about, however, is a matter requiring further study.

Cultural Influences

Another question concerns how much culture is an influence on the development of a hereditary tendency toward altruism. Some have suggested that after generations and generations of cultural emphasis on the need for altruism, it might come to have a genetic basis. There is little hard evidence that this would occur. On the other hand, humans have had a very rapid cultural evolution, and it possible that they may have had strong genetic propensities for altruism which have been culturally overlaid. Some argue that biology and culture evolve simultaneously—that the culture is formed as a result of the imposition of genetic factors while, at the same time, genetic traits are evolving in response to cultural change. In order to understand the source of altruism in humankind, one must study such behavior in the context of many factors in human development—biological as well as social, cultural, economic, and ecologic.

Studying Altruistic Behavior
Those investigating the sources of altruism usually begin with a thorough understanding of whatever organism is the subject of the study. When the insect or animal cannot be studied in the wild, the ethologist tries to simulate the important features of the natural habitat in a captive environment, at least in the beginning. Models are devised, based on observable data; formulas are employed; and projections are made, which provide a basis for speculative argument when absolutes cannot be assured.

By observing, it is possible to determine whether various evidences of altruism exist within a population. Altruism may be manifested in as simple a way as the sharing of food when there is a scarcity. In some populations, one might observe a division of labor in which some forfeit their reproductive possibilities in order to care for the offspring of others. This phenomenon introduces the question of how altruism can survive in a population in which the genetic traits favoring the behavior are most evident in the individuals that do not reproduce themselves. It has been shown that the tendency for altruism can be perpetuated only within the family unit where the same genetic tendency exists to some degree in members that engage in reproductive activity; this can be demonstrated by a mathematical formula.

Each individual bears the inheritance coefficient or relatedness coefficient r. Offspring share with each parent an average of half of the genetic traits of each ($r = \frac{1}{2}$). Offspring share with each grandparent one-fourth of the genetic traits of each of the older generation ($r = \frac{1}{4}$); the same coefficient exists with cousins. Were the altruists not to reproduce, it would be required, in order for the trait to be passed on, that the reproductive chances of their siblings more than double or that the reproductive chances of their cousins more than quadruple. For the sacrifice to be of value, the genetic relationship must be close, according to the demonstration. The case has been made that in societies having evolved according to this principle, there is a diminishing willingness to put the interests of others ahead of one's own as the degree of kinship decreases.

In societies where males are produced from unfertilized eggs and females from fertilized ones, female offspring of a mated pair have a high relatedness coefficient ($r = \frac{3}{4}$). The altruists among the female siblings will benefit more, regarding their genetic potential, by caring for their sisters than for their own offspring, and it can again be observed that sacrifice is more likely to be made on behalf of the member that is more closely related.

Voluntary vs. Involuntary Altruism

If altruism exists in nature, and if it has come about through natural selection, then one can argue that it must be a behavior with value. When applying the human connotation to the term altruism, however, one must consider the role of choice in the manifestation of the behavior. Humans claim to admire acts of unselfishness that are seemingly done with no expectation of reward. The admiration would diminish or become nonexistent, however, if there were to be proof that the act was performed because of some primitive biological predisposition rather than because of a decision on the part of the donor. Therefore, it is necessary to make the distinction, when discussing the importance of altruism, as to whether one is referring to the acts of human beings that are performed in the face of emergency or tragedy, where a sacrifice is made as a matter of choice, or whether the intent is to consider altruism as it occurs in other creatures and seems to be involuntary.

In the case of nonhuman forms, altruism as an act of voluntary sacrifice is infrequent—if indeed it exists at all. Altruism, however, as an act which is dictated by genetics, is observable, and it has been shown to have been necessary for the survival of certain species. Where animal societies have formed in which some members of the society have spent their lives caring for the offspring of others or performing other sacrificial behavior which benefits the group, there can be little doubt that such altruism has been dictated by nature for its own unique purposes.

Moreover, the fact that voluntary self-sacrifice on the part of human beings does exist does not automatically make it desirable human behavior any more than aggressive behavior is automatically undesirable. The case can be made that both types of behavior are important. Perhaps the larger question is when and under what circumstances certain types of human behavior should be acceptable or desirable for the individual and for the group, and, even more important, who is qualified to decide what type of behavior is appropriate.

P. R. Lannert

See also: Communication; Defense mechanisms; Displays; Ethology; Hierarchies; Insect societies; Mammalian social systems; Mimicry; Pheromones; Population genetics; Predation; Reproductive strategies; Territoriality and aggression.

Sources for Further Study

Boorman, Scott A., and Paul R. Levitt. *The Genetics of Altruism.* New York: Academic Press, 1980.

Bradie, Michael. *The Secret Chain: Evolution and Ethics.* Albany: State University of New York Press, 1994.

Wright, Robert. *Nonzero: The Logic of Human Destiny.* New York: Pantheon Books, 2000.

Zahn-Waxler, Carolyn, E. Mark Cummings, and Ronald Iannotti, eds. *Altruism and Aggression: Biological and Social Origins.* New York: Cambridge University Press, 1986.

ANIMAL-PLANT INTERACTIONS

Types of ecology: Community ecology

The ways in which certain animals and plants interact have evolved in some cases to make them interdependent for nutrition, respiration, reproduction, or other aspects of survival.

The realm of ecology involves a systematic analysis of plant-animal interactions through the considerations of nutrient flow in food chains and food webs, exchange of such important gases as oxygen and carbon dioxide between plants and animals, and strategies of mutual survival between plant and animal species through the processes of pollination and seed dispersal. Having the unique ability, by photosynthesis, to take carbon dioxide and incorporate it into organic molecules, green plants are classified as ecological producers. Animals are classified as consumers, taking the products of photosynthesis and chemically breaking them down at the cellular level to produce energy for life activities. Carbon dioxide is a waste product of this process. Given their respective status as producers and consumers, plants and animals have over the ages formed many ecological relationships.

Mutualism
Mutualism is an ecological interaction in which two different species of organisms beneficially reside together in close association, usually revolving around nutritional needs. One such example is a small aquatic flatworm that absorbs microscopic green algae into its tissues. The benefit to the animal is one of added food supply. The mutual adaptation is so complete that the flatworm does not actively feed as an adult. The algae, in turn, receive adequate supplies of nitrogen and carbon dioxide and are literally transported throughout tidal flats in marine habitats as the flatworm migrates, thus exposing the algae to increased sunlight. This type of mutualism, which verges on parasitism, is called symbiosis.

Coevolution
Coevolution is an evolutionary process wherein two organisms interact so closely that thy evolve together in response to shared or antagonistic selection pressure. A classic example of coevolution involves the yucca plant and a species of small, white moth (*Tegitecula*). The female moth collects pollen grains from the stamen of one flower on the plant and transports

these pollen loads to the pistil of another flower, thereby ensuring cross-pollination and fertilization. During this process, the moth will lay her own fertilized eggs in the flowers' undeveloped seed pods. The developing moth larvae have a secure residence for growth and a steady food supply. These larvae will rarely consume all the developing seeds; thus, both species (plant and animal) benefit.

Although this example represents a mutually positive relationship between plants and animals, other interactions are more antagonistic. Predator-prey relationships between plants and animals are common. Insects and larger herbivores consume large amounts of plant material. In response to this selection pressure, many plants have evolved secondary metabolites that make their tissues unpalatable, distasteful, or even poisonous. In response, herbivores have evolved ways to neutralize these plant defenses.

Mimicry

In mimicry, an animal or plant has evolved structures or behavior patterns that allow it to mimic either its surroundings or another organism as a defensive or offensive strategy. Certain types of insects, such as the leafhopper, walking stick, praying mantis, and katydid (a type of grasshopper), often duplicate plant structures in environments ranging from tropical rain forests to northern coniferous forests. Mimicry of their plant hosts affords these insects protection from their own predators as well as camouflage that enables them to capture their own prey readily. Certain species of ambush bugs and crab spiders have evolved coloration patterns that allow them to hide within flower heads of such common plants as goldenrod, enabling them to ambush the insects that visit these flowers.

Nonsymbiotic Mutualism

In nonsymbiotic mutualism, plants and animals coevolve morphological structures and behavior patterns by which they benefit each other but without living physically together. This type of mutualism can be demonstrated in the often unusual shapes, patterns, and colorations that more advanced flowering plants have developed to attract various insects, birds, and mammals for pollination and seed dispersal purposes. Accessory structures, called fruits, form around seeds and are usually tasty and brightly marked to attract animals for seed dispersal. Although the fruits themselves become biological bribes for animals to consume, often the seeds within these fruits are not easily digested and thus pass through the animals' digestive tracts unharmed, sometimes great distances from the parent plant. Some seeds must pass through the digestive plant of an animal to stimulate germination. Other types of seed dispersal mechanisms involve the evolution of hooks, barbs, and sticky substances on seeds that enable them to be easily transported by animal fur, feet, feathers, or beaks. Such strategies of dispersal reduce competition between the parent plant and its offspring.

Pollinators

Because structural specialization increases the possibility that a flower's pollen will be transferred to a plant of the same species, many plants have evolved a vast array of scents, colors, and nutritional products to attract pollinators. Not only does pollen include the plant's sperm cells; it also represents a food reward. Another source of animal nutrition is a substance called nectar, a sugar-rich fluid produced in specialized structures called nectaries within the flower or on adjacent stems and leaves. Assorted waxes and oils are also produced by plants to ensure plant-animal interactions. As species of bees, flies, wasps, butterflies, and hawkmoths are attracted to flower heads for these nutritional rewards, they unwittingly become agents of pollination by transferring pollen from stamens to pistils.

Some flowers have evolved distinctive, unpleasant odors reminiscent of rotting flesh or feces, thereby attracting carrion beetles and flesh flies in search of places to reproduce and deposit their own fertilized eggs. As these animals copulate, they often become agents of pollination for the plant itself. Some tropical plants, such as orchids, even mimic a female bee, wasp, or beetle, so that the insect's male counterpart will attempt to mate with them, thereby encouraging precise pollination.

Among birds, hummingbirds are the best examples of plant pollinators. Various types of flowers with bright, red colors, tubular shapes, and strong, sweet odors have evolved in tropical and temperate regions to take advantage of hummingbirds' long beaks and tongues as an aid to pollination. Because most mammals, such as small rodents and bats, do not detect colors as well as bees and butterflies do, some flowers instead focus upon the production of strong, fermenting, or fruitlike odors and abundant pollen rich in protein. In certain environments, bats and mice that are primarily nocturnal have replaced day-flying insects and birds as pollinators.

Thomas C. Moon, updated by Bryan Ness

See also: Allelopathy; Coevolution; Communities: ecosystem interactions; Communities: structure; Competition; Defense mechanisms; Food chains and webs; Herbivores; Hierarchies; Lichens; Mycorrhizae; Omnivores; Poisonous plants; Pollination; Predation; Symbiosis; Trophic levels and ecological niches.

Sources for Further Study

Abrahamson, Warren G., and Arthur E. Weis. *Evolutionary Ecology Across Three Trophic Levels: Goldenrods, Gallmakers, and Natural Enemies*. Princeton, N.J.: Princeton University Press, 1997.

Barth, Friedrich G. *Insects and Flowers: The Biology of a Partnership*. Princeton, N.J.: Princeton University Press, 1991.

Buchmann, Stephen L., and Gary Paul Nabhan. *The Forgotten Pollinators*. Washington, D.C.: Island Press/Shearwater Books, 1997.

Dickerman, Carolyn. "Pollination: Strategies for Survival." *Ward's Natural Science Bulletin*, Summer, 1986, 1-4.

Howe, Henry F., and Lynne C. Westley. *Ecological Relationships of Plants and Animals*. New York: Oxford University Press, 1990.

John, D. M., S. J. Hawkins, and J. H. Price, eds. *Plant-Animal Interactions in the Marine Benthos*. Oxford, England: Clarendon Press, 1992.

Lanner, Ronald M. *Made for Each Other: A Symbiosis of Birds and Pines*. New York: Oxford University Press, 1996.

Meeuse, Bastian, and Sean Morris. *The Sex Life of Flowers*. New York: Facts on File, 1984.

Price, Peter W., G. Wilson Fernandes, Thomas H. Lewinsohn, and Woodruff W. Benson, eds. *Plant-Animal Interactions: Evolutionary Ecology in Tropical and Temperate Regions*. New York: John Wiley & Sons, 1991.

Rudman, William B. "Solar-Powered Animals." *Natural History* 96 (October, 1987): 50-53.

BALANCE OF NATURE

Types of ecology: Ecoenergetics; Theoretical ecology

The ecological concept of the balance of nature—a view that proposes that nature, in its undisturbed state, is constant—has never been legitimized in science as either a hypothesis or a theory. However, it laid the groundwork for the science of ecology and persists as a designation for a healthy environment.

The notion of the "balance of nature" has a deep history that dates back to ancient times and has persisted into modern times. During the scientific revolution in the seventeenth century, John Graunt, a merchant, analyzed London's baptismal and death records in 1662 and discovered the balance in the sex ratio and the regularity of most causes of death (excluding epidemics). England's chief justice, Sir Matthew Hale, was interested in Graunt's discoveries, but he nevertheless decided that the human population, in contrast to animal populations, must have steadily increased throughout history. He surveyed the known causes of animal mortality and in 1677 published the earliest explicit account of the balance of nature.

English scientist Robert Hooke studied fossils and in 1665 concluded that they represented the remains of plants and animals, some of which were probably extinct. However, a clergyman-naturalist, John Ray, replied that the extinction of species would contradict the wisdom of the ages, by which he seems to have meant the balance of nature. Ray also studied the hydrologic cycle, the geochemical cycle of water. Antoni van Leeuwenhoek, one of the first investigators to make biological studies with a microscope, discovered that parasites are more prevalent than anyone had suspected and that they are often detrimental or even fatal to their hosts. Before that, it was commonly assumed that the relationship between host and parasite was mutually beneficial.

Richard Bradley, a botanist and popularizer of natural history, pointed out in 1718 that each species of plant has its own kind of insect and that there are even different insects that eat the leaves and bark of a tree. His book *A Philosophical Account of the Works of Nature* (1721) explored aspects of the balance of nature more thoroughly than had been done before. Ray's and Bradley's books may have inspired the comment in Alexander Pope's *Essay on Man* (1733) that all species are so closely interdependent that the extinction of one would lead to the destruction of all living nature.

Toward a Science of Ecology

Swedish naturalist Carolus Linnaeus was an important protoecologist. His essay *Oeconomia Naturae* (1749; *The Economy of Nature*, 1749) attempted to organize the aspect of natural history dealing with the balance of nature, but he realized that one must study not only ways that plants and animals interact but also their habitats. He knew that while balance had to exist, there occurred over time a succession of plants, beginning with a bare field and ending with a forest. In *Politia Naturae* (1760; *Governing Nature*, 1760) he discussed the checks on populations that prevent some species from becoming so numerous that they eliminate others. He noticed the competition among different species of plants in a meadow and concluded that feeding insects kept them in check. French naturalist Comte de Buffon developed a dynamic perspective on the balance of nature from his studies on rodents and their predators. Rodents can increase in numbers to plague proportions, but then predators and climate reduce their numbers. Buffon also suspected that humans had exterminated some large mammals, such as mammoths and mastodons.

Later another Frenchman, Jean-Baptiste Lamarck, published his book on evolution called *Philosophie zoologique* (1809; *Zoological Philosophy*, 1914), which cast doubt on extinction by arguing that fossils only represent early forms of living species: Mammoths and mastodons evolved into African and Indian elephants. In developing this idea, he minimized the importance of competition in nature. An English opponent, the geologist Charles Lyell, argued in 1833 that species do become extinct, primarily because of competition among species. Charles Darwin was inspired by his own investigations during a long voyage around the world and by his reading of the works of Linnaeus and Lyell. Darwin's revolutionary book, *On the Origin of Species by Means of Natural Selection* (1859), argued an intermediate position between Lamarck and Lyell: Species do evolve into different species, but in the process, some species do indeed become extinct.

Darwin's theory of evolution might have brought an end to the balance-of-nature concept, but it did not. Instead, American zoologist Stephen A. Forbes developed an evolutionary concept of the balance of nature in his essay, "The Lake as a Microcosm" (1887). Although the reproductive rate of aquatic species is enormous and the struggle for existence among them is severe, "the little community secluded here is as prosperous as if its state were one of profound and perpetual peace." He emphasized the stabilizing effects of natural selection.

The Science of Ecology

The science of ecology became formally organized between the 1890's and the 1910's. One of its important organizing concepts was that of "biotic communities." An American plant ecologist, Frederic E. Clements, wrote a large monograph titled *Plant Succession* (1916), in which he drew a morphological and developmental analogy between organisms and plant communities. Both the individual and the community have a life history during which each changes its anatomy and physiology. This superorganismic concept was an extreme version of the balance of nature that seemed plausible as long as one believed that a biotic community was a real entity rather than a convenient approximation of what one sees in a pond, a meadow, or a forest. However, the studies of Henry A. Gleason in 1917 and later indicated that plant species merely compete with one another in similar environments; he concluded that Clements's superorganism was poetry, not science.

While the balance-of-nature concept was giving way to ecological hypotheses and theories, Rachel Carson decided that she could not argue her case in *Silent Spring* (1962) without it. She admitted, "The balance of nature is not a *status quo*; it is fluid, ever shifting, in a constant state of adjustment." Nevertheless, to her the concept represented a healthy environment, which humans could upset. Her usage of the phrase has persisted within the environmental movement.

In 1972 English medical chemist James E. Lovelock developed a new balance-of-nature idea, which he calls Gaia, named for a Greek earth goddess. His reasoning owed virtually nothing to previous balance-of-nature notions that focused upon the interactions of plants and animals. His concept emphasized the chemical cycles that flow from the earth to the waters, atmosphere, and living organisms. He soon had the assistance of a zoologist named Lynn Margulis. Their studies convinced them that biogeochemical cycles are not random, but exhibit homeostasis, just as some animals exhibit homeostasis in body heat and blood concentrations of various substances. They believe that living beings, rather than inanimate forces, mainly control the earth's environment. In 1988 three scientific organizations sponsored a conference of 150 scientists from all over the world to evaluate their ideas. Although science more or less understands how homeostasis works when a brain within an animal controls it, no one has succeeded in satisfactorily explaining how homeostasis can work in a world "system" that lacks a brain. The Gaia hypothesis is as untestable as were earlier balance-of-nature concepts.

Frank N. Egerton

See also: Animal-plant interactions; Biomass related to energy; Biosphere concept; Deep ecology; Ecology: definition; Ecosystems: definition and history; Evolution: history; Food chains and webs; Geochemical cycles; Hydrologic cycle; Nutrient cycles; Trophic levels and ecological niches.

Sources for Further Study

Arthur, Wallace. *The Green Machine: Ecology and the Balance of Nature.* Cambridge, Mass.: Blackwell, 1990.

Egerton, Frank N. "Changing Concepts of the Balance of Nature." *Quarterly Review of Biology* (June, 1973).

Kirchner, James W. "The Gaia Hypotheses: Are They Testable? Are They Useful?" In *Scientists on Gaia*, edited by Stephen H. Schneider and Penelope J. Boston. Cambridge, Mass.: MIT Press, 1991.

Milne, Lorus Johnson. *The Balance of Nature.* New York: Knopf, 1960.

BIODIVERSITY

Types of ecology: Community ecology; Ecosystem ecology; Global ecology; Population ecology; Restoration and conservation ecology

The Wildlife Society has defined biodiversity as "the richness, abundance, and variability of plant and animal species and communities and the ecological process that link them with one another and with soil, air, and water."

Many kinds of specialists—including organismic biologists, population and evolutionary biologists, geneticists, and ecologists—investigate biological processes that are encompassed by the concept of biodiversity. Conservation biologists are concerned with the totality of biodiversity, including the process of speciation that forms new species, the measurement of biodiversity, and factors involved in the extinction process. However, the primary thrust of their efforts is the development of strategies to preserve biodiversity. The biodiversity paradigm connects classical taxonomic and morphological studies of organisms with modern techniques employed by those working at the molecular level.

It is generally accepted that biodiversity can be approached at three levels of organization, commonly identified as species diversity, ecosystem diversity, and genetic diversity. Some also recognize biological phenomena diversity.

Species Diversity

No one knows how many species inhabit the earth. Estimates range from five million to several times that number. Each species consists of individuals that are somewhat similar and capable of interbreeding with other members of their species but are not usually able to interbreed with individuals of other species. The species that occupy a particular ecosystem are a subset of the species as a whole. Ecosystems are generally considered to be local units of nature; ponds, forests, and prairies are common examples.

Conservation biologists measure the species diversity of a given ecosystem by first conducting a careful, quantitative inventory. From such data, scientists may determine the "richness" of the ecosystem, which is simply a reflection of the number of species present. Thus, an island with three hundred species would be 50 percent richer than another with only two hundred species. Some ecosystems, especially tropical rain forests and coral reefs, are much richer than others. Among the least rich are tundra regions and deserts.

A second aspect of species diversity is "evenness," defined as the degree to which each of the various elements are present in similar percentages of the total species. As an example, consider two forests, each of which has a total of twenty species of trees. Suppose that the first forest has a few tree species represented by rather high percentages and the remainder by low percentages. The second forest, with its species more evenly distributed, would rate higher on a evenness scale.

Species diversity, therefore, is a value that combines measures of both species richness and species evenness. Values obtained from a diversity index are used in comparing species diversity among ecosystems of both the same and different types. They also have implications for the preservation of ecosystems; other things being equal, it would be preferable to preserve ecosystems with a high diversity index, thus protecting a larger number of species.

Species Diversity and Geographical Areas

Considerable effort has been expended to predict species diversity as determined by the nature of the area involved. For example, island biogeography theory suggests that islands that are larger, nearer to other islands or continents, and have a more heterogeneous landscape would be expected to have a higher species diversity than those possessing alternate traits. Such predictions apply not only to literal islands but also to other discontinuous ecosystems. Examples would be the ecosystems of isolated mountaintops in alpine tundra or those of ponds several miles apart.

The application of island biogeography theory to designing nature preserves was proposed by Jared Diamond in 1975. His suggestion began the "single larger or several smaller," or SLOSS, area controversy. Although island biogeography theory would, in many instances, suggest selecting one large area for a nature preserve, it is often the case that several smaller areas, if carefully selected, could preserve more species.

The species diversity of a particular ecosystem is subject to change over time. Pollution, deforestation, and other types of habitat degradation invariably reduce diversity. Conversely, during the extended process of ecological succession that follows disturbances, species diversity typically increases until a permanent, climax ecosystem with a large index of diversity results. It is generally assumed by ecologists that more diverse ecosystems are more stable than are those with less diversity. Certainly, the more species present, the greater the opportunity for various interactions, both with other species and with the environment. Examples of interspecific reactions include mutualism, predation, and parasitism. Such interactions apparently help to integrate a community into a whole, thus increasing its stability.

Ecosystem Diversity

Ecology can be defined as the study of ecosystems. From a conservation standpoint, ecosystems are important because they sustain their particular assemblage of living species. Conservation biologists also consider ecosystems to have an intrinsic value beyond the species they harbor. Therefore, it would be ideal if representative global ecosystems could be preserved.

However, this is far from realization. Just deciding where to draw the line between interfacing ecosystems can be a problem. For example, the water level of a stream running through a forest is subject to seasonal fluctuation, causing a transitional zone characterized by the biota from both adjoining ecosystems. Such ubiquitous zones negate the view that ecosystems are discrete units with easily recognized boundaries.

The protection of diverse ecosystems is of utmost importance to the maintenance of biodiversity. However, ecosystems throughout the world are threatened by global warming, air and water pollution, acid deposition, ozone depletion, and other destructive forces. At the local level, deforestation, thermal pollution, urbanization, and poor agricultural practices are among the problems affecting ecosystems and therefore reducing biodiversity. Both global and local environmental problems are amplified by rapidly increasing world population pressures.

In the process of determining which ecosystems are most in need of protection, it has become apparent to many scientists that a system for naming and classifying ecosystems is highly desirable, if not imperative. Efforts are being made to establish a system similar to the hierarchical system applied to species that was developed by Swedish botanist Carl Linnaeus during the eighteenth century. However, a classification system for ecosystems is far from complete. Freshwater, marine, and terrestrial ecosystems are recognized as main categories, with each further divided into particular types. Though tentative, this has made possible the identification and preservation of a wide range of representative, threatened ecosystems.

In 1995 conservation biologist Reed F. Noss of Oregon State University and his colleagues identified more than 126 types of ecosystems in the United States that are threatened or critically endangered. The following list illustrates their diversity: southern Appalachian spruce-fir forests; eastern grasslands, savannas, and barrens; California native grasslands; Hawaiian dry forests; caves and Karst systems; old-growth forests of the Pacific Northwest; and southern forested wetlands.

Not all ecosystems can be saved. Establishing priorities involves many considerations, some of which are economic and political. Ideally, choices would be made on merit: rarity, size, number of endangered species they include, and other objective, scientific criteria.

Genetic Diversity

Most of the variation among individuals of the same species is caused by the different genotypes (combinations of genes) that they possess. Such genetic diversity is readily apparent in cultivated or domesticated species such as cats, dogs, and corn, but also exists, though usually to a lesser degree, in wild species. Genetic diversity can be measured only by exacting molecular laboratory procedures. The tests detect the amount of variation in the deoxyribonucleic acid (DNA) or isoenzymes (chemically distinct enzymes) possessed by various individuals of the species in question.

A significant degree of genetic diversity within a population or species confers a great advantage. This diversity is the raw material that allows evolutionary processes to occur. When a local population becomes too small, it is subject to a serious decline in vigor from increased inbreeding. This leads, in turn, to a downward, self-perpetuating spiral in genetic diversity and further reduction in population size. Extinction may be imminent. In the grand scheme of nature, this is a catastrophic event; never again will that particular genome (set of genes) exist anywhere on the earth. Extinction is the process by which global biodiversity is reduced.

Biological Phenomena Diversity

Biological phenomena diversity refers to the numerous unique biological events that occur in natural areas throughout the world. Examples include the congregation of thousands of monarch butterflies on tree limbs at Point Pelee in Ontario, Canada, as they await favorable conditions before continuing their migration, or the return of hundreds of loggerhead sea turtles each April to Padre Island in the Gulf of Mexico in order to lay their eggs.

Conservation Biology

Although biologists have been concerned with protecting plant and animal species for decades, only recently has conservation biology emerged as an identifiable discipline. Conceived in a perceived crisis of biological extinctions, conservation biology differs from related disciplines, such as ecology, because of its advocative nature and its insistence on maintaining biodiversity as intrinsically good. Conservation biology is a value-laden science, and some critics consider it akin to a religion with an accepted dogma.

The prospect of preserving global ecosystems and the life processes they make possible, all necessary for maintaining global diversity, is not promising. Western culture does not give environmental concerns a high priority. For those who do, there is more often a concern over issues relating to immediate health effects than concern over the loss of biodiversity.

Only when education in basic biology and ecology at all levels is extended to include an awareness of the importance of biodiversity will there develop the necessary impetus to save ecosystems and all their inhabitants, including humans.

Thomas E. Hemmerly

See also: Animal-plant interactions; Biogeography; Biomes: determinants; Biomes: types; Biosphere concept; Communities: ecosystem interactions; Communities: structure; Conservation biology; Ecosystems: definition and history; Ecosystems: studies; Endangered animal species; Endangered plant species; Extinctions and evolutionary explosions; Gene flow; Genetic diversity; Habitats and biomes; Restoration ecology; Species loss; Zoos.

Sources for Further Study

Baskin, Y. "Ecologists Dare to Ask: How Much Does Diversity Matter?" *Science* 264 (April 8, 1994): 202-203.

Burton, John, ed. *The Atlas of Endangered Species*. 2d ed. New York: Macmillan, 1999.

Cracraft, Joel, and Francesca T. Grifo, eds. *The Living Planet in Crisis: Biodiversity Science and Policy*. Foreword by Edward O. Wilson. New York: Columbia University Press, 1999.

DiSilvestro, Roger L. *The Endangered Kingdom: The Struggle to Save America's Wildlife*. New York: John Wiley & Sons, 1989.

Ehrlich, Paul, and Anne Ehrlich. *Extinction: The Causes and Consequences of the Disappearance of Species*. New York: Random House, 1981.

National Research Council. *Science and the Endangered Species Act*. Washington, D.C.: National Academy Press, 1995.

New, T. R. *An Introduction to Invertebrate Conservation*. New York: Oxford University Press, 1995.

Raven, Peter H., and George B. Johnson. *Biology*. 4th ed. Boston: McGraw-Hill, 1996.

Reaka-Kudla, Marjorie L., Don E. Wilson, and Edward O. Wilson, eds. *Biodiversity II: Understanding and Protecting Our Biological Resources*. Washington, D.C.: Joseph Henry Press, 1997.

Ricketts, Taylor H., et al. *Terrestrial Ecoregions of North America: A Conservation Assessment*. Washington, D.C.: Island Press, 1999.

Stefoff, Rebecca. *Extinction*. New York: Chelsea House, 1992.

Tudge, Colin. *The Variety of Life: The Meaning of Biodiversity*. New York: Oxford University Press, 2000.

Wilson, Edward O., ed. *Biodiversity*. Washington, D.C.: National Academy Press, 1988.

BIOGEOGRAPHY

Types of ecology: Community ecology; Population ecology

To understand the underlying geography of plant and animal distributions, biogeographers integrate considerations of historical and current events and conditions.

Biogeography is the science that seeks to understand spatial patterns of biodiversity. By examining past and present distributions, biogeographers attempt to explain why certain groups of organisms occur where they do, what enables them to live there, -and what factors prevent them from moving or living elsewhere. To address these issues, biogeographic investigations must consider the effects of climate, topography, and other kinds of organisms, as well as historical events such as tectonic effects and glaciation. Such historical considerations are frequently important and conditions are constantly changing, so a group's closest relatives and their distribution must be taken into account in order to accommodate evolutionary events.

Because biogeography is such a broad discipline, individuals can rarely address an entire spectrum of relevant questions. Consequently, most biogeographers become specialists. For example, phytogeographers study plant distributions and zoogeographers examine those of animals. Historical biogeographers reconstruct origins, dispersal events, and extinctions through time. Ecological biogeographers concentrate on interactions between organisms and their environments to explain distribution patterns, and paleoecologists try to bridge the gap between historical and current conditions. Also related to the breadth of the discipline is the necessity for biogeographers to be conversant in one or more related fields. A broad knowledge of biology is obviously fundamental, as is an understanding of physical geography, but geology, paleontology, and climatology, among others, may be equally important.

Trends in Biogeography

Several consistent trends have emerged from biogeographic studies. Communities in isolated regions (especially large islands that have not been in contact with continents for long periods of time) tend to be unlike those found anywhere else. More types of organisms are found in tropical than in temperate or arctic regions. Fewer types of organisms are found on oceanic islands, although the organisms that are present may be found in phenomenal densities. On the other hand, some inconsistencies are striking. For example, some groups of related organisms are found throughout the world whereas other groups have very restricted ranges.

Generally speaking, groups of organisms that are broadly distributed are either very old (their ancestors were in place before the continents drifted apart) or very mobile. Mobile organisms may be able to disperse actively, that is, on their own power (some species of birds and large marine mammals are good examples), whereas others are dispersed passively. Seeds of many plants, microscopic planktonic organisms, and insects are often transported by wind, currents, or other organisms. Limits to the dispersal of organisms are myriad. Size limits the ability to be blown by the wind; for example, dandelion thistles are more readily dispersed by even mild breezes than walnuts. If water is a factor, buoyancy is critical. Coconuts are found on tropical shorelines around the world in large part because they do not sink, nor are their hard shells easily penetrated by salt water. Physical barriers often limit dispersal. Deserts effectively block organisms that require moisture, land halts movement of aquatic forms, and mountains prevent the passage of plants and animals that cannot tolerate the conditions associated with high elevations. Some barriers even appear to be "psychological"; birds that could easily fly cross a stream or lake, for example, often will not.

Island Biogeography

In 1967, Robert H. MacArthur and Edward O. Wilson published a classic volume titled *The Theory of Island Biogeography*. Although islands have figured prominently in modern biology (Charles Darwin and Alfred Russel Wallace both relied heavily on evidence from islands when formulating their theories of natural selection), only during the last third of the twentieth century was island biogeography recognized as a distinct discipline. Many of the principles that form the foundation of biogeography emanated from studies of islands. Among these are relationships between biodiversity and island size, ecological heterogeneity, and proximity to continents; between isolation and endemism (species evolving in a given area and found nowhere else in the world); and between island size and location and rates of immigration, colonization, and extinction. In addition, island biogeographers frequently have been at the center of debates arguing the relevance of dispersal versus vicariance. Central to these disputes is whether disjunct distributions of organisms are attributable to movement over barriers (dispersal) or to the creation of a barrier that separated a previously contiguous range (vicariance). Although both have undoubtedly played important roles, the debate rages over which was primarily respon-

sible for the current distributions of many faunas, both on islands surrounded by water or on terrestrial "islands" surrounded by other inhospitable habitats.

Robert Powell

See also: Adaptations and their mechanisms; Adaptive radiation; Biodiversity; Biological invasions; Clines, hybrid zones, and introgression;
Communities: ecosystem interactions; Communities: structure; Evolution:
definition and theories; Food chains and webs; Gene flow; Genetic diversity; Genetic drift; Isolating mechanisms; Landscape ecology; Migration;
Nonrandom mating, genetic drift, and mutation; Population genetics; Speciation; Species loss.

Sources for Further Study

Cox, C. Barry, and Peter D. Moore. *Biogeography: An Ecological and Evolutionary Approach.* 6th ed. Malden, Mass.: Blackwell Science, 2000.

Groombridge, Brian, and M. K. Jenkins, eds. *Global Biodiversity: Earth's Living Resources in the Twenty-first Century.* Cambridge, England: World
Conservation Monitoring Centre, 2000.

Myers, Alan A., and Paul S. Giller, eds. *Analytical Biogeography: An Integrated Approach to the Study of Animal and Plant Distributions.* New York:
Chapman & Hall, 1988.

Whittaker, Robert J. *Island Biogeography: Ecology, Evolution, and Conservation.* New York: Oxford University Press, 1998.

Wilson, Edward O., ed. *Biodiversity.* Washington, D.C.: National Academic
Press, 1988.

BIOLOGICAL INVASIONS

Types of ecology: Community ecology; Ecosystem ecology;
Ecotoxicology

Biological invasions are the entry of a type of organism into an ecosystem outside its historic range. In a biological invasion, the "invading" organism may be an infectious virus, a bacterium, a plant, an insect, or an animal.

Species introduced to an area from somewhere else are referred to as alien or exotic species or as invaders. Because an exotic species is not native to the new area, it is often unsuccessful in establishing a viable population and disappears. The fossil record, as well as historical documentation, indicates that this is the fate of many species in new environments as they move from their native habitats. Occasionally, however, an invading species finds the new environment to its liking. In this case, the invader may become so successful in exploiting its new habitat that it can completely alter the ecological balance of an ecosystem, decreasing biodiversity and altering the local biological hierarchy. Because of this ability to alter ecosystems, exotic invaders are considered major agents in driving native species to extinction.

Biological invasions by notorious species constitute a significant component of earth's history. In general, large-scale climatic changes and geological crises are at the origin of massive exchanges of flora and fauna. On a geologic time scale, migrations of invading species from one continent to another are true evolutionary processes, just as speciation and extinction are. On a smaller scale, physical barriers such as oceans, mountains, and deserts can be overcome by many organisms as their populations expand. Organisms can be carried by water in rivers or ocean currents, transported by wind, or carried by other species as they migrate seasonally or to escape environmental pressures. Humans have transplanted plants since the beginning of plant cultivation in pre-Columbian times. The geological and historical records of the earth suggest that biological invasions contribute substantially to an increase in the rate of extinction within ecosystems.

Invasive Plants

In modern times, most people are not aware of the distinction between native plants and exotic species growing in their region. Recent increases in intercontinental invasion rates by exotic species, brought about primarily by human activity, create important ecological problems for the recipient lands. Invasive plants in North America include eucalyptus trees, morning glory, and pampas grass.

It would seem logical to assume that invading species might add to the biodiversity of a region, but many invaders have the opposite effect. In all ecosystems the new species are often opportunistic, driving out native species by competing with them for resources. For example, *Pueraria lobata*, or kudzu, is a vine native to Japan. Introduced in the United States at the 1876 Philadelphia Exposition, kudzu was planted to control erosion on hillsides and for livestock forage. By the end of the twentieth century, it could be found from Connecticut to Missouri, extending south to Texas and Florida. Kudzu covers everything in its path and grows as much as 1 foot (0.3 meter) per day. Similarly, English ivy (*Hedera helix*), a native of Eurasia, is considered a serious problem in West Coast states. It forms "ivy deserts" in forests and crowds out native trees and shrubs that make up essential wildlife habitat.

The invasion of an ecosystem by an exotic species can effectively alter ecosystem processes. An invading species does not simply consume or compete with native species but can actually change the rules of existence within the ecosystem by altering processes such as primary productivity, decomposition, hydrology, geomorphology, nutrient cycling, and natural disturbance regimes.

Invasive Insects and Microorganisms

The invasion of native forests alone by nonnative insects and microorganisms has been devastating on many continents. The white pine blister rust and the balsam woolly adelgid have invaded both commercial and preserved forest lands in North America. Both exotics were brought to North America in the late 1800's on nursery stock from Europe. The balsam woolly adelgid attacks fir trees and causes their death within two to seven years from chemical damage and by feeding on the tree's vascular tissue. The adelgid has killed nearly every adult cone-bearing fir tree in the southern Appalachian Mountains. The white pine blister rust attacks five-needle pines; in the western United States fewer than 10 pine trees in 100,000 are resistant. Because white pine seeds are an essential food source for bears and other animals, the loss of the trees is having severe consequences across the food chain.

Since the 1800's the deciduous trees of eastern North America have been attacked numerous times by waves of invading exotic species and diseases. One of the most notable invaders is the gypsy moth, which consumes a variety of tree species. Other invaders have virtually eliminated the once-dominant American chestnut and the American elm. Tree species that continue to decline because of new invaders include the American beech, mountain ash, white birch, butternut, sugar maple, flowering dogwood, and eastern hemlock. It is widely accepted that the invasion of exotic species is the single greatest threat to the diversity of deciduous forests in North America.

Effects on Humans and Humans as Invaders

Some introduced exotic species are beneficial to humanity. It would be impossible to support the present world human population entirely on species native to their regions. Humans, the ultimate biological invaders, have been responsible for the extinction of many species and will continue to be in the future. At the beginning of the twenty-first century, the United States was spending $4 billion annually to eradicate invasive plant species, a figure that does not take into account loss of biodiversity or wildlife habitat.

Randall L. Milstein, updated by Elizabeth Slocum

See also: Biodiversity; Biogeography; Biomagnification; Communities: ecosystem interactions; Deforestation; Endangered animal species; Endangered plant species; Eutrophication; Invasive plants; Succession.

Sources for Further Study

Bright, Chris. *Life out of Bounds: Bioinvasion in a Borderless World*. New York: Norton, 1998.

Cox, George W. *Alien Species in North America and Hawaii: Impacts on Natural Ecosystems*. Washington, D.C.: Island Press, 1999.

Crosby, Alfred. *Ecological Imperialism: The Biological Expansion of Europe, 900-1900*. New York: Cambridge University Press, 1994.

Drake, J. A., et al. *Biological Invasions: A Global Perspective*. New York: Wiley, 1989.

Hengeveld, Rob. *Dynamics of Biological Invasions*. New York: Chapman and Hall, 1989.

BIOLUMINESCENCE

Types of ecology: Chemical ecology; Physiological ecology

Bioluminescence is visible light emitted from living organisms. Half of the orders of animals include luminescent species. These organisms are widespread, occurring in marine, terrestrial, and freshwater habitats. Bioluminescence is used for defense, predation, and communication.

Bioluminescence is the visible light produced by luminous animals, plants, fungi, protists, and bacteria that results from a biochemical reaction with oxygen. Unlike incandescent light from electric light bulbs, bioluminescence is produced without accompanying heat. Bioluminescence was first described in 500 B.C.E., but the chemical mechanism of bioluminescence was not elucidated until the beginning of the twentieth century. The ability to luminesce appears to have arisen as many as thirty times during evolution. The chemical systems used by luminescent organisms are similar but not exactly the same. Most organisms use a luciferin/luciferase system. The luciferin molecules are oxidized through catalysis by an oxidase enzyme (luciferase). The oxidized form of luciferin is in an excited electronic state that relaxes to the ground state through light emission.

Types of Bioluminescence

Animals may produce light in one of three ways. The bioluminescence may be intracellular: Chemical reactions within specialized cells result in the emission of visible light. These specialized cells are often found within photophores. These light-producing organs may be arranged in symmetrical rows along the animal's body, in a single unit overhanging the mouth, or in patches under the eyes, and are connected to the nervous system. Alternatively, the bioluminescence may be extracellular: The animals secrete chemicals that react in their surroundings to produce light. The third option involves a symbiotic relationship between an animal and bioluminescent bacteria. Several species of fish and squid harbor bioluminescent bacteria in specialized light organs. The symbiotic relationship is specific: Each type of fish or squid associates with a certain type of bacteria. The bacteria-filled organ is continuously luminous. The animal regulates the emission of light either by melanophores scattered over the surface of the organ or by a black membrane that may be mechanically drawn over the organ.

A female Photuris *firefly devours a male firefly that she has lured by imitating the bioluminescent flash pattern of his species' mating signal.* (AP/Wide World Photos)

Although bioluminescence is widespread in animals, its occurrence is sporadic. Most of the bioluminescent animal species are invertebrates. Among the vertebrates, only fish exhibit bioluminescence; there are no known luminous amphibians, reptiles, birds, or mammals. Although bioluminescence is found in terrestrial and freshwater environments, the majority of luminous organisms are marine. Scientists estimate that 96 percent of all creatures in the deep sea possess some form of light generation.

Functions of Bioluminescence
There appear to be three main uses of bioluminescence: finding or attracting prey, defense against predators, and communication. Although visible light penetrates into the ocean to one thousand meters at most, most fish living below one thousand meters possess eyes or other photoreceptors. Many deep-sea fish have dangling luminous light organs to attract prey. Terrestrial flies have also exploited bioluminescence for predation. The glow of glowworms (fly larvae) living in caves serves to attract insect prey, which get snared in the glowworms' sticky mucous threads. Fungus gnats (carnivorous flies) attract small arthropods through light emission and capture the prey in webs of mucus and silk.

Bioluminescence can serve as a decoy or camouflage. For example, jel-lyfish, such as comb jellies, produce bright flashes to startle a predator, while siphonophores can release thousands of glowing particles into the water as a mimic of small plankton to confuse the predator. Other jellyfish produce a glowing slime that can stick to a potential predator and make it vulnerable to its predators. Many squid and some fish possess photo-phores that project light downward, regardless of the orientation of the an-imal's body. The emitted light matches that of ambient light when viewed from below, rendering the squid invisible to both predators and prey.

The best-known example of bioluminescence used as communication is in fireflies, the common name for any of a large family of luminescent bee-tles. Luminescent glands are located on the undersides of the rear abdomi-nal segments. There is an exchange of flashes between males and females. Females respond to the flashes of flying males, with the result that the male eventually approaches the female for the purpose of mating. To avoid con-fusion between members of different types of fireflies, the signals of each species are coded in a unique temporal sequence of flashing, the timing of which is controlled by the abundant nerves in the insect's light-making or-gan. Females of one genus of fireflies (*Photuris*) take advantage of this by mimicking the response of females of another genus (*Photinus*) to lure *Photinus* males that the *Photuris* females then kill and eat.

Some marine animals also utilize bioluminescence for communication. For example, lantern fish and hatchetfish (the most abundant vertebrate on earth) possess distinct arrangements of light organs on their bodies that can serve as species- and sex-recognition patterns; female fire worms re-lease luminescent chemicals into the water during mating, beginning one hour after sundown on the three nights following the full moon; and deep-sea dragonfish emit red light that is undetectable except by other drag-onfish.

Lisa M. Sardinia

See also: Adaptations and their mechanisms; Camouflage; Communica-tion; Defense mechanisms; Marine biomes; Metabolites; Mimicry; Phero-mones; Poisonous animals; Poisonous plants; Pollination; Predation; Tro-pisms.

Sources for Further Study
Presnall, Judith Janda. *Animals That Glow*. Salem, Mass.: Franklin Watts, 1993.
Robison, Bruce H. "Light in the Ocean's Midwaters: Bioluminescent Ma-rine Animals." *Scientific American* 273 (July, 1995): 60-64.

Silverstein, Alvin, and Virginia Silverstein. *Nature's Living Lights: Fireflies and Other Bioluminescent Creatures*. Boston: Little, Brown, 1988.

Toner, Mike. "When Squid Shine and Mushrooms Glow, Fish Twinkle, and Worms Turn into Stars." *International Wildlife* 24 (May/June, 1994): 30-37.

Tweit, Susan J. "Dance of the Fireflies." *Audubon* 101 (July/August, 1999): 26-30.

BIOMAGNIFICATION

Type of ecology: Ecotoxicology

Biomagnification is the accumulation of toxic contaminants in the environment as they move up through the food chain. As members of each level of the food chain are progressively eaten by those organisms found in higher levels of the chain, the concentration of toxic chemicals within the tissues of the higher organisms increases.

Not all chemicals, potentially toxic or not, are equally likely to undergo biomagnification. However, molecules susceptible to biomagnification have certain characteristics in common. They are resistant to natural microbial degradation and therefore persist in the environment. They are also lipophilic, tending to accumulate in the fatty tissue of organisms. In addition, the chemical must be biologically active in order to have an effect on the organism in which it is found. Such compounds are likely to be absorbed from food or water in the environment and stored within the membranes or fatty tissues.

Pesticides

The process usually begins with the spraying of pesticides for the purpose of controlling insect populations. Industrial contamination, including the release of heavy metals, can be an additional cause of such pollution. Biomagnification results when these chemicals contaminate the water supply and are absorbed into the lipid membranes of microbial organisms. This process, often referred to as bioaccumulation, results in the initial concentration of the chemical in an organism in a form that is not naturally excreted with normal waste material. Levels of the chemical may reach anywhere from one to three times that found in the surrounding environment. Since the nature of the chemical is such that it is neither degraded nor excreted, it remains within the organism.

As organisms on the bottom of the food chain are eaten and digested by members of the next level in the chain, the concentration of the accumulated material significantly increases; at each subsequent level, the concentration may reach one order of magnitude (a tenfold increase) higher. Consequently, the levels of the pollutant at the top of the environmental food chain for the ecosystem in question—such as fish, carnivorous birds, or humans—may be as much as one million times more concentrated than the original, presumably safe, levels in the environment.

DDT

For example, studies of dichloro-diphenyl-trichloroethane (DDT) levels in the 1960's found that zooplankton at the bottom of the food chain had accumulated nearly one thousand times the level of the pollutant in the surrounding water. Ingestion of the plankton by fish resulted in concentration by another factor of several hundred. By the time the fish were eaten by predatory birds, the level of DDT was concentrated by a factor of more than two hundred thousand. DDT is characteristic of most pollutants subject to potential biomagnification. It is relatively stable in the environment, persisting for decades. It is soluble in lipids and readily incorporated into the membranes of organisms.

Since pesticides are, by their nature, biologically active compounds, which reflects their ability to control insects, they are of particular concern if subject to biomagnification. DDT remains the classical example of how bioaccumulation and biomagnification may have an effect on the environment. Initially introduced as a pesticide for control of insects and insect-borne disease, DDT was not thought to be particularly toxic. However, biomagnification of the chemical was found to result in the deaths of birds and other wildlife. In addition, DDT contamination was found to result in formation of thin egg shells that greatly reduced the birthrate among birds. Before the use of DDT was banned in the 1960's, the population levels of predatory birds such as eagles and falcons had fallen to a fraction of the levels found prior to use of the insecticide. Though it was unclear whether there was any direct effect on the human population in the United States, the discovery of elevated levels of DDT in human tissue contributed to the decision to ban the use of the chemical.

Other Toxic Pesticides

While DDT represents the classic example of biomagnification of a toxic chemical, it is by no means the only representative of potential environmental pollutants. Other pesticides with similar characteristics include pesticides such as aldrin, chlordane, parathion, and toxaphene. In addition, cyanide, polychlorinated biphenyls (PCBs), and heavy metals—such as selenium, mercury, copper, lead, and zinc—have also been found to concentrate within the food chain.

Some heavy metals are inherently toxic or may undergo microbial modification to increase their toxic potential. For example, mercury does not naturally accumulate in membranes and was therefore not originally viewed as a significant danger to the environment. However, some microorganisms are capable of adding a methyl group to the metal and producing methyl mercury, a highly toxic material that does accumulate in fatty tissue and membranes.

Prevention

Several procedures have been adopted since the 1960's to prevent the biomagnification of toxic materials. In addition to outright bans, pesticides are often modified to prevent their accumulation in the environment. Most synthetic pesticides contain chemical structures that are easily degraded by microorganisms found in the environment. Ideally, the pesticide should survive no longer than a single growing season before being rendered harmless by the environmental flora. Often such chemical changes require only simple modification of the basic structure.

Richard Adler

See also: Acid deposition; Biological invasions; Biopesticides; Endangered animal species; Food chains and webs; Genetically modified foods; Integrated pest management; Ocean pollution and oil spills; Pesticides; Phytoplankton; Pollution effects; Waste management.

Sources for Further Study

Atlas, Ronald, and Richard Bartha. *Microbial Ecology: Fundamentals and Applications.* Redwood City, Calif.: Benjamin/Cummings, 1993.

Carson, Rachel. *Silent Spring.* Boston: Houghton Mifflin, 1962.

Colborn, Theo, Dianne Dumanoski, and John Peterson Myers. *Our Stolen Future: Are We Threatening Our Fertility, Intelligence, and Survival? A Scientific Detective Story.* New York: Dutton, 1996.

BIOMASS RELATED TO ENERGY

Type of ecology: Ecoenergetics

The relationship between the accumulation of living matter resulting from the primary production of plants or the secondary production of animals (biomass) and the energy potentially available to other organisms in an ecosystem forms the basis of the study of biomass related to energy.

Biomass is the amount of organic matter, such as animal and plant tissue, found at a particular time and place. The rate of accumulation of biomass is termed productivity. Primary production is the rate at which plants produce new organic matter through photosynthesis. Secondary production is the rate at which animals produce their organic matter by feeding on other organisms. Biomass is an instantaneous measure of the amount of organic matter, while primary and secondary production give measures of the rates at which biomass increases. Plant and animal biomass consists mostly of carbon-rich molecules, such as sugars, starches, proteins, and lipids, and other substances, such as minerals, bone, and shell. The carbon-rich organic molecules are not only the building blocks of life but also the energy-rich molecules used by organisms to fuel their activities.

Primary Production: Photosynthesis

Ultimately, all energy used by organisms to produce the building blocks of life and to drive life processes originated as solar energy captured by plants. Only a small fraction, less than 2 percent, of the total solar light energy received by a plant is absorbed and transformed by photosynthesis into energy-containing organic molecules. The rest of the sun's energy passes out of the plant as heat. The rate at which plants capture light energy and transform it into chemical energy is called primary production. Because plants do not rely on other organisms to provide their energy needs, they are referred to as primary producers, or autotrophs (meaning "self-feeding"). In addition to light energy, plants must absorb water, carbon dioxide gas, and simple nutrients, such as nitrate and phosphate, to produce various organic molecules during photosynthesis. Oxygen gas is also produced.

Sugars are the first energy-containing organic molecules produced in photosynthesis, and they can be changed to other, more complex, molecules, such as starches, proteins, and fats. The energy in the sugar molecules can be used immediately by the plants to maintain their own respiration needs, stored as starches and fats, or can be converted to new plant tissue. It is the stored organic matter plus new tissue that contributes to the growth of plants and to biomass.

Because the energy-containing products of photosynthesis can be used either immediately in respiration or in the formation of new plant biomass, two types of primary production can be distinguished. Gross production refers to the total amount of energy produced by photosynthesis. It includes both the energy used by the plant for respiration and the energy that goes into new biomass. Net production refers only to the amount of energy that accumulates as new biomass. It is only the energy in net production that is potentially available to animal consumers as food.

The rate of primary production varies directly with the rate of photosynthesis; therefore, factors in the environment that affect the rate of photosynthesis affect the rate of primary production. These factors most often

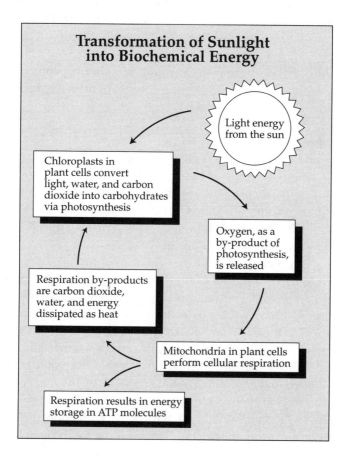

Transformation of Sunlight into Biochemical Energy

Light energy from the sun

Chloroplasts in plant cells convert light, water, and carbon dioxide into carbohydrates via photosynthesis

Oxygen, as a by-product of photosynthesis, is released

Respiration by-products are carbon dioxide, water, and energy dissipated as heat

Mitochondria in plant cells perform cellular respiration

Respiration results in energy storage in ATP molecules

include light intensity, temperature, nutrient concentrations, and moisture conditions. Each species of plant has a specific combination of these factors that promotes maximum rates of primary production. If one or more of these factors is in excess or is in short supply, then the rate of primary production is slowed.

On land, the rate of primary production by plants is determined largely by light, temperature, and rainfall. The favorable combination of intense sunlight for twelve hours per day, warm temperatures throughout the year, and considerable rainfall make the tropical rain forests the most productive ecosystem on land. In contrast, Arctic tundra vegetation is exposed to reduced light intensity, very cold winters, and cool summers. Primary production there is very low. In deserts, the lack of water severely limits primary production even though light and temperature are otherwise favorable.

In aquatic habitats, rates of primary production by algae, such as phytoplankton, are determined by nutrient concentration and light intensity. As sunlight penetrates water, it is quickly absorbed by the water molecules and by small suspended particles. Thus, all primary production occurs near the surface, as long as nutrients are available. Although the waters of the open ocean are very clear, and sunlight can penetrate to great depths, the scarcity of nutrients reduces the rate of primary production to less than one-tenth that of coastal bays.

Secondary Production
The energy and material needs of some organisms are met by consuming the organic materials produced by others. These consumer organisms are called heterotrophs; there are two types. Those that obtain their food from other living organisms are called consumers and include all animals. Those that obtain their energy from dead organisms are called decomposers and include mostly the fungi and bacteria.

The energy available to each type of consumer becomes progressively less at each level of the food chain. Each consumer level uses most of its food energy, about 90 percent, to fuel its respiratory activities. In this energy-releasing process, most of the food energy is actually converted to heat and is lost to the environment. Only 10 percent or less of the original food energy is used to form new biomass. It is only this small amount of energy that is available for the next consumer level. The result is that food chains are limited in their number of links or levels by the reduced amount of energy available at each higher level.

Generally, the greater the amount of primary production, the larger the number of consumer organisms and the longer the food chain. Most food chains consist of three levels; rarely are there examples of up to five levels. It should be noted that the food chain concept is a simplified view of a more complex network of energy pathways, known as food webs, that occur in nature. Another outcome of the reduction in energy flow up the food chain is a progressive decrease in production and biomass. The most productive level, and the one with the greatest biomass, is therefore the primary producers, or plants.

Human Threats to Primary Production

The total natural primary production of the earth is limited, and human efforts to increase total world primary production much beyond its present levels may be futile. One reason for this is that much of the earth's surface lacks optimal conditions for plant growth. The open ocean, which covers about 71 percent of the earth's surface, has very little plant growth. On land, the Arctic, subarctic, and Antarctic regions are very unproductive most of the year. Human attempts to increase primary production in the form of food or fuel crops usually involve changing the characteristics of the land, converting forests into croplands, for example, and adding large quantities of nutrients and water. It has been estimated that humans are currently utilizing most of the easily workable croplands and that the development of additional lands for agriculture would require major changes to currently unworkable habitats, changes that would be expensive and demand much fuel energy.

The study of production processes is vitally important in understanding the ecology of natural ecosystems. Such information is necessary to manage and conserve habitats and their organisms in the face of human pressures. These processes provide insight into the general health of ecosystems. Pollutants, such as acid rain or industrial toxic wastes, are known to reduce the primary and secondary productivity of forests and lakes.

Throughout the world, humans are reducing the biomass of the world's primary producers through deforestation. This is particularly true in the tropics, where high population pressures have necessitated that land be cleared for agriculture and development. There is a worldwide demand for lumber. One obvious consequence is the dramatic reduction in the primary and secondary production of these areas. The clear-cutting (removal of all the trees) of tropical forests allows unprotected soils to wash away quickly during the heavy tropical rains. It will take hundreds of years for new soils to develop and for the forest to return—if it can return at all.

Deforestation is also harmful in that tropical forests form a major part of the world's life-support system. For millions of years these forests have buffered the earth's atmosphere by producing the oxygen gas needed by animals and by removing carbon dioxide and other toxic gases. The low level of carbon dioxide in the atmosphere is believed to have moderated the earth's temperature, counteracting the so-called greenhouse effect. It is therefore of great importance to understand and preserve these forests and other primary producers of the world.

Ray P. Gerber

See also: Balance of nature; Communities: structure; Deforestation; Ecology: definition; Ecosystems: studies; Food chains and webs; Geochemical cycles; Herbivores; Hydrologic cycle; Nutrient cycles; Omnivores; Phytoplankton; Predation; Rain forests and the atmosphere; Trophic levels and ecological niches.

Sources for Further Study

Brower, James E., and Jerrold H. Zar. *Field and Laboratory Methods for General Ecology.* 4th ed. Boston, Mass.: WCB McGraw-Hill, 1998.

Nybakken, James W. *Marine Biology: An Ecological Approach.* 5th ed. Menlo Park, Calif.: Benjamin Cummings, 2001.

Odum, Howard T. *Ecological and General Systems: An Introduction to Systems Ecology.* Rev. ed. Niwot: University Press of Colorado, 1994.

Pasztor, Janos, and Lars A. Kristoferson, eds. *Bioenergy and the Environment.* Boulder, Colo.: Westview Press, 1990.

Ricklefs, Robert E. *Ecology.* 4th ed. New York: W. H. Freeman, 1999.

Smith, Robert L. *Ecology and Field Biology.* 5th ed. Menlo Park, Calif.: Addison Wesley Longman, 1996.

BIOMES: DETERMINANTS

Types of ecology: Biomes; Ecosystem ecology; Global ecology;
Theoretical ecology

*Biomes are large-scale ecological communities of both plants and animals, deter-
mined primarily by geography and climate. Worldwide, there are six major types
of biomes on land: forest, grassland, woodland, shrubland, semidesert scrub, and
desert.*

One who travels latitudinally from the equator to the Arctic will cross
tropical forests, deserts, grasslands, temperate forests, coniferous forest, tundra, and ice fields. Those major types of natural vegetation at regional scales are called biomes. A biome occurs wherever a particular set of
climatic and edaphic (soil-related) conditions prevail with similar physiognomy. For example, prairies and other grasslands in the North American Middle West and West form a biome of temperate grasslands, where
moderately dry climate prevails. Tropical rain forests in the humid tropical
areas of South and Central America, Africa, and Southeast Asia create a
biome where rainfall is abundant and well-distributed through the year.

In general, biomes are delineated by both physiognomy and environment. There are six major physiognomic types on land: forest, grassland,
woodland, shrubland, semidesert scrub, and desert. Each of the six types
occurs in a wide range of environments. Therefore, more than one biome
may be defined within each physiognomic type according to major differences in climate. Tropical forests, temperate deciduous forest, and coniferous forests are, for example, separate biomes, although forests dominate
all of them. On the other hand, some biome types, such as the tundra, are
dominated by a range of physiognomic types and are in one prevailing environmental region.

Classification of Biomes

There are many ways to classify biomes. One system, which designates a
small number of broadly defined biomes, divides global vegetation into
nine major terrestrial biomes: tundra, taiga, temperate forest, temperate
rain forest, tropical rain forest, savanna, temperate grasslands, chaparral,
and desert. Other systems more narrowly define biomes, designating a
larger total number. In those cases, some of the broadly defined biomes are
divided into two or more biomes. For example, the biome called temperate
forest in a broad classification may be separated into temperature decidu-

Biomes and Their Features

Biome	Annual Mean Rainfall[1]	Climate and Temperature[2]
Desert	250 mm or less	Arid, with extremes of heat and cold
Grasslands	250-750 mm	Cold winters, warm summers; dry periods
Mediterranean scrub	Low to moderate	Cool winters, hot summers; latitudes 30° to 45°; includes chaparral, maquis
Rain forest (tropical)	2,500-4,500 mm	20-30°
Savanna, deciduous tropics	1,500-2,500 mm	Hot summers; 3-6 months dry; seasonal fires
Taiga (boreal forest)	1,000 mm	Cold, long winters; mild, short summers; seasonal fires
Tundra	Very low year-round	Very cold (3° or less); soil characerized by permafrost; Arctic tundra occurs in Arctic Circle; alpine tundra in other high elevations

[1] In millimeters
[2] Degrees Celsius

ous forest and temperate evergreen forest in a fine classification. The biome of desert in the broad classification may be broken into warm semidesert, cool semidesert, Arctic-alpine semidesert, Arctic-pine desert, and true desert in the fine classification.

Description of Biome Distributions

Naturalists, geographers, and ecologists have tried to correlate world major types of biomes to climatic patterns in both descriptive and quantitative approaches. For example, in northern North America, the tundra and boreal forests are two broad belts of vegetation that stretch from east to west. The distribution of the two biomes is primarily influenced by temperature. South of those two belts are biome types that are mostly controlled by precipitation and evaporation. From east to west in North America, available moisture decreases, influencing biome distribution. Humid regions along the East Coast support forest biomes, including temperate coniferous forests and temperate deciduous forest. West of the eastern forests is a biome type of grasslands, including tall-grass prairie and short-grass steppe. In this zone, there is less precipitation than evaporation. The ratio of precipitation to evaporation is about 0.6 to 0.8 in the land that supports a tall-grass prairie and 0.2 to 0.4 farther west, where a short-grass steppe is supported. Beyond the short-grass steppe are shrubland and the deserts of the West. Western Northern America is a mountainous country in which vegetation zones reflect climatic changes on an altitudinal gradient. The vegetation in the lowlands is characteristic of the regions (short-grass steppe in the east side of Rocky Mountains, sagebrush cold semideserts in the Great Basin between the Rocky Mountains and the Sierra Nevada, and grasslands in California's Central Valley west of the Sierra Nevada). Above the base regions, the vegetation changes from shrub, woodland, or deciduous forest to montane coniferous forests or alpine tundra. In Central America, from Mexico to Panama where precipitation becomes ample and temperatures are high, tropical rain forests and tropical seasonal forests occur.

Similar distributions of biomes along latitude and altitude can be found in South America, Africa, and Eurasia. In general, the climate-induced patterns of vegetation are influenced by latitude; the location of regions within a continent, which affects the amount of moisture they receive; and altitude, in which mountains modify the climate patterns. In addition, other factors, such as fire and human disturbance, may influence distributions of biomes. For example, most grasslands require periodic fires for maintenance, renewal, and elimination of incoming woody growth. Grasslands at one time covered about 42 percent of the land surface of the world. Humans have converted much of that area into croplands.

Quantitative Relationships
Descriptive relationships can provide pictures of world vegetation distributions along latitudinal and altitudinal gradients of temperature and moisture. Ecologists in the past several decades have also sought quantitative relationships between distributions of biomes and environmental factors. For example, when R. H. Whittaker plotted various types of biomes on gradients of mean annual temperature and mean annual precipitation in 1975, a global pattern emerged relating biomes to climatic variables. It was shown that tropical rain-forest biomes are distributed in regions with annual mean precipitation of 2,500 to 4,500 millimeters and annual mean temperatures of 20 to 30 degrees Celsius. Tropical seasonal forest and sa-

vannas also occur in warm regions with precipitation of 1,500-2,500 millimeters and 500-1,500 millimeters per year, respectively. Temperate forests occupy regions with annual temperature of 5 to 20 degrees Celsius and precipitation exceeding 1,000 millimeters per year. This thermal zone can support temperate rain forest when annual precipitation is more than 2,500 millimeters and temperate grassland when annual precipitation is below 750 millimeters. Temperate woodland occurs between temperate forests and grasslands. Tundra and taiga are distributed in regions with annual mean temperature below 3 degrees Celsius, whereas deserts occupy areas with annual precipitation below 250 millimeters.

These relationships between climatic variables and biomes provide a reasonable approximation of global vegetation patterns. Many types of biomes intergrade with one another. Soil, exposure to fire, and regional climate can influence distributions of biomes in a given area.

Yiqi Luo

See also: Biomes: types; Chaparral; Deserts; Forests; Grasslands and prairies; Habitats and biomes; Lakes and limnology; Marine biomes; Mediterranean scrub; Mountain ecosystems; Old-growth forests; Rain forests; Rain forests and the atmosphere; Rangeland; Reefs; Savannas and deciduous tropical forests; Taiga; Tundra and high-altitude biomes; Wetlands.

Sources for Further Study

Archibold, O. W. *Ecology of World Vegetation*. London: Chapman & Hall, 1995.

Smith, R. L., and T. M. Smith. *Ecology and Field Biology*. 6th ed. San Francisco: Benjamin Cummings, 2001.

Whittaker, R. H. *Communities and Ecosystems*. 2d ed. New York: Macmillan, 1975.

BIOMES: TYPES

Types of ecology: Biomes; Ecosystem ecology; Global ecology; Theoretical ecology

Biomes are classified by geography, climate, temperature, precipitation, soil types, and typical aggregates of flora and fauna. Although classification systems vary, most agree that there are half a dozen major world biomes, which can be further classified into minor biomes.

Temperature, precipitation, soil, and length of day affect the survival and distribution of biome species. Species diversity within a biome may increase its stability and capability to deliver natural services, including enhancing the quality of the atmosphere, forming and protecting the soil, controlling pests, and providing clean water, fuel, food, and drugs. Major biomes include the temperate, tropical, and boreal forests; tundra; deserts; grasslands; chaparral; and oceans.

Temperate Forests
The temperate forest biome occupies the so-called temperate zones in the midlatitudes (from about 30 to 60 degrees north and south of the equator). Temperate forests are found mainly in Europe, eastern North America, and eastern China, and in narrow zones on the coasts of Australia, New Zealand, Tasmania, and the Pacific coasts of North and South America. Their climates are characterized by high rainfall and temperatures that vary from cold to mild.

Temperate forests contain primarily deciduous trees—including maple, oak, hickory, and beechwood—and, secondarily, evergreen trees—including pine, spruce, fir, and hemlock. Evergreen forests in some parts of the Southern Hemisphere contain eucalyptus trees. The root systems of forest trees help keep the soil rich. The soil quality and color are due to the action of earthworms. Where these forests are frequently logged, soil runoff pollutes streams, which reduces spawning habitat for fish. Raccoons, opossums, bats, and squirrels are found in the trees. Deer and black bears roam forest floors. During winter, small animals such as marmots and squirrels burrow in the ground.

Tropical Forests
Tropical forests exist in frost-free areas between the Tropic of Cancer and the Tropic of Capricorn. Temperatures range from warm to hot year-round.

These forests are found in northern Australia, the East Indies, Southeast Asia, equatorial Africa, and parts of Central America and northern South America.

Tropical forests have high biological diversity and contain about 15 percent of the world's plant species. Animal life lives at each layer of tropical forests. Nuts and fruits on the trees provide food for birds, monkeys, squirrels, and bats. Monkeys and sloths feed on tree leaves. Roots, seeds, leaves, and fruit on the forest floor feed larger animals. Tropical forest trees produce rubber and hardwood, such as mahogany and rosewood. Deforestation for agriculture and pastures has caused reduction in plant and animal diversity in these forests.

Boreal Forests
The boreal forest is a circumpolar Northern Hemisphere biome spread across Russia, Scandinavia, Canada, and Alaska. The region is very cold. Evergreen trees such as white spruce and black spruce dominate this zone, which also contains larch, balsam, pine, fir, and some deciduous hardwoods such as birch and aspen. The acidic needles from the evergreens make the leaf litter that is changed into soil humus. The acidic soil limits the plants that develop.

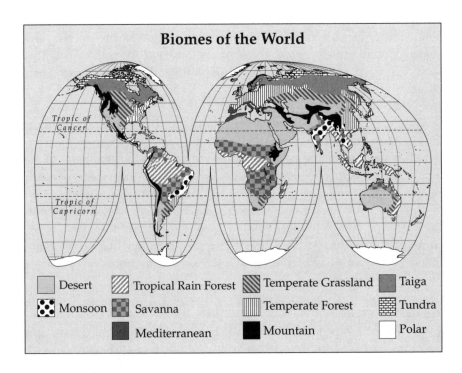

Animals in boreal forests include deer, bears, and wolves. Birds in this zone include red-tailed hawks, sapsuckers, grouse, and nuthatches. Relatively few animals emigrate from this habitat during winter. Conifer seeds are the basic winter food.

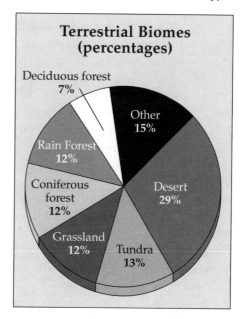

Terrestrial Biomes (percentages)

Deciduous forest 7%
Other 15%
Rain Forest 12%
Coniferous forest 12%
Desert 29%
Grassland 12%
Tundra 13%

Tundra

About 5 percent of the earth's surface is covered with Arctic tundra, and 3 percent with alpine tundra. The Arctic tundra is the area of Europe, Asia, and North America north of the boreal coniferous forest zone, where the soils remain frozen most of the year. Arctic tundra has a permanent frozen subsoil, called permafrost. Deep snow and low temperatures slow the soil-forming process. The area is bounded by a 50 degrees Fahrenheit (122 degrees Celsius) circumpolar isotherm, known as the summer isotherm. The cold temperature north of this line prevents normal tree growth.

The tundra landscape is covered by mosses, lichens, and low shrubs, which are eaten by caribou, reindeer, and musk oxen. Wolves eat these herbivores. Bears, foxes, and lemmings also live there. The most common Arctic bird is the old squaw duck. Ptarmigans and eider ducks are also very common. Geese, falcons, and loons are some of the nesting birds of the area.

The alpine tundra, which exists at high altitude in all latitudes, is acted upon by winds, cold temperatures, and snow. The plant growth is mostly cushion- and mat-forming plants.

Deserts

The desert biome covers about one-seventh of the earth's surface. Deserts typically receive no more than 10 inches (25 centimeters) of rainfall per year, and evaporation generally exceeds rainfall. Deserts are found around the Tropic of Cancer and the Tropic of Capricorn. As warm air rises over the equator, it cools and loses its water content. The dry air descends in the two subtropical zones on each side of the equator; as it warms, it picks up moisture, resulting in drying the land.

Rainfall is a key agent in shaping the desert. The lack of sufficient plant cover contributes to soil erosion during wind- and rainstorms. Some desert plants—for example, the mesquite tree, which has roots that grow 40 feet (13 meters) deep—obtain water from deep below the earth's surface. Other plants, such as the barrel cactus, store large amounts of water in their leaves, roots, or stems. Some plants slow the loss of water by having tiny leaves or shedding their leaves. Desert plants have very short growth periods, because they cannot grow during the long drought periods.

Grasslands

Grasslands cover about one-quarter of the earth's surface and can be found between forests and deserts. Treeless grasslands exist in parts of central North America, Central America, and eastern South America that have between 10 and 40 inches (250-1,000 millimeters) of erratic rainfall per year. The climate has a high rate of evaporation and periodic major droughts. Grasslands are subject to fire.

Some grassland plants survive droughts by growing deep roots, while others survive by being dormant. Grass seeds feed the lizards and rodents that become the food for hawks and eagles. Large animals in this biome include bison, coyotes, mule deer, and wolves. The grasslands produce more food than any other biome. Overgrazing, inefficient agricultural practices, and mining destroy the natural stability and fertility of these lands, resulting in reduced carrying capacity, water pollution, and soil erosion. Diverse natural grasslands appear to be more capable of surviving drought than are simplified manipulated grass systems. This may be due to slower soil mineralization and nitrogen turnover of plant residues in the simplified system.

Savannas are open grasslands containing deciduous trees and shrubs. They are near the equator and are associated with deserts. Grasses there grow in clumps and do not form a continuous layer.

Chaparral

The chaparral, or Mediterranean, biome is found in the Mediterranean Basin, California, parts of Australia, middle Chile, and the Cape Province of South America. This region has a climate of wet winters and summer drought. The plants have tough, leathery leaves and may have thorns. Regional fires clear the area of dense and dead vegetation. The seeds from some plants, such as the California manzanita and South African fire lily, are protected by the soil during a fire and later germinate and rapidly grow to form new plants. Vegetation dwarfing occurs as a result of the severe summer drought and extreme climate changes.

Oceans

The ocean biome covers more than 70 percent of the earth's surface and includes 90 percent of its volume. Oceans have four zones. The intertidal zone is shallow and lies at the land's edge. The continental shelf, which begins where the intertidal zone ends, is a plain that slopes gently seaward. The neritic zone (continental slope) begins at a depth of about 600 feet (180 meters), where the gradual slant of the continental shelf becomes a sharp tilt toward the ocean floor, plunging about 12,000 feet (3,660 meters) to the ocean bottom. This abyssal zone is so deep that it does not have light.

Plankton are animals that float in the ocean. They include algae and copepods, which are microscopic crustaceans. Jellyfish and animal larva are also considered plankton. The nekton are animals that move freely through the water by means of their muscles. These include fish, whales, and squid. The benthos are animals that are attached to or crawl along the ocean's floor. Clams are examples of benthos. Bacteria decompose the dead organic materials on the ocean floor.

The circulation of materials from the ocean's floor to the surface is caused by winds and water temperature. Runoff from the land contains pollutants such as pesticides, nitrogen fertilizers, and animal wastes. Rivers carry loose soil to the ocean, where it builds up the bottom areas. Overfishing has caused fisheries to collapse in every world sector.

Human Impact on Biomes

Human interaction with biomes has increased biological invasions, reduced species biodiversity, changed the quality of land and water resources, and caused the proliferation of toxic compounds. Managed care of biomes may not be capable of undoing these problems.

Ronald J. Raven

See also: Biomes: determinants; Chaparral; Deserts; Forests; Grasslands and prairies; Habitats and biomes; Lakes and limnology; Marine biomes; Mediterranean scrub; Mountain ecosystems; Old-growth forests; Rain forests; Rain forests and the atmosphere; Rangeland; Reefs; Savannas and deciduous tropical forests; Taiga; Tundra and high-altitude biomes; Wetlands.

Sources for Further Study

Food and Agriculture Organization of the United Nations. *State of the World's Forests, 2001.* Rome: Author, 2001.

Gawthorp, Daniel, and David Suzuki. *Vanishing Halo: Saving the Boreal Forest.* Seattle: Mountaineers, 1999.

Linsenmair, K. E., ed. *Tropical Forest Canopies: Ecology and Management*. London: Kluwer Academic, 2001.

Prager, Ellen J., with Cynthia A. Earle. *The Oceans*. New York: McGraw-Hill, 2000.

Solbrig, Otto Thomas, E. Medina, and J. F. Silva, eds. *Biodiversity and Savanna Ecosystem Processes: A Global Perspective*. New York: Springer, 1996.

BIOPESTICIDES

Types of ecology: Agricultural ecology; Ecotoxicology

Biopesticides are biological agents, such as viruses, bacteria, fungi, mites, and other organisms used to control insect and weed pests in an environmentally and ecologically friendly manner.

Biopesticides allow biologically based, rather than chemically based, control of pests. Pests are any unwanted animal, plant, or microorganism. When the environment provides no natural resistance to a pest and when no natural antagonists are present, pests can run rampant. For example, spread of the fungus *Endothia parasitica*, which entered New York in 1904, caused the nearly complete destruction of the American chestnut tree because no natural control was present. Viruses, bacteria, fungi, protozoa, mites, insects, and flowers have all been used as biopesticides.

Advantages and Disadvantages

Many plants and animals are protected from pests by passive means. For example, plant rotation is a traditional method of insect and disease protection that is achieved by removing the host plant long enough to reduce a region's pathogen and pest populations. Biopesticides have several significant advantages over commercial pesticides. They appear to be ecologically safer than commercial pesticides because they do not accumulate in the food chain. Some biopesticides provide persistent control, as more than a single mutation is required to adapt to them and because they can become an integral part of a pest's life cycle. In addition, biopesticides have slight effects on ecological balances because they do not affect nontarget species. Finally, biopesticides are compatible with other control agents.

The major drawbacks to using biopesticides are the time required for them to kill their targets and the inefficiency with which they work. Also, if the organism being used as a biopesticide is a nonnative species, it may cause unforeseen damage to the local ecosystem.

Viruses and Bacteria

Viruses have been developed against insect pests such as *Lepidoptera* (butterflies and moths), *Hymenoptera* (bees, wasps, and ants), and *Dipterans* (flies). Gypsy moths and tent caterpillars, for example, periodically suffer from epidemic virus infestations, which could be exploited and encouraged.

Many commensal microorganisms (microorganisms that live on or in other organisms causing no direct benefit or harm) that occur on plant roots and leaves can passively protect plants against microbial pests by competitive exclusion (that is, simply crowding them out). *Bacillus cereus* has been used as an inoculum on soybean seeds to prevent infection by fungal pathogens in the genus *Cercospora*. Some microorganisms used as biopesticides produce antibiotics, but the major mechanism in most cases seems to be competitive exclusion. For example, *Agrobacterium radiobacter* antagonizes *Agrobacterium tumefaciens*, which causes the disease crown gall. Species of two bacterial genera–*Bacillus* and *Streptomyces*—when added as biopesticides to soil, help control the damping-off disease of cucumbers, peas, and lettuce caused by *Rhizoctonia solani*. *Bacillus subtilis* added to plant tissue also controls stem rot and wilt rot caused by species of the fungus *Fusarium*. *Mycobacteria* species produce cellulose-degrading enzymes, and their addition to young seedlings helps control fungal infection by species of *Pythium*, *Rhizoctonia*, and *Fusarium*. Species of *Bacillus* and *Pseudomonas* produce enzymes that dissolve fungal cell walls.

Bacillus thuringiensis Toxins
The best examples of microbial insecticides are *Bacillus thuringiensis* (*B.t.*) toxins, which were first used in 1901. They have had widespread commercial production and use since the 1960's and have been successfully tested on 140 insects, including mosquitoes. Insecticidal endotoxins are produced by *B.t.* during sporulation, and exotoxins are contained in crystalline parasporal protein bodies. These protein crystals are insoluble in water but readily dissolve in an insect's gut. Once dissolved, the proteolytic enzymes paralyze the gut. Spores that have been consumed germinate and kill the insect. *Bacillus popilliae* is a related bacterium that produces an insecticidal spore that has been used to control Japanese beetles, a corn pest.

The gene for the *B.t.* toxin has also been inserted into the genomes of cotton and corn, producing genetically modified, or GM, plants that produce their own *B.t.* toxin. GM cotton and *B.t.* corn both express the gene in their roots, which provides them with protection from root worms. Ecologists and environmentalists have expressed concern that constantly exposing pests to *B.t.* will cause insects to develop resistance to the toxin. In such a scenario, the effectiveness of traditionally applied *B.t.* would decrease.

Fungi and Protozoa
Saprophytic fungi can compete with pathogenic fungi. There are several examples of fungi used as biopesticides, such as *Gliocladium virens*, *Tri-*

choderma hamatum, Trichoderma harzianum, Trichoderma viride, and *Talaromyces flavus.* For example, *Trichoderma* species compete with pathogenic species of *Verticillium* and *Fusarium. Peniophora gigantea* antagonizes the pine pathogen *Heterobasidion annosum* by three mechanisms: It prevents the pathogen from colonizing stumps and traveling down into the root zone, it prevents the pathogen from traveling between infected and uninfected trees along interconnected roots, and it prevents the pathogen from growing up to stump surfaces and sporulating.

Nematodes are pests that interfere with commercial button mushroom (*Agaricus bisporus*) production. Several types of nematode-trapping fungi can be used as biopesticides to trap, kill, and digest the nematode pests. The fungi produce constricting and nonconstricting rings, sticky appendages, and spores, which attach to the nematodes. The most common of the nematode-trapping fungi are *Arthrobotrys oligospora, Arthrobotrys conoides, Dactylaria candida,* and *Meria coniospora.*

Protozoa have occasionally been used as biopesticide agents, but their use has suffered because of slow growth and the complex culture conditions associated with their commercial production.

Mites, Insects, and Flowers

Well-known "terminator" bugs include praying mantis and ladybugs as well as decollate snails, which eat the common brown garden snail. Fleas, grubs, beetles, and grasshoppers often have natural nematode species that prey on them, which can be used as biocontrol agents. Predaceous mites are used as a biopesticide to protect cotton from other insect pests such as the boll weevil. Parasitic wasps of the genus *Encarsia,* especially *E. formosa,* prey on whiteflies, as does *Delphastus pusillus,* a small, black ladybird beetle.

Dalmatian and Persian insect powders contain pyrethrins, which are toxic insecticidal compounds produced in chrysanthemum flowers. Synthetic versions of these naturally occurring compounds are found in products used to control head lice.

Mark S. Coyne, updated by Elizabeth Slocum

See also: Biomagnification; Food chains and webs; Genetically modified foods; Integrated pest management; Multiple-use approach; Pesticides; Pollution effects; Soil; Soil contamination.

Sources for Further Study

Carozzi, Nadine, and Michael Koziel, eds. *Advances in Insect Control: The Role of Transgenic Plants.* Bristol, Pa.: Taylor & Francis, 1997.

Deacon, J. W. *Microbial Control of Plant Pests and Diseases.* Washington, D.C.: American Society for Microbiology, 1983.

Hall, Franklin R., and Julius J. Menn, eds. *Biopesticides: Use and Delivery.* Totowa, N.J.: Humana Press, 1999.

Hokkanen, Heikki M. T., and James M. Lynch, eds. *Biological Control: Benefits and Risks.* New York: Cambridge University Press, 1995.

BIOSPHERE CONCEPT

Types of ecology: Global ecology; Theoretical ecology

The term "biosphere" was coined in the nineteenth century by Austrian geologist Eduard Suess in reference to the 20-kilometer-thick zone extending from the floor of the oceans to the top of mountains, within which all life on earth exists. Thought to be more than 3.5 billion years old, the biosphere supports nearly one dozen biomes, regions of climatic conditions within which distinct biotic communities reside.

Compounds of hydrogen, oxygen, carbon, nitrogen, potassium, and sulfur are cycled among the four major spheres, one of which is the biosphere, to make the materials that are essential to the existence of life. The other spheres are the lithosphere, the outer part of the earth; the atmosphere, the whole mass of air surrounding the earth; and the hydrosphere, the aqueous vapor of the atmosphere, sometimes defined as including the earth's bodies of water.

The Water Cycle
The most critical of these compounds is water, and its movement among the spheres is called the hydrologic cycle. Dissolved water in the atmosphere condenses to form clouds, rain, and snow. The annual precipitation for any region is one of the major factors in determining the terrestrial biome that can exist. The precipitation takes various paths leading to the formation of lakes and rivers. These flowing waters interact with the lithosphere (the outer part of the earth's crust) to dissolve chemicals as they flow to the oceans. Evaporation of water from the oceans then supplies most of the moisture in the atmosphere. This cycle continually moves water among the various terrestrial and oceanic biomes.

Solar Energy
The biosphere is also dependent upon the energy that is transferred from the various spheres. Solar energy is the basis for almost all life. Light enters the biosphere as the essential energy source for photosynthesis. Plants take in carbon dioxide, water, and light energy, which is converted via photosynthesis into chemical energy in the form of sugars and other organic molecules. Oxygen is generated as a by-product. Most animal life reverses

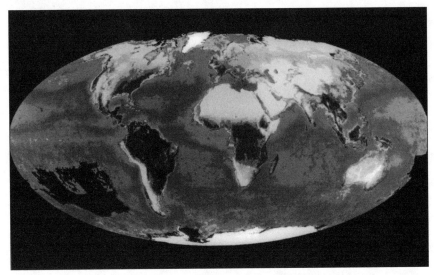

This composite image of the earth's biosphere shows the planet's heaviest vegetative biomass in the dark sections, known to be rain forests. The combination of intense sunlight for twelve hours per day, warm temperatures throughout the year, and considerable rainfall make tropical rain forests the most productive ecosystems on land. (NASA)

this process during respiration, as chemical energy is released to do work by the oxidation of organic molecules to produce carbon dioxide and water.

Incoming solar energy also interacts dramatically with the water cycle and the worldwide distribution of biomes. Because of the earth's curvature, the equatorial regions receive a greater amount of solar heat than the polar regions. Convective movements in the atmosphere—such as winds, high- and low-pressure systems, and weather fronts—and the hydrosphere—such as water currents—are generated during the redistribution of this heat. The weather patterns and climates of earth are a response to these energy shifts. Earth's various climates are defined by the mean annual temperature and the mean annual precipitation.

Toby R. Stewart and Dion Stewart

See also: Biodiversity; Biomes: determinants; Biomes: types; Ecology: definition; Ecosystems: definition and history; Ecosystems: studies; Geochemical cycles; Global warming; Greenhouse effect; Habitats and biomes; Hydrologic cycle; Ozone depletion and ozone holes; Rain forests and the atmosphere.

Sources for Further Study

McNeely, Jeffrey A. *Conserving the World's Biological Diversity*. Washington, D.C.: International Union for Conservation of Nature and Natural Resources, 1990.

Smith, Vaclav. *Cycles of Life: Civilization and the Biosphere*. New York: W. H. Freeman, 2000.

Vernadskii, V. I. *The Biosphere*. Translated by Mark A. S. McMenamin. New York: Copernicus, 1998.

Weiner, Jonathon. *The Next One Hundred Years: Shaping the Fate of Our Living Earth*. New York: Bantam Books, 1991.

Wilson, Edward O., ed. *Biodiversity*. Washington, D.C.: National Academy Press, 1988.

CAMOUFLAGE

Type of ecology: Physiological ecology

Both predatory and prey species use camouflage to minimize the chance that their presence will be detected. Although camouflage is often thought of as being exclusively a visual phenomenon, for it to work, it must occur in all sensory modalities.

Crypsis is the art of remaining hidden. Camouflage is usually thought of as color matching: a green aphid, for example, is likely to go unnoticed while feeding on a green leaf. Background matching, or cryptic coloration, is indeed the most common form of camouflage, but most crypsis involves far more than matching a single color. Very small animals such as aphids can get away with using a single camouflage color because they are much smaller than the plants on which they spend their entire lives: They need to match only one thing. Most animals, however—even most insects—are significantly larger than aphids and are likely to spend time in more than one place. Their camouflage must be more sophisticated if it is to be useful.

Disruptive Coloration

If a large organism is to remain undetected, it must be camouflaged with respect to an entire scene. One way to do this is with the use of disruptive coloration, that is, the use of stripes, spots, or patches of color for camouflage. Disruptive coloration can involve large color patches, as on a pinto pony, a tabby cat, or a diamond-backed rattlesnake, or may involve tiny variations of color on each scale, feather, or hair. Many brownish or grayish mammals actually have what is called agouti coloring, with three different colors appearing on each hair.

The irregular borders of multiple color patches on an animal's body help to obscure its outline against an irregular and multicolored background, just like the blotchy greens and browns on military uniforms. An animal that has a mix of browns in its fur, feathers, skin or scales, for example, will blend into a forest or even an open desert or tundra much better than one that is a single solid color. Even the black-and-white stripes of zebras, which seem so striking, act as a form of disruptive coloration: From far away, and especially to an animal such as a lion, which does not have good color vision, the stripes of zebras help them blend into the tall, wavy grasses of the savannah.

Countershading is another form of crypsis, involving differently colored patches. Countershaded animals appear dark when viewed from above and light when viewed from underneath. Animals with counter-shading include orca whales with their black backs and white bellies, penguins, blue jays, bullfrogs, and weasels. Countershading works and is found as camouflage in so many kinds of animals because no matter where one lives—a desert, a forest, a meadow, or an ocean—the sun shines from above. When looking up toward the sun and sky, dark things stand out and light colors blend in; when looking down toward the ground or the ocean floor, light colors stand out and dark colors blend in. Predators that are countershaded can thus approach their prey with equal stealth from either above or below; likewise, prey species that are countershaded will be equally hard to find whether a predator is searching from on high or from underneath. Countershading and other forms of disruptive coloration can occur in the same organism, so that dark spots, blotches, or stripes appear on top while paler ones appear below.

Mimicry

Another way of remaining undetected in a complex scene is by using protective mimicry, the ability to mimic an inanimate object in both color and form. Some insects look like thorns on plant stems; others look like leaves or twigs or flowers. Some insects, frogs, and fish look like rocks, lichens, or corals. Sea lions, sea dragons, and even eels can look like floating kelp or other forms of seaweed.

Zebras' stripes not only help them blend into their grassland habitats but also make it difficult for predators to pick out a single individual for attack. (Digital Stock)

Some animals may not look much like the objects around them but will disguise themselves by attaching pieces of plants or sand or other debris to their body. Some caterpillars use silk to tie bits of flowers and leaves to their bodies; others use saliva as a glue. Some crabs glue broken bits of shell and coral to their exoskeletons. By using local materials to camouflage itself, an animal can ensure that it matches the background and can even change its disguise as it moves from one area into another.

Transparency

Being transparent is another way to match whatever background happens to be present. Many marine invertebrates such as worms, jellyfish, and shrimp, are completely transparent. Complete transparency is less common among land animals, but some land invertebrates have transparent body parts, such as their wings, allowing them to break up the outline of their body and blend into whatever happens to be in the immediate background.

Predation and Prey

Behavior is an important factor in the success or lack of success of any form of crypsis. For example, not even disruptive camouflage can hide something that is moving quickly with respect to its background. Predatory species that rely on speed or stamina to outrun, outswim, or outfly their prey therefore have little use for camouflage. On the other hand, so-called sit-and-wait predators (such as boa constrictors or praying mantises) must be virtually perfectly camouflaged in order to remain undetected while their prey approach to within grabbing distance. In between are the stealth hunters, which sneak up on their prey before making a final high-speed attack; such animals must be camouflaged and slow-moving when out of attack range but do not have to be camouflaged or slow when at close range.

Like predators, prey species that rely on rapid escape maneuvers do not often exhibit camouflage coloration, while prey species that cannot rely on efficient escape tactics must, instead, rely on camouflage to avoid being seen in the first place. Prey species that can move quickly but not as quickly as their predators must detect their predators before their predators detect them, and then they must remain absolutely still until the danger is passed.

Life Stages

Some species use different strategies as they go through different stages in life. In many altricial species (species with dependent young that require extended parental care of the offspring), the eggs and/or young are camouflaged, even though the adults are not; the temporary spots on deer fawns and mountain lion cubs are examples. In other species, nesting or brooding females may be camouflaged while the adult males retain their gaudy plumage or attention-getting behaviors; the changing seasonal patterns of color and behavior of ducks and songbirds provide examples here. Some species may be toxic and gaudy during one stage of life, yet tasty and cryptic during another.

Multisensory Camouflage

Finally, although camouflage is usually thought of as a visual phenomenon, crypsis is important in every sensory modality. If a prey animal is virtually invisible to its predators but puts out a sound, a scent, or a vibration that makes it easy to locate, visual crypsis alone is useless. For successful protection, prey species must be cryptic in whatever sensory modalities their predators use for hunting. Likewise, for successful hunting, predatory species must be cryptic in whatever sensory modalities their prey use to detect danger. For most species of both predator and prey, this means being camouflaged or blending into the background in several sensory modalities all at once.

Linda Mealey

See also: Adaptations and their mechanisms; Bioluminescence; Defense mechanisms; Displays; Metabolites; Mimicry; Pollination.

Sources for Further Study

Dettner, K., and C. Liepert "Chemical Mimicry and Camouflage." *Annual Review of Entomology* 39 (1994): 129-154.

Ortolani, Alessia. "Spots, Stripes, Tail Tips, and Dark Eyes: Predicting the Function of Carnivore Colour Patterns Using the Comparative Method." *Biological Journal of the Linnean Society* 67, no. 4 (August, 1999): 433-476.

Owen, Denis. *Camouflage and Mimicry.* Chicago: University of Chicago Press, 1980.

Ramachandran, V. S., et.al. "Rapid Adaptive Camouflage in Tropical Flounders." *Nature* 379, no. 6568 (1996): 815-818.

Wicksten, Mary K. "Decorator Crabs." *Scientific American* 242, no. 2 (February, 1980): 146-154.

CHAPARRAL

Types of ecology: Biomes; Ecosystem ecology

Chaparral is the name of a major ecosystem (or biome) found in areas with moist, cool to cold winters and long, dry hot summers (Mediterranean climate).

Chaparral ecosystems with different names occur in the Mediterranean, South Africa, Chile, Australia, and Mexico. The word "chaparral" is a colloquial adaptation of the original Mexican name, *chaparro*. Chaparral communities in other parts of the world have the same basic characteristics and very similar adaptations; this article focuses on the chaparral of California and the American Southwest.

Chaparral is an interesting and unique ecosystem. It is an elfin (stunted) forest dependent on a fire ecology, and its adaptations to a harsh and variable climate are remarkable. The chaparral's geology, latitude, altitude, and climate are all related and have played a role in its formation.

Chaparral is an ecological community composed of shrubby plants adapted to dry summers and moist winters that occurs especially in Southern California. Shrubs well adapted to fire are widespread throughout chaparral and all such mediterranean scrub ecosystems. (AP/Wide World Photos)

Location

In California, the chaparral is located mainly along the central and southern coastal areas, primarily between elevations of 500 and 2,500 feet. It is also found in some areas of the Sierra Nevada foothills, one hundred miles or more inland, and in the lower elevations of some other interior mountains. The geology of most of the areas where this ecosystem occurs is believed to have started with massive upheavals from half a million to ten thousand years ago. The most common substrate was granite. Relentless disintegration resulted in rocky and sandy debris which would allow increasing amounts of plant life to grow. As the organic matter became more abundant, its debris (leaves, twigs, and decaying dead plants) became more and more mixed into the materials of the granite decomposition. This resulted in a rich, sandy loam.

Chaparral Flora

Because of the cool, moist winters and dry, hot summers, plants evolved to survive these marked changes. In most of the chaparral, there are quite extreme diurnal temperature changes, with fluctations of fifty to sixty degrees or more. Compounding these harsh conditions are frequent strong, dry winds, often reaching forty miles per hour.

The plants that have become the residents of the chaparral are mainly shrubby, small-leaved evergreens with leathery, thick stems. Shrubs predominate, but there are also small trees and in many areas abundant wildflowers. All plants are adapted to conserve water. There is little humus in the soil, which is relatively nutrient-poor. The sandy nature of the soil and the variable periods of dry and sudden rain can cause a quick runoff.

The predominant plants are between three and nine feet in height, with some trees being taller. They hug close to the ground to provide shade. The ratio of the surface area of the leaves and stems to their body mass is reduced, and they tend to have thick, heat-resistant surfaces. Some of the plants are capable of turning their leaves so the edges face the sun, which cuts down the warming effect on their surfaces. All these mechanisms greatly reduce evaporative water loss. Most of the bushes and trees also have unusually long tap roots. A three-foot plant might have a tap root that goes ten or more feet below the surface, enabling it to get more water and nutrients.

The most common plant in the chaparral is the greasewood-chamise bush (*Adenostoma fasciculatum*). Others that predominate are the christmasberry-toyon bush (*Photinia arbutifolia*), the scrub oak (*Quercus dumosa*), the yucca (*Yucca whipplei*), and the hoary manzanita (*Arctostaphylos canescens*).

The chamise is characterized by numerous small, club-shaped leaves with a waxy substance that protects them from drying. When there is a fire, the chamise burns with an intense heat and creates a very black smoke (hence the name "greasewood").

The Role of Fire

The role of fire in the maintenance and regeneration of the chaparral is of paramount importance. The hot, dry summer weather, often fanned by winds, makes the chaparral very prone to fires. Because of the high fuel content in the dense plants, with their waxy and oily components, these fires are very intense and can spread rapidly. Fire is necessary to clear excess growth and allow new seeds to germinate. Indeed, several of the key species need fire to release their seeds or they will not germinate. The amount and distribution of the canopy fuel can have a marked effect on regrowth and even spatial variation. Naturally recurring fires are usually good for germination.

However, unusually intense fires, often from years of fire suppression, may do harm by damaging the plants severely. In much of the chaparral, human interference has caused this suppression, creating a difficult paradox, since many people now reside in the chaparral and chaparral fires can spread very rapidly, especially with strong winds. Conversely, in some areas where fire has been controlled, the chaparral has been retreating. A good example of this retreat has occurred on the southern slopes of Mount Tamalpais, a mountain just north of San Francisco.

Chaparral Fauna

A wide variety of reptiles, birds, and mammals make the chaparral their home. They have developed adaptations to survive and thrive in this harsh environment. Several species of skinks, lizards, and a variety of snakes, including gopher snakes, the California king snake, and both the red diamond and western rattlesnakes are residents. There are dozens of birds, from several species of hummingbirds to the large birds: the turkey vulture, barn owl, roadrunner, and golden eagle. There are many species of rodents, including kangaroo rats, chipmunks and gophers. The variety of medium to large mammals of common interest is impressive and includes the coyote, gray fox, badger, lynx, bobcat, mountain lion, and mule deer.

C. Mervyn Rasmussen

See also: Biomes: determinants; Biomes: types; Deserts; Forests; Grasslands and prairies; Habitats and biomes; Lakes and limnology; Marine biomes; Mediterranean scrub; Mountain ecosystems; Old-growth forests;

Rain forests; Rain forests and the atmosphere; Rangeland; Reefs; Savannas and deciduous tropical forests; Taiga; Tundra and high-altitude biomes; Wetlands.

Sources for Further Study

Brown, O. E., and R. A. Minnich. "Fire and Changes in the Creosote Bush Scrub of the Western Sonoran Desert, California." *American Midland Naturalist* 116, no. 2 (1986): 411-422.

Collis, P. H., ed. *Dictionary of Ecology and the Environment*. 3d ed. Chicago: Fitzroy Dearborn, 1998.

Head, W. S. *The California Chaparral: An Elfin Forest*. 1972. Reprint. Happy Camp, Calif.: Naturegraph, 1998.

Jasson-Holt, Sophie. *Unfold the Chaparral*. San Francisco: San Francisco State University Chapbook, 1996.

Odion, Dennis C., and Frank W. Davis. "Fire, Soil Heating, and the Formation of Vegetation Patterns in Chaparral." *Ecological Monographs* 70, no. 1 (2000): 149-169.

CLINES, HYBRID ZONES, AND INTROGRESSION

Types of ecology: Population ecology; Speciation

A cline is a genetic variation in the characteristics of populations of the same species that results from a variation in the geographical area that it occupies. Hybrid zones are areas with populations of a species composed of individuals with characteristics of one or more species that have interbred. Introgression is speciation that occurs when the genes of one species are incorporated into the gene pool of another as the result of successful hybridization.

Gene flow among populations tends to increase the similarity of characters among all the demes (local populations) of a species. Natural selection has the opposite effect: It tends to make every deme uniquely specialized for its specific habitat. Clines are one possible result of these two opposing forces; a cline is a phenomenon in which a genetic variation occurs that is caused by a difference in geographical habitat. Each species is continuously adjusting its gene pool to ensure that the species survives in the face of an environment that is continuously changing.

Comparing the characteristics of the demes of a single species usually will reveal that they are not identical. The greater the distance between the demes, the greater the differences between them will be. The grass frogs in Wisconsin differ from the grass frogs in Texas more than they differ from those in Michigan. On the average, the song sparrows of Alaska are heavier and have darker coloration than those in California. These phenomena, in which a single character shows a gradient of change across a geographical area, are called clines.

North-South Clines

Many birds and mammals exhibit north-south clines in average body size and weight, being larger and heavier in the colder climate farther north and smaller and lighter in warmer climates to the south. In the same way, many mammalian species show north-south clines in the sizes of body extremities such as tails and ears, these parts being smaller in northern demes and larger in southern demes. Increase in average body size with increasing cold is such a common observation that it has been codified as Bergmann's rule. The tendency toward shorter and smaller extremities in colder climates and longer and larger ones in warmer climates is known as

Allen's rule. The trend toward lighter colors in southern climates and darker shades in northern climates has been designated Gloger's rule. The zebra, for example, shows a cline in the amount of striping on the legs. The northernmost races are fully leg-striped, and the striping diminishes toward the southern latitudes of Africa; this appears to be an example of Gloger's rule.

Another example of a cline, which does not fit any of the biogeographical rules mentioned, is the number of eggs laid per reproductive effort (the clutch size) by the European robin: This number is larger in northern Europe than it is for the same species in northern Africa. Other birds, such as the crossbill and raven, which have wide distribution in the Holarctic realm, show a clutch-size cline that reveals a larger clutch size in lower latitudes. The manifestation of such clines in clutch size is a consequence of the interplay of two different reproductive strategies that may give a species a competitive advantage in a given environment. The stability of the environment is what elicits the appropriate strategy.

In unstable environments, such as those in the temperate zone, where there may occur sudden variations in weather and extremes between seasons, a species needs to reproduce rapidly and build its numbers quickly to take advantage of the favorable warm seasons to ensure survival of the species during the harsh, unfavorable conditions of winter. This strategy is known as r strategy (r stands for the rate of increase). In the tropics, the climate is more equable throughout the year. The environment, however, can only support a limited number of individuals throughout the year. This number is called the carrying capacity. When carrying capacity is reached, competition for resources increases, and the reproductive effort is reduced to maintain the population at the carrying capacity. This is called K strategy, with K standing for carrying capacity.

In birds, clutch size tends to be inversely proportional to the climatic stability of the habitat: In temperate climates, more energy is directed to increase the reproductive rate. In the tropics, the carrying capacity is more important, resulting in a reduced reproductive rate. In the apparent contradiction of the crossbills and ravens, it may be the harshness of the habitat at higher latitudes that limits the resources available for successfully fledging a larger number of young.

Frog Clines

The cline exhibited by the common grass frog is one of the best known of all the examples of this phenomenon. It has the greatest range, occupies the widest array of habitats, and possesses the greatest amount of morphological variability of any frog species. This variability and adaptation are not haphazard. The species includes a number of temperature-adapted demes, varying from north to south. These adaptations involve the departmental processes from egg to larva. The northernmost demes have larger eggs that develop faster at lower temperatures than those of the southernmost demes. These physiological differences are so marked that matings between individuals from the extreme ends of the cline result in abnormal larvae or offspring that are inviable (cannot survive) even at a temperature that is average for the cline region.

Leopard frogs from Vermont can interbreed readily with ones from New Jersey. Those in New Jersey can hybridize readily with those in the Carolinas, and those in turn with those in Georgia. Yet, hybrids of Vermont demes and Florida demes are usually abnormal and inviable. Thus, it appears that the Vermont gene pool has been selected for a rate of development that corresponds to a lower environmental temperature. The gene pool of the Florida race has a rate of development that is slower at a higher average temperature. The mixture of the genetic makeup of the northern and southern races is so discordant that it fails to regulate characteristic rates of development at any sublethal temperature, so the resulting embryo dies before it becomes a tadpole.

There are two primary reasons why characters within a species may show clinal variation. First, if gene flow occurs between nearby demes of a population, the gene pools of demes that are close to one another will share more alleles than the gene pools of populations that are far apart. Second, environmental factors, such as annual climate, vary along gradients that can be defined longitudinally, latitudinally, or altitudinally. Because these environmental components act as selective pressures, the phenotypic characters that are best adapted to such pressures will also vary in a gradient.

Hybridization, Hybrid Swards, Introgressive Demes

Hybridization is the process whereby individuals of different species produce offspring. A hybrid zone is an area occupied by interbreeding species. Partial species can and do develop on the way to becoming new species as products of hybridization. Natural hybridization and gene flow can take place between biological species no matter how sterile most of the hybrid offspring may be. As long as the mechanisms that prevent free exchange of genes between populations can be penetrated, there is the potential for a new species to develop. Because the parental species has a tendency to be replaced by the hybrid types if natural selection favors them, hybridization can be a threat to the integrity of the parental species as a distinct entity.

Hybridization between different species leads to various and unpredictable results. Any time that hybridization occurs, the isolation mechanisms of populations are overcome, forming bridging populations. Such connecting demes of hybrid origin fall into one of two general categories: hybrid swarms or introgressive demes. The formation of these types of demes reverses the process of speciation and changes the formerly distinct species into a complex mixture of highly variable individuals that are the products of the segregation and independent assortment of traits. This is the primary advantage of sexual reproduction: to produce variation in the population that is acted upon by natural selection over time. It cannot be overemphasized that hybrid swarms and introgressive demes are highly variable.

The environmental conditions that contour animal communities have endured for a very long time. In long-lived communities, every available niche has been filled by well-adapted species. When populations with new adaptive characteristics occur, there is no niche for them to occupy, so they usually die out. In contrast, when such communities are disturbed, the parity among their component species is upset, which gives new variants an opportunity to become established.

Hybrid swarms can be observed in nature by the careful investigator. The hybrid swarm forms in a disturbed habitat, where hybrid individuals backcross with the parental types to form a third population, which result from the migration of the genes of one population into the other. Such a population is designated an introgressive population. The progeny of such populations resemble the parent species, but the variations are in the direction of one parental species or the other. If introgression is extensive enough, it may eradicate the morphological and ecological distinctions of the parental types. The parental types become rarer and rarer, until they are no longer the representatives of the species.

There appear to be three reasons that first-generation hybrids occurring naturally are more likely to form offspring by backcrossing to one of the parental species than by mating with each other. Primarily, the hybrids are always rarer than the parents. Second, the parental individuals are so much more fertile than the hybrids that many more parental gametes are available than hybrid ones. Finally, backcross progeny, since they contain primarily parentally derived genes, are more likely to be well adapted to the habitat in which they originated than are the purely hybrid individuals.

Introgression

Thus, the most likely result of hybridization is backcrossing to one of the parental species. Genotypes containing the most parental genes usually have the selective advantage, and the fact that they contain a few chromosomal segments from another species gives them unique characteristics that may also be advantageous. This sequence of events—hybridization, backcrossing, and stabilization of backcross types—is known as introgression. Hybrid swarms are interesting phenomena, but they are unlikely to be of evolutionary significance except through introgression.

There are many examples of introgression among plants, but examples of introgression in animals are not common. Those that have been demonstrated are usually associated with the domestication of livestock. In the Himalayan region of Asia, there exists a relative of cattle, the yak, which is also domesticated. Many of the herds of cattle found along the western edge of the Himalayas, in central Asia, contain characteristics that clearly are derived from the gene pool of the yak. Many of these characteristics are manifested as adaptations to the harsh climatic conditions in this region.

In western Canada, there has been a modest introgression of the genes of the American bison into the gene pool of strains of range cattle. The bisonlike characters incorporated into beef cattle created a new breed called the beefalo, which exhibits such characteristics as greater body musculature, lower fat content of the flesh, and great efficiency in the utilization of range forage. A beefalo steer is ready for market in only eight months, while the same live weight is not obtained in the standard beef breed until eighteen months.

These examples serve to illustrate the concept that, as an evolutionary force, introgression is rather insignificant in natural biomes. It is almost always in the wake of human activity or the activities of their domesticated animals that the process of introgression can and does result in new combinations of gene pools from different species.

Edward N. Nelson

See also: Adaptive radiation; Biodiversity; Biogeography; Convergence and divergence; Demographics; Extinctions and evolutionary explosions; Gene flow; Genetic diversity; Genetic drift; Isolating mechanisms; Nonrandom mating, genetic drift, and mutation; Population analysis; Population fluctuations; Population genetics; Population growth; Punctuated equilibrium vs. gradualism; Reproductive strategies; Speciation.

Sources for Further Study
Anderson, Edgar. *Introgressive Hybridization.* New York: Hafner Press, 1968.
Arnold, M. L. *Natural Hybridization and Evolution.* New York: Oxford University Press, 1997.
Briggs, D., and S. M. Walters. *Plant Variation and Evolution.* New York: Cambridge University Press, 1997.

Dobzhansky, Theodosius G. *Genetics and the Origin of Species*. 1951. 3d rev. ed. Reprint. New York: Columbia University Press, 1982.

Endler, John A. *Geographic Variation: Speciation and Clines*. Princeton, N.J.: Princeton University Press, 1977.

Grant, V. *The Origin of Adaptations*. New York: Columbia University Press, 1963.

Kimbel, William H., and Lawrence B. Martin, eds. *Species, Species Concepts, and Primate Evolution*. New York: Plenum Press, 1993.

Ridley, M. *Evolution*. Boston: Blackwell Scientific Publications, 1993.

Roughgarden, Jonathan. *Theory of Population Genetics and Evolutionary Ecology: An Introduction*. Upper Saddle River, N.J.: Prentice Hall, 1996.

Stebbins, G. L., Jr. *Variation and Evolution in Plants*. New York: Columbia University Press, 1950.

COEVOLUTION

Types of ecology: Community ecology; Evolutionary ecology

Coevolution is the interactive evolution of two or more species that results in a mutualistic or antagonistic relationship.

When two or more different species evolve in a way that affects one another's evolution, coevolution is taking place. This interactive type of evolution is characterized by the fact that the participant life-forms are acting as a strong selective pressure upon one another over a period of time. The coevolution of plants and animals, whether animals are considered strictly in their plant-eating role or also as pollinators, is abundantly represented in every terrestrial ecosystem throughout the world where flora has established itself. Moreover, the overall history of some of the multitude of present and past plant and animal relationships is displayed (although fragmentally) in the fossil record found in the earth's crust.

Beginnings

The most common coevolutionary relationships between plants and animals concern plants as a food source. Microscopic, unicellular plants were the earth's first autotrophs (organisms that can produce their own organic energy through photosynthesis, that is, from basic chemical ingredients derived from the environment). In conjunction with the appearance of autotrophs, microscopic, unicellular heterotrophs (organisms, such as animals, that must derive food from other sources, such as autotrophs) evolved to exploit the autotrophs.

Sometime during the later part of the Mesozoic era, angiosperms, the flowering plants, evolved and replaced most of the previously dominant land plants, such as the gymnosperms and the ferns. New species of herbivores evolved to exploit these new food sources. At some point, probably during the Cretaceous period of the late Mesozoic era, animals became unintentional aids in the angiosperm pollination process. As this coevolution proceeded, the first animal pollinators became more and more indispensable as partners to the plants.

Eventually, highly coevolved plants and animals developed relationships of extreme interdependence, exemplified by the honeybees and their coevolved flowers. This angiosperm-insect relationship is thought to have arisen in the Mesozoic era by way of beetle predation, possibly on early, magnolia-like angiosperms. The fossil record gives some support to this theory. Whatever the exact route along which plant-animal pollination partnerships coevolved, the end result was a number of plant and animal species that gained mutual benefit from the new type of relationship.

Coevolutionary Relationships

Coevolved relationships include an immense number of relationships between plants and animals, and even between plants and other plants. Among these coevolved situations can be found commensalisms, in which different species have coevolved to live intimately with one another without injury to any participant, and symbioses, in which species have coevolved to literally "live together."

Such intertwined relationships can take the form of mutualism, in which neither partner is harmed and indeed one or both benefit—as in the relationships between fungi and algae in lichens, fungi and roots in mycorrhizae, and ants and acacia trees in a symbiotic mutualism in which the ants protect the acacias from herbivores. In parasitism, however, one partner benefits at the expense of the other; a classic example is the relationship between the mistletoe parasite and the oak tree. Another coevolu-

A nonsymbiotic mutualism has coevolved between this hummingbird and the funnel shape of the flower from which it extracts nectar with its long, pointed bill. The plant has achieved greater evolutionary fitness through its ability to attract the bird, which helps the plant propagate by facilitating pollination and seed dispersal. At the same time, the hummingbird benefits from a source of nutrition tailored to its anatomy and protected from competitors. (Corbis)

tionary relationship, predation, is restricted primarily to animal-animal relationships (vertebrate carnivores eating other animals, most obviously), although some plants, such as Venus's flytrap, mimic predation in having evolved means of trapping and ingesting insects as a source of food. Some highly evolved fungi, such as the oyster mushroom, have evolved anesthetizing compounds and other means of trapping protozoa, nematodes, and other small animals.

One of the most obvious and complex coevolutionary relationships are the mutualisms that have evolved between plants bearing fleshy fruits and vertebrate animals, which serve to disperse the seeds in these fruits. Over time, plants that produce these fruits have benefited from natural selection because their seeds have enjoyed a high degree of survival and germination: Animals eat the fruits, whose seeds are passed through their digestive systems (or regurgitated to feed offspring) unharmed; at times the seeds are even encouraged toward germination as digestion helps break down the seed coat. Furthermore, dispersal through the animals' mobility allows the seeds to enjoy more widely distributed propagation. The coevolutionary process works on the animals as well: Birds and animals that eat the fruits enjoy a higher degree of survival, and so natural selection favors both fleshy-fruit-producing plants and fleshy-fruit-eating animals. Similar selection has favored the coevolution of flowers with colors and smells that attract pollinators such as bees.

Eventually some plant-animal mutualisms became so intertwined that one or both participants reached a point at which they could not exist without the aid of the other. These obligatory mutualisms ultimately involve other types of animal partners besides insects. Vertebrate partners such as birds, reptiles, and mammals became involved in mutualisms with plants. In the southwestern United States, for example, bats and the agave and saguaro cactus have a special coevolutionary relationship: The bats, nectar drinkers and pollen eaters, have evolved specialized feeding structures such as erectile tongues similar to those found among moths and other insects with similar lifestyles. In turn, angiosperms coevolutionarily involved with bats have developed such specializations as bat-attractive scents, flower structures that match the bats' feeding habits and minimize the chance of injuring the animals, and petal openings timed to the nocturnal activity of bats.

Defense Mechanisms
Coevolution is manifested in defense mechanisms as well as attractants: Botanical structures and chemicals (secondary metabolites) have evolved to discourage or to prevent the attention of plant eaters. These include the

development of spines, barbs, thorns, bristles, and hooks on plant leaves, stems, and trunk surfaces. Cacti, hollies, and rose bushes illustrate this form of plant strategy. Some plants produce chemical compounds that are bitter to the taste or poisonous. Plants that contain organic tannins, such as trees and shrubs, can partially inactivate animals' digestive juices and create cumulative toxic effects that have been correlated with cancer. Grasses with a high silica content act to wear down the teeth of plant eaters. Animals have counteradapted to these defensive innovations by evolving a higher degree of resistance to plant toxins or by developing more efficient and tougher teeth with features such as harder enamel surfaces or the capacity of grinding with batteries of teeth.

Frederick M. Surowiec, updated by Christina J. Moose

See also: Adaptations and their mechanisms; Adaptive radiation; Colonization of the land; Convergence and divergence; Defense mechanisms; Evolution: definition and theories; Evolution of plants and climates; Grazing and overgrazing; Metabolites; Natural selection; Paleoecology; Pollination; Symbiosis.

Sources for Further Study

Bakker, Robert T. *The Dinosaur Heresies*. Reprint. Secaucus, N.J.: Citadel, 2001.

Barth, Friedrich G. *Insects and Flowers: The Biology of a Partnership*. Translated by M. A. Biederman-Thorson. Princeton, N.J.: Princeton University Press, 1991.

Gilbert, Lawrence E., and Peter H. Raven, eds. *Coevolution of Animals and Plants*. Austin: University of Texas Press, 1980.

Gould, Stephen Jay. *The Panda's Thumb*. Reprint. New York: W. W. Norton, 1992.

Grant, Susan. *Beauty and the Beast: The Coevolution of Plants and Animals*. New York: Charles Scribner's Sons, 1984.

Hughes, Norman F. *Paleobiology of Angiosperm Origins*. Reprint. New York: Cambridge University Press, 1994.

Thompson, John N. *The Coevolutionary Process*. Chicago: University of Chicago Press, 1994.

COLONIZATION OF THE LAND

Types of ecology: Evolutionary ecology; Paleoecology

The advent of animals and plants on land during the Ordovician period added new complexity to preexisting ecosystems and paved the way for land ecosystems. The newly increased mass of vegetation on land served to stabilize soils against erosion and promoted the weathering of their nutrient minerals. Arthropods, too, found a place in this early ecosystem of nonvascular plants on land.

The appearance of animals and plants on land by the Middle Ordovician period, some 450 million years ago, was a major event in the evolution of terrestrial ecosystems. Nevertheless, they probably were not the earth's first inhabitants; there is a fossil record of blue-green algae and other microscopic life well back into Precambrian time, as much as 3.5 billion years ago. Indeed, it is doubtful that plants and animals visible to the naked eye could have lived on land without preexisting microbial ecosystems, which served to stabilize minerals in soils, to decompose and circulate organic matter of dead organisms, and to oxygenate the atmosphere by photosynthesis.

The increased mass of more complex animals and plants on land during Ordovician time further stabilized soils, invigorated the recycling of organic matter, and boosted atmospheric oxygenation. In addition, large plants provided greater depth and structure to terrestrial ecosystems than was possible with microbes and so may have promoted photosynthetic efficiency, biological diversity, and perhaps also resistance to disturbance by floods and storms. This self-reinforcing boost to terrestrial productivity firmly established life on land.

Invasion from the Sea
Because there are marine fossils of plants and animals visible to the naked eye in Precambrian rocks (at least 600 million years old), it has commonly been assumed that the earliest creatures on land during the Ordovician and Silurian periods invaded from the sea. Reasons advanced to explain why the land was unavailable for marine creatures for more than 200 million years include the lack of available oxygen, the poverty of terrestrial microbial photosynthetic productivity, and an unpredictable land surface of flash floods and erosional badlands. This view of an invasion from the sea has been used to explain the origins of earliest land animals, which probably were arthropods, such as millipedes and spiders. A tremendous

variety of fossil arthropods have been found in Cambrian, Ordovician, and Silurian deposits of shallow seas and estuaries. Like modern marine crabs, these creatures may have ventured out to a limited extent on land, and some may have become more fully adapted to more difficult conditions there. The external skeletons of arthropods, important for defense in the sea, also are effective for support, movement, and preventing desiccation on land.

On the other hand, millipedes and spiders are not very closely related either to any known fossil or to living aquatic arthropods. A reassessment of the earliest fossil scorpions, formerly regarded as possible early land animals, has shown that they had a breathing apparatus that would have been effective only in water. Substantial evolution on land must have occurred to produce the earliest spiders and millipedes, perhaps from microscopic early microbial feeders that have left no fossil record.

Immigrant vs. Indigenous Evolution

The idea of invasion of the land by marine and freshwater algae is supported to some extent by the close biochemical similarities between modern land plants and charophytes (a kind of pond weed commonly called stonewort because of its calcified egg cells). Charophytes, however, are very different from land plants, and it is unlikely that such soft-bodied aquatic algae in the geological past were any more successful in colonizing the land than are the mounds of rotting seaweed now thrown up on beaches by storms.

Land plants differ from stoneworts and seaweeds in many ways: They have a waxy and proteinaceous outer coating (cuticle) to prevent desiccation and to allow the plant body to remain turgid through internal water pressure; they have small openings (stomates) surrounded by cells that can open and close the opening in order to control loss of water and oxygen and intake of carbon dioxide; they have internal systems of support and water transport, which include tubular thick-walled cells (hydroids) in nonvascular plants, such as mosses and liverworts, and elongate cells with helical or banded woody thickenings (tracheids) in vascular plants; they have roots, unicellular root hairs, or rootlike organs (rhizoids) that gather water and nutrients from soil; and they have propagules (spores) protected from desiccation and abrasion by proteinaceous envelopes. To some botanists, the coordinated evolution of all these features from aquatic algae is extremely unlikely, notwithstanding the impressive diversity of algae today. This consideration, plus the simple nature of the earliest fossil land plants, has led to the argument that land plants evolved on land from microscopic algae already accustomed to such conditions.

The Earliest Land Ecosystems

While immigrant versus indigenous evolutionary origins of the earliest land creatures remains a theoretical problem, there is fossil evidence of very early land ecosystems. In Late Ordovician rocks are found the earliest spores of land plants. Most of them are smooth and closely appressed in groups of four, somewhat similar to spores of liverworts and mosses today. This is not to say that they belonged to liverworts and mosses; no clear fossils of land plants visible to the naked eye have yet been found in rocks of this age. Early moss and liverwort ancestors are found in Silurian rocks, but so are extinct nonvascular plants, such as nematophytes. These early experiments in the evolution of land plants had tissues supported by densely interwoven proteinaceous tubes. In life, they had the rubbery texture of a mushroom and a variety of bladelike and elongate forms similar to those of some living algae.

Although the botanical affinities of the earliest spores of nonvascular land plants remain unclear, there is evidence that they grew in clumps. Buried soils of Late Ordovician age have been found with surficial erosion scours of the kind formed by wind around clumps of vegetation. The clumps are represented by gray spots from the reducing effect of remnant organic matter. Burrows also have been found in Late Ordovician buried soils as an indication of animals in these early land ecosystems. The fossil burrows are quite large (2 to 21 millimeters). They are similar in their clayey linings, backfill structures, and fecal pellets to the burrows of modern roundback millipedes. The buried soils are calcareous and strongly ferruginized—indications that they were nutrient-rich, periodically dry, and well drained, as are modern soils preferred by millipedes. Actual fossils of millipedes have not yet been found in rocks older than Late Silurian, so all that can be said at present is that these very early animals on land were in some ways like millipedes.

Diversification of Life on Land

By Silurian time (some 438 million years ago) there was a considerable diversification of life on land. Spores of fungi and of vascular land plants have been found fossilized in Early Silurian rocks. During Mid-Silurian time, there were small, leafless plants with bifurcating rhizomes and photosynthetic stems terminated by globular, spore-bearing organs. These matchstick-sized fossil plants have been called *Cooksonia*. Although not so well preserved as to show their water-conducting cells, they have been regarded as the earliest representatives of the extinct group of vascular plants called rhyniophytes. In Devonian rocks (some 408 million years old), some well-preserved rhyniophytes are known to have been true vascular plants, but there are other plants similar in general appearance that had simpler thick-walled conducting cells like those of nonvascular plants. By Devonian time, there were also vascular plants with spore-bearing organs borne above lateral branches (zosterophylls), plants with true roots and spore-bearing organs borne in clusters (trimerophytes), and spore-producing plants with woody roots and tree trunks (progymnosperms). The evolution of the earliest vascular plant cover on land, and of the first forests, involved different kinds of plants now extinct.

To fossil millipedes of Silurian age were added during Devonian time spiders, centipedes, springtails (*Collembola*), and bristletails (*Thysanura*). The earliest vertebrates on land are known from bones of extinct amphibians (*Ichthyostegalia*) and from footprints of Devonian age, some 370 million years old.

This great Silurian and Devonian evolutionary radiation promoted environmental changes similar to those initiated by the first colonization of land by plants and animals, as well as some new changes. For example, the formation of charcoal from wildfires in woodlands and the accumulation of peat in swamps were ways of burying carbon that otherwise might have decayed or digested into carbon dioxide in the atmosphere. Removal of carbon dioxide in this way allowed increased oxygenation of the atmosphere. Oxygenation was kept within bounds by increased flammability of woodlands when oxygen reached amounts much in excess of the present atmospheric level.

Late Devonian ecosystems were very different from modern ones. Major ecological roles, such as insect-eating large animals on land, were still being added. More changes were to come, but the world at that time would have seemed a much more familiar place than the meadows of *Cooksonia* during the Silurian, the patchy cover of Ordovician nonvascular plants, and the red and green microbial earths of earlier times.

Gregory J. Retallack

See also: Coevolution; Convergence and divergence; Evolution: definition and theories; Evolution: history; Evolution of plants and climates; Extinctions and evolutionary explosions; Mycorrhizae; Natural selection; Paleoecology; Punctuated equilibrium vs. gradualism.

Sources for Further Study

Gordon, M. S., and E. C. Olson. *Invasions of the Land: The Transition of Organisms from the Aquatic to Terrestrial Life*. New York: Columbia University Press, 1995.

Little, C. *The Colonization of the Land*. Cambridge, England: Cambridge University Press, 1983.

Schopf, J. William, ed. *Major Events in the History of Life*. Boston: Jones and Bartlett, 1992.

Schumm, Stanley A. *The Fluvial System*. New York: Wiley-Interscience, 1977.

Stanley, Steven M. *Earth and Life Through Time*. 2d ed. New York: W. H. Freeman, 1989.

Stebbins, G. L., and G. J. C. Hill. "Did Multicellular Plants Invade the Land?" *American Naturalist* 115 (1980): 342-353.

Wright, V. P., and Alfred Fischer, eds. *Paleosols: Their Recognition and Interpretation*. Princeton, N.J.: Princeton University Press, 1986.

Zimmer, Carl. *At the Water's Edge: Macroevolution and the Transformation of Life*. New York: Free Press, 1998.

COMMUNICATION

Types of ecology: Behavioral ecology; Chemical ecology

In animal communication, information is exchanged through signals. Such signals are vital for survival, finding mates, and rearing young.

A simple definition of animal communication is the transmission of information between animals by means of signals. Developing a more precise definition is difficult because of the broad array of behaviors that are considered messages or signals and the variety of contexts in which these behaviors may occur. Animal signals can be chemical, visual, auditory, tactile, or electrical. The primary means of communication used within a species will depend upon its sensory capacities and its ecology.

Pheromones
Of the modes of communication available, chemical signals, or pheromones, are assumed to have been the earliest signals used by animals. Transmission of chemical signals is not affected by darkness or by obstacles. One special advantage is that the sender of a chemical message can leave the message behind when it moves. The persistence of the signal may also be a disadvantage when it interferes with transmission of newer information. Another disadvantage is that the transmission is relatively slow.

The speed at which a chemical message affects the recipient varies. Some messages have an immediate effect on the behavior of recipients. Alarm and sex-attractant pheromones of many insects, aggregation pheromones in cockroaches, or trail substances in ants are examples. Other chemical messages, primers, affect recipients more slowly, through changes in their physiology. Examples of primers include pheromones that control social structure in hive insects such as termites. Reproductive members of the colony secrete a substance that inhibits the development of reproductive capacity in other hive members. The chemicals important for controlling the hive are spread through grooming and food sharing (trophallaxis). Chemical communication is important not only among social and semisocial insects but also among animals, both vertebrate and invertebrate. Particularly common is the use of a pheromone to indicate that an animal is sexually receptive.

Visual Signals
Visual communication holds forth the advantage of immediate transmission. A visual signal or display is also able to encode a large amount of in-

One of the most interesting forms of communication is a "bee dance" in which one bee, after finding a food source, returns to the hive to perform a waggle dance that indicates the location of the food to its hivemates. (Jeff J. Daly/Photo Agora)

formation, including the location of the sender. Postures and movements of parts of an animal's body are typical elements of visual communication. Color and timing are additional means of providing information. Some visual signals are discrete; that is, the signal shows no significant variation from performance to performance. Other displays are graded so that the information content of the signal can be varied. An example of a graded display is found in many of the threat or aggressive postures of birds. Threat postures of the chaffinch vary between low-intensity and high-intensity postures. The elevation of the crest varies in ways that indicate the bird's relative readiness for combat. The song spreads of red-winged blackbirds and cowbirds show variation in intensity. In red-winged blackbirds, the red epaulets, or shoulder patches, are exposed to heighten the effect of the display. Discrete and graded signals may be used together to increase the information provided by the signal. In zebras, ears back indicates a threat and ears up indicates greeting. The intensity of either message is shown by the degree to which the mouth is held open. A widely open mouth indicates a heightened greeting or threat.

Visual displays depend upon the presence of light or the production of light. The ability to produce light, bioluminescence, is found most frequently in aquatic organisms, but its use in communication is probably best documented in fireflies, beetles belonging to the family *Lampyridae*.

Firefly males advertise their presence by producing flashes of light in a species-specific pattern. Females respond with simple flashes, precisely timed, to indicate that they belong to the appropriate species. This communication system is used to advantage by females in a few predatory species of the genus *Photuris*. After females of predatory species have mated with males of their own species, they attract males of other species by mimicking the responses of the appropriate females. The males that are tricked are promptly eaten. The luminescence of fireflies does not attract a wide variety of nocturnal predators, because their bodies contain a chemical that makes them unpalatable.

Visual displays are limited in the distance over which they can be used and are easily blocked by obstacles. Visual communication is important in primates, birds, and some insects, but can be dispensed with by many species that do not have the necessary sensory capacities.

Auditory Communication

The limitation of visual communication is frequently offset by the coupling of visual displays with other modes of communication. Visual displays can be coupled with auditory communication, for example. There are many advantages to using sound: It can be used in the dark, and it can go around obstacles and provide directional information. Because pitch, volume, and temporal patterns of sound can be varied, extremely complex messages can be communicated. The auditory communication of many bird species has been studied intensively. Bird vocalizations are usually classified into two groups, calls and songs. Calls are usually brief sounds, whereas songs are longer, more complex, and often more suited to transmission over distances.

The call repertoire of a species serves a broad array of functions. Many young birds use both a visual signal, gaping, and calling in their food begging. Individuals that call more may receive more food. Begging calls and postures may also be used by females in some species to solicit food from mates. One call type that has been intensively studied is the alarm call. Alarm calls of many species are similar, and response is frequently interspecific (that is, interpretable by more than one species). Alarm calls are likely to be difficult to locate, a definite advantage to the individual giving the call. Calls used to gather individuals for mobbing predators are also similar in different species. Unlike alarm calls, mobbing calls provide good directional information, so that recruitment to the mobbing effort can be rapid.

Call repertoires serve birds in a great variety of contexts important for survival of the individual. Song, on the other hand, most often serves a reproductive function, that of helping a male hold a territory and attract a mate. Songs are species-specific, like the distinctive markings of a species. In some cases, songs are more distinctive than physical appearance. The chiffchaff and willow warbler were not recognized as separate species until an English naturalist named Gilbert White discovered, by examining their distinctive songs, that they are separate. The North American wood and hermit thrushes can also be distinguished more readily by song than by appearance. Birdsong can communicate not only the species of the individual singing but also information about motivational state. Most singing is done by males during the breeding season. In many species, only the male sings. In some species, females sing as well. Their songs may be similar to the songs of the males of their species or they may be distinctive. If the songs are similar to those of the males, the female may sing songs infrequently and with less volume. In some instances, the female song serves to notify her mate of her location. An interesting phenomenon found in some species is duetting, in which the male and female develop a duet. Mates may sing in alternate and perfectly timed phrases, as is done by the African boubou shrike, *Laniarius aethiopicus*. An individual shrike can recall its mate by singing the entire song alone.

Individuals in some bird species have a single song, and individuals of other species have repertoires of songs. Average repertoire size of the individual is characteristic of a species. Whether songs in repertoires are shared with neighbors or unique to the individual is also characteristic of a species. Sharing songs with neighbors permits song matching in countersinging. Cardinals and tufted titmice are species that frequently match songs in countersinging. Possible uses for matching are facilitating the recognition of intruders and indicating which neighbor has the attention of a singer. Some species of birds have dialects. The species-specific songs of one geographic region can be differentiated from the song of another geographic region. The development of dialects may be useful in maintaining local adaptations within a species, provided that females select mates of the same dialect as their fathers.

Although auditory signals of birds have received a disproportionate share of attention in the study of animal communication, auditory communication is used by a broad spectrum of animals. Crickets have species-specific songs to attract females and courtship songs to encourage an approaching female. The ears of most insects can hear only one pitch, so the temporal pattern of sound pulses is the feature by which a species can be identified. Vervet monkeys use three different alarm calls, depending upon the kind of threat present; they respond to the calls appropriately by looking up, looking down, or climbing a tree, depending upon the kind of call given.

Tactile Communication

Tactile communication differs significantly from other forms of communication in that it cannot occur over a distance. This form of communication is important in many insects, equipped as they are with antennae rich in receptors. Shortly after a termite molts, for example, it strokes the end of the abdomen of another individual with its antennae and mouthparts. The individual receiving this signal responds by extruding a fluid from its hindgut. Tactile signals are frequently used in eliciting trophallaxis (food sharing) in social insects. Tactile signals are also important in the copulatory activity of a number of vertebrates.

Additional channels of communication available in animals are electrical and surface vibration. Many modes of communication are used in combination with other modes. The channels used will depend in part on the sensory equipment of the species, its ecology, and the particular context. Most messages will be important either for the survival of the individual or the group or for the individual's ability to transmit its genes to the next generation.

Donna Janet Schroeder

See also: Altruism; Defense mechanisms; Displays; Ethology; Hierarchies; Insect societies; Mammalian social systems; Mimicry; Pheromones; Poisonous animals; Predation; Reproductive strategies; Territoriality and aggression.

Sources for Further Study

De Waal, Frans. *Chimpanzee Politics: Power and Sex Among Apes*. New York: Harper & Row, 1982.

Goodall, Jane. *In the Shadow of Man*. Rev. ed. Boston: Houghton Mifflin, 2000.

Gould, James L. *Ethology*. New York: W. W. Norton, 1982.

Grier, James W. *Biology of Animal Behavior*. 2d ed. St. Louis: Times Mirror/Mosby, 1992.

Hart, Stephen. *The Language of Animals*. New York: H. Holt, 1996.

Hauser, Marc D. *The Evolution of Communication*. Cambridge, Mass.: MIT Press, 1996.

Peters, Roger. *Mammalian Communication: A Behavioral Analysis of Meaning*. Monterey, Calif.: Brooks/Cole, 1980.

Roitblat, Herbert L., Louis M. Herman, and Paul E. Nachtigall, eds. *Language and Communication: Comparative Perspectives*. Hillsdale, N.J.: Lawrence Erlbaum, 1993.

Wilson, Edward O. *Insect Societies*. Cambridge, Mass.: Harvard University Press, 1971.

COMMUNITIES: ECOSYSTEM INTERACTIONS

Types of ecology: Community ecology; Ecosystem ecology

Ecosystems are complex organizations of living and nonliving components. They are frequently named for their dominant biotic or physical features (such as marine kelp beds or coniferous forests). Communities are groups of species usually classified according to their most prominent members (such as grassland communities or shrub communities). The interactions between species and their ecosystems have lasting impacts on both.

In an ecological sense, a community consists of all populations residing in a particular area. Examples of communities range in scale from all the trees in a given watershed, all soil microbes on an agricultural plot, or all phytoplankton in a particular harbor to all plants, animals, and microbes in vast areas, such as the Amazon River basin or the Chesapeake Bay.

An ecosystem consists of the community of species as well as the environment of a given site. A forest ecosystem would include all living plants and animals, along with climate, soils, disturbance, and other abiotic (nonliving) factors. An estuarine ecosystem, likewise, would include all the living things present in addition to climate, currents, salinity, nutrients, and more.

Interactions between species in communities and ecosystems range from mutually beneficial to mutually harmful. One such category of interaction is mutualism, which usually involves two species. Both species derive benefit from a mutualism. Commensalism is used to describe a situation in which one species benefits without harming the other. If the two species are neither helped nor harmed, a neutralism is said to occur, and an amensalism happens when one species is harmed while the other remains unaffected. During competition, both species involved are negatively affected. A number of terms are used to describe a relationship in which one species benefits at another's expense, including herbivory, predation, parasitism, and pathogenicity. The choice of term more often than not depends on the relative sizes of the species involved.

Competition
Plants typically compete for resources, such as light, space, nutrients, and water. One way an individual may outcompete its neighbors is to outgrow

them, thus capturing more sunlight for itself (and thus producing more sugars and other organic molecules for itself). Another way is to be more fecund than the neighbors, flooding the surroundings with one's progeny and thereby being more likely to occupy favorable sites for reproduction. For example, in closed forests treefall gaps are quickly filled with growth from the canopy, thus shading the ground and making it more difficult for competing seedlings and saplings to survive.

Plants compete in the root zone as well, as plants with a more extensive root network can acquire more of the water and other inorganic nutrients necessary for growth and reproduction than can their competitors. In semiarid areas, for example, trees often have trouble colonizing grasslands because the extensive root systems of grasses are much more effective in capturing available rainwater.

Sometimes plants resort to chemical "warfare," known as allelopathy, in order to inhibit the growth of competitors in the surrounding area. The existence of allelopathy remains a controversial topic, and simpler explanations have been offered for many previously alleged instances of the phenomenon. Allelopathy cannot be rejected outright; however, the controversy most likely proves only that many aspects of nature cannot be pigeonholed into narrow explanations.

Competition involves a cost in resources devoted to outgrowing or outreproducing the neighbors. Because of the cost, closely related, competing species will diverge in their ecological requirements. This principle is known as competitive exclusion.

Mutualism, Commensalism, and Parasitism

Many flowering plants could not exist without one of the most important mutualisms of all: pollination. In concept, pollination is simple: In exchange for carrying out the physical work of exchanging genetic material (in pollen form) between individual plants (thus enabling sexual reproduction), the carrier is rewarded with nutrients in the form of nectar or other materials. Many types of animals are involved in pollination: insects such as bees, flies, and beetles; birds, particularly the hummingbirds; and mammals such as bats.

Another highly important mutualism is that between plant roots and fungal hyphae, or mycorrhizae. Mycorrhizae protect plant roots from pathogenic fungi and bacteria; their most important role, however, is to enhance water and nutrient uptake by the plant. In fact, regeneration of some plants is impossible in the absence of appropriate mycorrhizae. Mycorrhizae benefit, in turn, by receiving nutrients and other materials synthesized by the host plant. There are two types of mycorrhizae: ectomycorrhizae, whose hyphae may fill the space between plant roots but do not penetrate the roots themselves; and vesicular-arbuscular mycorrhizae, whose hyphae penetrate and develop within root cells.

Few people can envision a swamp in the southeastern United States without thinking of bald cypress trees (*Taxodium distichum*) draped in ethereal nets of Spanish moss (*Tillandsia usneoides*), which is actually not a moss but a flowering plant in the monocot family Bromeliaceae. *Tillandsia* is an epiphyte, a plant that grows on the stems and branches of a tree. Epiphytism is one of the most common examples of a commensalism, in which one organism, for instance the epiphyte, benefits without any demonstrable cost to the other, in this case the host tree. Epiphytes are common in tropical rain forests and include orchids, bromeliads, cacti, and ferns. In temperate regions more primitive plants, such as lichens, are more likely to become epiphytes.

Not all epiphytes are commensal, however. In the tropics, strangler figs, such as *Ficus* or *Clusia*, begin life as epiphytes but send down roots that in time completely encircle and kill the host. Mistletoes, such as *Phoradendron* or *Arceuthobium*, may draw off the photosynthetic production of the host, thus severely depleting its resources.

Herbivory and Pathogenicity

Plants, because of their ability to harvest light energy from the sun to produce the organic nutrients and building blocks necessary for life, are the primary producers of most of the earth's ecosystems. Thus, they face an onslaught of macroscopic and microscopic consumers. If macroscopic, the consumers are generally regarded as herbivores (plant-eating animals); if microscopic, they are pathogens. Either way, herbivores and pathogens generally devour the tissues of the host.

Plants have evolved a number of defense mechanisms in response to pressure from herbivores and pathogens. Some responses may be mechanical. For example, trees on an African savanna may evolve greater height to escape grazing pressures from large herbivores, but some large herbivores, specifically giraffes, may evolve to grow to greater heights as well. Plants may encase themselves in nearly indestructible outer coatings or arm themselves with spines in order to discourage grazers.

Other responses may be chemical. Cellulose, one of the important chemical components of plant tissues such as wood, is virtually indigestible—unless the herbivore itself hosts a bacterial symbiont in its stomach that can manage the job of breaking down cellulose. Other chemicals, such as phenols and tannins—the class of compounds that gives tea its brown color—are likewise indigestible, thus discouraging feeding by insects.

Plants produce a wide range of toxins, such as alkaloids, which poison or kill herbivores. A number of hallucinogenic drugs are made from plant alkaloids.

Phytoalexins are another group of defensive compounds produced by plants in response to bacterial and fungal pathogens. Substances present in the cell walls of bacteria and fungi are released via the action of plant enzymes and spread throughout the plant. The bacterial and fungal substances function as hormones to stimulate, or elicit, phytoalexin production. Hence, these substances are referred to as elicitors. The phytoalexins act as antibiotics, killing the infective agents. Tannins, phenols, and other compounds also serve to defend against pathogen attack.

David M. Lawrence

See also: Allelopathy; Animal-plant interactions; Biodiversity; Biogeography; Biological invasions; Coevolution; Communities: structure; Competition; Defense mechanisms; Food chains and webs; Lichens; Metabolites; Mycorrhizae; Pollination; Predation; Speciation; Succession; Symbiosis; Trophic levels and ecological niches.

Sources for Further Study

Barbour, Michael G., et al. *Terrestrial Plant Ecology.* 3d ed. Menlo Park, Calif.: Benjamin Cummings, 1999.

Barnes, Burton V., Donald R. Zak, Shirley R. Denton, and Stephen H. Spurr. *Forest Ecology.* 4th ed. New York: John Wiley & Sons, 1998.

Mitch, William J., and James G. Gosselink. *Wetlands.* 3d ed. New York: John Wiley & Sons, 2000.

Morin, Peter J. *Community Ecology.* Oxford, England: Blackwell Science, 1999.

COMMUNITIES: STRUCTURE

Type of ecology: Community ecology

An ecological community is the assemblage of species found in a given time and place. The species composition of different ecosystems and the ways in which they maintain equilibrium and react to disturbances are manifestations of the community's stability.

The populations that form a community interact through the processes of competition, predation, parasitism, and mutualism. The structures of communities are determined, in part, by the nature and strength of these biotic factors. Abiotic factors (physical factors such as temperature, rainfall, and soil fertility) are the other set of important influences determining community structure. An ecological community together with its physical environment is called an ecosystem. No ecosystem can be properly understood without a careful study of the biotic and abiotic factors that shape it.

Energy Flow

The most common way to characterize a community functionally is by describing the flow of energy through it. Based on the dynamics of energy flow, organisms can be classified into three groups: those that obtain energy through photosynthesis (called producers), those that obtain energy by consuming other organisms (consumers), and those that decompose dead organisms (decomposers). The pathway through which energy travels from producer through one or more consumers and finally to decomposer is called a food chain. Each link in a food chain is called a trophic level. Interconnected food chains in a community constitute a food web.

Very few communities are so simple that they can be readily described by a food web. Most communities are compartmentalized: A given set of producers tends to be consumed by a limited number of consumers, which in turn are preyed upon by a smaller number of predators, and so on. Alternatively, consumers may obtain energy by specializing on one part of their prey (for example, some birds may eat only seeds of plants) but utilize a wide range of prey species. Compartmentalization is an important feature of community structure; it influences the formation, organization, and persistence of a community.

Dominant and Keystone Species

Some species, called dominant species, can exert powerful control over the abundance of other species because of the dominant species' large size, extended life span, or ability to monopolize energy or other resources. Communities are named according to their dominant species: for example, oak-hickory forest, redwood forest, sagebrush desert, and tall-grass prairie. Some species, called keystone species, have a disproportionately large effect on community structure. These interact with other members of the community in such a way that loss of the keystone species can lead to the loss of many other species. Keystone species may also be the dominate species, but they may also appear insignificant to the community until they are gone. For example, cordgrass (*Spartina*) is the dominant plant in many tidal estuaries, and it is also a keystone species because so many members of the community depend on it for food and shelter.

The species that make up a community are seldom distributed uniformly across the landscape; rather, some degree of patchiness is characteristic of virtually all species. There has been conflicting evidence as to the nature of this patchiness. Moving across an environmental gradient (for example, from wet to dry conditions or from low to high elevations), there is a corresponding change in species and community composition. Some studies have suggested that changes in species composition usually occur along relatively sharp boundaries and that these boundaries mark the border between adjacent communities. Other studies have indicated that species tend to respond individually to environmental gradients and that community boundaries are not sharply defined; rather, most communities broadly intergrade into one another, forming what is often called an ecotone.

Degrees of Species Interaction
These conflicting results have fueled a continuing debate as to the underlying nature of communities. Some communities seem to behave in a coordinated manner. For example, if a prairie is consumed by fire, it regenerates in a predictable sequence, ultimately returning to the same structure and composition it had before the fire. This process, called ecological succession, is to be expected if the species in a community have evolved together with one another. In this case, the community is behaving like an organism, maintaining its structure and function in the face of environmental disturbances and fluctuations (as long as the disturbances and fluctuations are not too extreme). The existence of relatively sharp boundaries between adjacent communities supports this explanation of the nature of the community.

In other communities, it appears that the response to environmental fluctuation or disturbance is determined by the evolved adaptations of the species available. There is no coordinated community response but rather a coincidental assembly of community structure over time. Some sets of species interact together so strongly that they enter a community together, but there is no evidence of an evolved community tendency to resist or accommodate environmental change. In this case, the community is formed primarily of species that happen to share similar environmental requirements.

Competition and Predation
Disagreement as to the underlying nature of communities usually reflects disagreement about the relative importance of the underlying mechanisms that determine community structure. Interspecific competition has long been invoked as the primary agent structuring communities. Competition is certainly important in some communities, but there is insufficient evidence to indicate how widespread and important it is in determining community structure. Much of the difficulty occurs because ecologists must infer the existence of past competition from present patterns in communities. It appears that competition has been important in many vertebrate communities and in communities dominated by sessile organisms, such as plants. It does not appear to have been important in structuring communities of plant-eating insects. Furthermore, the effects of competition typically affect individuals that use identical resources, so that only a small percentage of species in a community may be experiencing significant competition at any time.

The effects of predation on community structure depend on the nature of the predation. Keystone predators usually exert their influence by preying on species that are competitively dominant, thus giving less competitive species a chance. Predators that do not specialize on one or a few species may also have a major effect on community structure, if they attack prey in proportion to their abundance. This frequency-dependent predation prevents any prey species from achieving dominance. If a predator is too efficient, it can drive its prey to extinction, which may cause a selective predator to become extinct as well. Predation appears to be most important in determining community structure in environments that are predictable or unchanging.

Disasters and Catastrophes
Chance events can also influence the structure of a community. No environment is completely uniform. Seasonal or longer-term environmental fluctuations affect community structure by limiting opportunities for colonization, by causing direct mortality, or by hindering or exacerbating the

The checkerboard pattern of clear-cutting in forests of the Pacific Northwest threatens the survival of the northern spotted owl, the marbled murrelet, Vaux's swift, and the red tree vole, even though fragments of the community remain. Such disruptions of community structure can be mitigated by thinning to sustain mixed-age, mixed-species trees. (PhotoDisc)

effects of competition and predation. Furthermore, all communities experience at least occasional disturbance: unpredictable, seemingly random environmental changes that may be quite severe.

It is useful in this regard to distinguish between regular disturbances and rarer, more frequent catastrophic events. For example, fire occurs so often in tall-grass prairies that most of the plant species have become fire-adapted—they have become efficient at acquiring nutrients left in the ash and at sprouting or germinating quickly after a fire. In contrast, the 1980 eruption of Mount St. Helens, a volcanic peak in Washington State, was so violent and so unexpected that no members of the nearby community were adequately adapted to such conditions.

Natural disturbances occur at a variety of scales. Small-scale disturbances may simply create small openings in a community. In forests, for example, wind, lightning, and fungi cause single mature trees to die and fall, creating gaps that are typically colonized by species requiring such openings. Large disturbances are qualitatively different from small disturbances in that large portions of a community may be destroyed, including some of the ability to recover from the disturbance. For example, following a large, intense forest fire, some tree species may not return for decades or centuries, because their seeds were consumed by the fire, and colonizers must travel a long distance.

Early ecologists almost always saw disturbances as destructive and disruptive for communities. Under this assumption, most mathematical models portrayed communities as generally being in some stable state; if a disturbance occurred, the community inevitably returned to the same (or some alternative) equilibrium. It later became clear, however, that natural disturbance is a part of almost all natural communities. Ecologists now recognize that few communities exhibit an equilibrium; instead, communities are dynamic, always responding to the last disturbance.

Long-Term Community Dynamics

The evidence suggests that three conclusions can be drawn about the long-term dynamics of communities. First, it can no longer be assumed that all communities remain at equilibrium until changed by outside forces. Disturbances are so common, at so many different scales and frequencies, that the community must be viewed as an entity that is constantly changing as its constituent species readjust to disturbance and to one another.

Second, communities respond in different ways to disturbance. A community may exhibit resistance, not markedly changing when disturbance occurs, until it reaches a threshold and suddenly and rapidly shifts to a new state. Alternatively, a community may exhibit resilience by quickly returning to its former state after a disturbance. Resilience may occur over a wide range of conditions and scales of disturbance in a dynamically robust system. On the other hand, a community that exhibits resilience only within a narrow range of conditions is said to be dynamically fragile.

Finally, there is no simple way to predict the stability of a community. At the end of the 1970's, many ecologists predicted that complex communities would be more stable than simple communities. It appeared that stability was conferred by more intricate food webs, greater structural complexity, and greater species richness. On the basis of numerous field studies and theoretical models, many ecologists now conclude that no such relationship exists. Both very complex communities, such as tropical rain forests, and very simple communities, such as Arctic tundra, may be very fragile.

Studying Communities

Most communities consist of thousands of species, and their complexity makes them very difficult to study. Most community ecologists specialize in taxonomically restricted subsets of communities (such as plant communities, bird communities, insect communities, or moss communities) or in functionally restricted subsets of communities (such as soil communities, tree-hole communities, pond communities, or detrivore communities).

The type of community under investigation and the questions of interest determine the appropriate methods of study. The central questions in most community studies are how many species are present and what is the abundance of each. The answers to these questions can be estimated using mark-recapture methods or any other enumeration method.

Often the aim is to compare communities (or to compare the same community at different times). A specialized parameter called similarity is used to compare and classify communities; more than two dozen measures of similarity are available. Measures of similarity are typically subjected to cluster analysis, a set of techniques that groups communities on the basis of their similarity.

Many multivariate techniques are used to search for patterns in community data. Direct gradient analysis is the simplest of these techniques; it is used to study the distribution of species along an environmental gradient. Ordination includes several methods for collapsing community data for many species in many communities along several environmental gradients onto a single graph that summarizes their relationships and patterns.

Community Disturbance

At the most basic level, destruction of a community eliminates the species that make up the community. If the community is restricted in its extent, and if its constituent species are found nowhere else, those species become extinct. If the community covers a large area or is found in several areas, local extinction of species may occur without causing global extinction.

Destruction of a community can cause unexpected changes in environmental conditions that were modified by the intact community. Even partial destruction of an extensive community can eliminate species. For example, the checkerboard pattern of clear-cutting in Douglas fir forests of the Pacific Northwest threatens the survival of the northern spotted owl, the marbled murrelet, Vaux's swift, and the red tree vole, even though fragments of the community remain. Many fragments are simply too small to support these species. A Douglas fir forest is regenerated following cutting, but this young, even-aged stand is so different from an old, mixed-age forest that it functions as a different type of community.

Altering the population of one species can affect others in a community. The black-footed ferret was once found widely throughout central North America as a predator of prairie dogs. As prairie dogs were poisoned, drowned, and shot throughout their range, the number of black-footed ferrets declined. The species nearly became extinct, and an attempt to increase their numbers and preserve the species was instituted in the late 1980's in a Wyoming breeding program.

Introducing a new species into a community can severely alter the interactions in the community. The introduction of the European rabbit into Australia led to a population explosion of rabbits, excessive predation on vegetation, and resulting declines in many native marsupials.

Finally, it appears that many communities exhibit stability thresholds; if a community is disturbed beyond its threshold, its structure is permanently changed. For example, acid deposition in lakes is initially buffered by natural processes. As acid deposition exceeds the buffering capacity of a lake, it causes insoluble aluminum in the lake bottom to become soluble, and this soluble aluminum kills aquatic organisms directly or by making them more susceptible to disease. The lesson is clear: It is far easier to disrupt or destroy natural systems (even accidentally) than it is to restore or reconstruct them.

Alan D. Copsey, updated by Bryan Ness

See also: Animal-plant interactions; Biodiversity; Biogeography; Biological invasions; Coevolution; Communities: ecosystem interactions; Competition; Defense mechanisms; Eutrophication; Food chains and webs; Invasive plants; Predation; Speciation; Species loss; Succession; Trophic levels and ecological niches.

Sources for Further Study

Aber, John D., and Jerry M. Melillo. *Terrestrial Ecosystems.* 2d ed. San Diego: Harcourt, 2001.

Begon, Michael, John L. Harper, and Colin R. Townsend. *Ecology: Individuals, Populations, and Communities.* 3d ed. Cambridge, Mass.: Blackwell Science, 1996.

Bormann, Frank H., and Gene E. Likens. "Catastrophic Disturbance and the Steady State in Northern Hardwood Forests." *American Scientist* 67 (1979): 660-669.

Goldammer, J. G., ed. *Tropical Forests in Transition: Ecology of Natural and Anthropogenic Disturbance Processes.* Boston: Springer-Verlag, 1992.

Krebs, Charles J. *Ecology: The Experimental Analysis of Distribution and Abundance.* 5th ed. San Francisco: Benjamin Cummings, 2001.

Pickett, S. T. A., and P. S. White, eds. *The Ecology of Natural Disturbance and Patch Dynamics.* Orlando, Fla.: Academic Press, 1985.

Pielou, E. C. *The Interpretation of Ecological Data: A Primer on Classification and Ordination.* New York: John Wiley & Sons, 1984.

COMPETITION

Types of ecology: Behavioral ecology; Community ecology

Competition is the conflict between different organisms for control of food, natural resources, territories, mates, and other aspects of survival. Competition can occur between individuals of the same species or between individuals of different species. In either case, it is natural selection for the fittest organisms and species; therefore, it is a major driving force in evolution.

The science of ecology can best be defined as the experimental analysis of the distribution and abundance of organisms. Natural selection influences the distribution and abundance of organisms from place to place. The possible selecting factors include physical factors (temperature and light, for example), chemical factors such as water and salt, and species interactions. Any of these factors can influence the survivability of organisms in any particular environment. According to ecologist Charles Krebs, species interactions include four principal types: mutualism, which is the living together of two species that benefit each other (for example, humans and their pets); commensalism, which is the living together of two species that results in a distinct benefit (or number of benefits) to one species while the other remains unhurt (commensalism is shown in the relationship of birds and trees); predation, which is the hunting, killing, and eating of one species by another (examples: cats and mice; dogs and deer); and competition, which is defined as an active struggle for survival among all the species in a given environment.

This struggle involves the acquisition of various resources: food, territory, and mates. Food is an obvious target of competition. All organisms must have energy in order to conduct the cellular chemical reactions (such as respiration) that keep them alive. Photoautotrophic organisms (plants, phytoplankton, photobacteria) obtain this energy by converting sunlight, carbon dioxide, and water into sugar, a process called photosynthesis. Photoautotrophs, also called producers, compete for light and water. For example, oak and hickory trees grow taller than most pines, thereby shading out smaller species and eventually dominating a forest. All other organisms—animals, zooplankton, and fungi—are heterotrophs; they must consume other organisms to obtain energy. Heterotrophs include herbivores, carnivores, omnivores, and saprotrophs. Herbivores (plant eaters such as rabbits and cattle) obtain the sugar manufactured by plants. Carnivores (meat eaters such as cats and dogs) eat other heterotrophs in order to get the sugar that these heterotrophs received from other organisms. Omnivores, such as humans, eat plants and animals for the same reason. Saprotrophs (such as fungi and bacteria) decompose dead organisms for the same reason. Life on earth functions by intricately complex food chains in which organisms consume other organisms in order to obtain energy. Each human being is composed of molecules that were once part of other living organisms, even other humans. Ultimately, the earth's energy comes from the sun.

Territoriality is equally important for two reasons: An organism needs a place to live, and this place must contain adequate food and water reserves. A strong, well-adapted organism will fight and drive away weaker individuals of the same or different species in order to maintain exclusive rights to an area containing a large food and water supply. Species that are less well adapted will be relegated to areas where food and water are scarce. The stronger species will have more food and will tend to produce more offspring, since they will easily attract mates. Being stronger or more adapted does not necessarily mean being physically stronger. A physically strong organism can be overwhelmed easily by numerous weak individuals. In general, adaptability is defined by an organism's ability to prosper in a hostile environment and leave many viable offspring.

Animals compete for territory, social status, food, and access to mates. Although competition may be violent and result in injury, competitive behavior (unlike predation) rarely results in death. (Corbis)

Types of Competition
Intraspecific competition occurs among individual members of the same population, for example, when sprouts from plants grow from seeds scattered closely together on the ground. Some seedlings will be able to grow faster than others and will inhibit the growth of less vigorous seedlings by overshadowing or overcrowding them.

Within animal species, males attempt to attract females to their territory, or vice versa, by courtship dances and displays, often including bright colors such as red and blue and exaggerated body size. Mating displays are very similar to the threat displays used to drive away competitors, although there is no hostility involved. Generally, females are attracted to dominant males having the best, not necessarily the largest, territories.

Interspecific competition involves two or more different species trying to use the same resources. All green plants, for example, depend on photosynthesis to derive the energy and carbon they need. Different areas or communities favor different growth characteristics. For plants with high light requirements, a taller-growing plant (or one with more or broader leaves) will have a competitive advantage if its leaves receive more direct sunlight than competitors. If, on the other hand, the species cannot tolerate much sun, a shorter-growing species that can benefit from sheltering shadows of larger plants nearby will have the competitive advantage over other shade-loving plants.

Competition for food and territory is both interspecific and intraspecific. Competition for mates is intraspecific. In an environment, the place where an organism lives (such as a eucalyptus tree or in rotting logs) is referred to as its habitat. Simultaneously, each species has its own unique niche, or occupation, in the environment (such as decomposer or carnivore). More than one species can occupy a habitat if they have different ecological niches. When two or more different species occupy the same habitat and niche, competition arises. One species will outcompete and dominate, while the losing competitors may become reduced in numbers and may be driven away from the habitat.

Pecking Orders
In vertebrate organisms, intraspecific competition occurs between males as a group and between females as a group. Rarely is there male-versus-female competition, except in species having high social bonding—primates, for example. Competition begins when individuals are young. During play fighting, individuals nip or peck at each other while exhibiting threat displays. Dominant individuals exert their authority, while weaker individuals submit. The net result is a very ordered ranking of individuals from top

to bottom, called a dominance hierarchy or pecking order. The top individual can threaten and force into submission any individual below it. The number two individual can threaten anyone except number one, and so on. The lowest-ranked individual can threaten no one and must submit to everyone. The lowest individual will have the least food, worst territory, and fewest (if any) mates. The number one individual will have the most food, best territory, and most mates. The pecking order changes over time because of continued group competition that is shown by challenges, aging, and accidents.

Pecking orders are evident in hens. A very dominant individual will peck other hens many times but will rarely be pecked. A less dominant individual will peck less but be pecked more. A correct ranking can be obtained easily by counting the pecking rate for each hen.

In the Netherlands, male black grouse contend with one another in an area called a "lek," which may be occupied by as many as twenty males. The males establish their territories by pecking, wing-beating, and threat displays. The most dominant males occupy small territories (several hundred meters) at the center of the lek, where the food supply is greatest. Less dominant males occupy larger territories with less food reserves to the exterior of the lek. Established territories are maintained at measurable distances by crowing and flutter-jumping, with the home territory owner nearly always winning. Females, which nest in an adjoining meadow, are attracted to dominant males in the heavily contested small central territories.

A baboon troop can range in size from ten to two hundred members, but usually averages about forty. Larger, dominant males and their many female mates move centrally within the troop. Less dominant males, with fewer females, lie toward the outside of the troop. Weak individuals at the troop periphery are more susceptible to predator attacks. Dominant males exert their authority by threat displays, such as the baring of the teeth or charging; weaker males submit by presenting their hindquarters. Conflicts are usually peacefully resolved.

Female lions maintain an organized pride with a single ruling male. Young males are expelled and wander alone in the wilderness. Upon reaching adulthood, males attempt to take over a pride in order to gain access to females. If a male is successful in capturing a pride and expelling his rival, he will often kill the cubs of the pride, simultaneously eliminating his rival's descendants and stimulating the females to enter estrus for mating.

Competition Within Niches
Interspecific competition occurs between different species over food and

water reserves and territories. Two or more species occupying the same niche and habitat will struggle for the available resources until either one species dominates and the others are excluded from the habitat or the different species evolve into separate niches by targeting different food reserves, thus enabling all to survive in the same habitat. Numerous interspecific studies have been conducted—on crossbills, warblers, blackbirds, and insects, to mention a few.

Crossbills are small birds that live in Europe and Asia. Three crossbill species inhabit similar habitats and nearly similar niches. Each species has evolved a slightly modified beak, however, for retrieving and eating seeds from three different cone-bearing (coniferous) trees. The white-winged crossbill has a slender beak for feeding from small larch cones, the common crossbill has a thicker beak for feeding from larger spruce cones, and the parrot crossbill appropriately has a very thick beak for feeding from pine cones. The evolution of different niches has enabled these three competitors to survive.

Another example of this phenomenon is shown by five species of warblers that inhabit the coniferous forests of the American northeast. The myrtle warbler eats insects from all parts of trees up to seven meters high. The bay-breasted warbler eats insects from tree trunks six to twelve meters above the ground. The black-throated green, blackburnian, and cape may warblers all feed near the treetops, according to elaborate studies by Robert H. MacArthur. The coexistence of five different species is probably the result of the warblers occupying different parts of the trees, with some warblers developing different feeding habits so that all survive.

G. H. Orians and G. Collier studied competitive exclusion between redwing and tricolored blackbirds. Introduction of tricolored blackbirds into redwing territories results in heavy redwing aggression, although the tricolored blackbirds nearly always prevail.

Two species of African ants, *Anoplolepis longipes* and *Oecophylla longinoda*, fight aggressively for territorial space. M. J. Way found that *Anoplolepis* prevails in sandy environments, whereas *Oecophylla* dominates in areas having thick vegetation.

Interspecific competition therefore results in the evolution of new traits and niches and the exclusion of certain species. Mathematical models of competition are based upon the work of A. J. Lotka and V. Volterra. The Lotka-Volterra equations attempt to measure competition between species for food and territory based upon the population size of each species, the density of each species within the defined area, the rate of population increase of each species, and time.

Observing Competiton
Studies of competition between individuals of the same or different species generally follow one basic method: observation. Interactions between organisms are observed and carefully measured to determine if the situation is competition, predation, parasitism, or mutualism. More detailed analyses of environmental chemical and physical conditions are used to determine the existence of additional influences. Observations of competition between organisms involve direct visual contact in the wild, mark-recapture experiments, transplant experiments, measurements of population sizes in given areas, and competition experiments in artificial environments.

Direct visual contact involves the scientist entering the field, finding a neutral, nonthreatening position, and watching and recording the actions of the subject organisms. The observer must be familiar with the habits of the subject organism and must be keen to detect subtle cues such as facial gestures, vocalizations, colors, and patterns of movement from individual to individual. Useful instruments include binoculars, telescopes, cameras, and sound recorders. The observer must be capable of tracking individuals over long distances so that territorial boundaries and all relevant actions are recorded. The observer may have to endure long periods of time in the field under uncomfortable conditions.

Mark-recapture experiments involve the capture of many organisms, tagging them, releasing them into an area, and then recapturing them (both tagged and untagged) at a later time. Repeated collections (recaptures) over time can give the experimenter an estimate of how well the species is faring in a particular environment. This technique is used in conjunction with other experiments, including transplants and population size measurements.

In transplant experiments, individuals of a given species are marked and released into a specific environmental situation, such as a new habitat or another species' territory. The objective of the experiment is to see how well the introduced species fares in the new situation, as well as the responses of the various species which normally inhabit the area. The tricolored blackbird takeover of redwing blackbird territories is a prime example. Another example is the red wolf, a species that was extinct in the wild until several dozen captive wolves were released at the Alligator River Wildlife Refuge in eastern North Carolina. Their survival is uncertain. Accidental transplants have had disastrous results for certain species; for example, the African honeybee poses a threat to the honey industry in Latin America and the southern United States because it is aggressive and produces poorly.

Measurements of population sizes rely upon the point-quarter technique, in which numerous rectangular areas of equal size are marked in the field. The number of organisms of each species in the habitat is counted for a given area; an averaging of all areas is then made to obtain a relatively accurate measure of each population's size. In combination with mark-recapture experiments, population measurements can provide information for birthrates, death rates, immigration, and emigration over time for a given habitat.

Laboratory experiments involve confrontations between different species or individuals of the same species within an artificial environment. For example, male mouse (*Mus musculus*) territoriality can be studied by introducing an intruder into another male's home territory. Generally, the winner of the confrontation is the individual that nips its opponent more times. Usually, home court advantage prevails; the intruder is driven away. Similar studies have been performed with other mammalian, reptile, fish, insect, and bird species.

Interactions between different species are subtle and intricate. Seeing how organisms associate enables scientists to understand evolution and to model various environments. Competition is a major driving force in evolution. The stronger species outcompete weaker species for the available ecological niches. Mutations in organisms create new traits and, therefore, new organisms (more species), which are selected by the environment for adaptability.

All environments consist of a complex array of species, each dependent on the others for survival. The area in which they live is their habitat. Each species' contribution to the habitat is that species' niche. More than one species in a given habitat causes competition. Two species will struggle for available territory and food resources until either one species drives the other away or they adapt to each other and evolve different feeding habits and living arrangements. Competition can be interspecific (between individuals of different species) or intraspecific (between individuals of the same species). The environment benefits because the most adapted species survive, whereas weaker species are excluded.

David Wason Hollar, Jr.

See also: Allelopathy; Animal-plant interactions; Biodiversity; Biogeography; Biological invasions; Coevolution; Communities: ecosystem interactions; Communities: structure; Defense mechanisms; Food chains and webs; Gene flow; Genetic diversity; Lichens; Mycorrhizae; Pollination; Predation; Speciation; Succession; Symbiosis; Trophic levels and ecological niches.

Sources for Further Study

Andrewartha, H. G. *Introduction to the Study of Animal Populations.* Chicago: University of Chicago Press, 1967.

Arthur, Wallace. *The Niche in Competition and Evolution.* New York: John Wiley & Sons, 1987.

Hartl, Daniel L. *Principles of Population Genetics.* 3d ed. Sunderland, Mass.: Sinauer Associates, 1997.

Keddy, Paul A. *Competition.* 2d ed. New York: Kluwer, 2000.

Krebs, Charles J. *Ecology: The Experimental Analysis of Distribution and Abundance.* 5th ed. San Francisco: Benjamin Cummings, 2001.

Lorenz, Konrad. *On Aggression.* New York: Harcourt, Brace & World, 1963.

Raven, Peter H., and George B. Johnson. *Biology.* 5th ed. Boston: WCB/McGraw-Hill, 1999.

Wilson, Edward O. *Sociobiology: The New Synthesis.* Cambridge, Mass.: Belknap Press of Harvard University Press, 1975.

CONSERVATION BIOLOGY

Type of ecology: Restoration and conservation ecology

Conservation biology is a multidisciplinary field that uses knowledge and skills from all aspects of biological science to design and implement methods to ensure the survival of species, ecosystems, and ecological processes.

The overriding mission of conservation biology is to ensure the continued survival of all life-forms and to maintain the structure and function of all ecosystems and ecological processes. Attempting to achieve this goal requires biological knowledge from disciplines such as genetics, physiology, systematics, and ecology in combination with skills and practices from applied fields such as forestry, fisheries, and wildlife management. In addition to biological knowledge, social and economic issues must be addressed in the development of natural resource policies designed to preserve and protect species and ecosystems.

Conservation biology, a relatively new discipline, arose from questions about how to develop methods and policies to maintain global biodiversity. Policy development and implementation must consider the role of evolutionary processes in order to preserve genetic diversity and enable the continued survival of threatened or endangered species. Conservation biologists are also concerned with practical issues such as the design of parks and nature reserves.

David D. Reed

See also: Biodiversity; Deforestation; Endangered animal species; Endangered plant species; Erosion and erosion control; Forest management; Genetic diversity; Grazing and overgrazing; Integrated pest management; Landscape ecology; Multiple-use approach; Old-growth forests; Reforestation; Restoration ecology; Species loss; Sustainable development; Urban and suburban wildlife; Waste management; Wildlife management; Zoos.

Sources for Further Study

Cox, G. W. *Conservation Biology: Concepts and Applications*. Dubuque, Iowa: William C. Brown, 1997.

Hunter, Malcolm, Jr. *Fundamentals of Conservation Biology*. Cambridge, Mass.: Blackwell Science, 1996.

Primack, Richard. *Essentials of Conservation Biology*. Sunderland, Mass.: Sinauer Associates, 1993.

CONVERGENCE AND DIVERGENCE

Types of ecology: Evolutionary ecology; Population ecology; Speciation

Some of the most dramatic examples of natural selection are the result of adaptation in response to stressful climatic conditions. Such selection may cause unrelated species to resemble one another in appearance and function, a phenomenon known as convergence. In other situations, subpopulations of a single species may split into separate species as the result of natural selection. Such divergence is best seen on isolated islands.

Convergent Evolution

Convergent evolution occurs when organisms from different evolutionary lineages evolve similar adaptations to similar environmental conditions. This can happen even when the organisms are widely separated geographically. A classic example of convergent evolution occurred with *Cactaceae*, the cactus family, of the Americas and with the spurge or euphorbs (*Euphorbiaceae*) family of South Africa, both of which have evolved succulent (water-storing) stems in response to desert conditions.

The most primitive cacti are vinelike, tropical plants of the genus *Pereskia*. These cacti, which grow on the islands of the West Indies and in tropical Central and South America, have somewhat woody stems and broad, flat leaves. As deserts developed in North and South America, members of the cactus family began to undergo selection for features that were adaptive to hotter, dryer conditions.

The stems became greatly enlarged and succulent as extensive water-storage tissues formed in the pith or cortex. The leaves became much reduced. In some cactus species, such as the common prickly pear (*Opuntia*), the leaves are small, cylindrical pegs that shrivel and fall off after a month or so of growth. In most cacti, only the leaf base forms and remains as a small hump of tissue associated with an axillary bud. In some cacti this hump is enlarged and is known as a tubercle. Axillary buds in cacti are highly specialized and are known as areoles. The "leaves" of an areole are reduced to one or more spines. Particularly in columnar cacti, the areoles are arranged in longitudinal rows along a multiple-ridged stem.

With the possible exception of the genus *Rhipsalis*, which has one species reported to occur naturally in Africa, all cacti are native to the Americas. As deserts formed in Africa, Eurasia, and Australia, different plant families evolved adaptations similar to those in cacti. The most notable examples are the candelabra euphorbs of South Africa. Desert-dwelling members of the *Euphorbiaceae* frequently have succulent, ridged, cylindrical stems resembling those of cacti. The leaves are typically reduced in size and are present only during the rainy season. They are arranged in rows along each of several ridges of the stem. Associated with each leaf are one or two spines. As a result, when the leaves shrivel and fall off during the dry season, a spiny, cactuslike stem remains.

The succulent euphorbs of Africa take on all the forms characteristic of American cacti, from pincushions and barrels to branched and unbranched columns. Other plant families that show convergence with the cacti, in having succulent stems or leaves, are the stem succulents of the milkweed family, sunflower family, stonecrop family, purslane family, grape family, leaf succulents of the ice plant family, daffodil family, pineapple family, geranium family, and lily family.

Divergent Evolution

Some of the most famous examples of divergent evolution have occurred in the Galápagos Islands. The Galápagos comprise fourteen volcanic islands located about 600 miles west of South America. A total of 543 species of vascular plants are found on the islands, 231 of which are endemic, found nowhere else on earth. Seeds of various species arrived on the islands by floating in the air or on the water or being carried by birds or humans.

With few competitors and many different open habitats, variant forms of each species could adapt to specific conditions, a process known as

An example of convergent evolution is the way that sharks (left), which are fish, and dolphins (right), which are mammals, have evolved similar body shapes to adapt to their marine ecological niches. Although the two animals look very much alike, their differences in evolutionary terms are vast. (Digital Stock)

121

adaptive radiation. Those forms of a species best suited to each particular habitat were continually selected for and produced progeny in that habitat. Over time, this natural selection resulted in multiple new species sharing the same ancestor. The best examples of divergent evolution in the Galápagos have occurred in the *Cactaceae* and *Euphorbiaceae*. Eighteen species and variety of cacti are found on the islands, and all are endemic. Of the twenty-seven species and varieties of euphorbs, twenty are endemic.

An interesting example of the outcome of divergent evolution can be seen in the artificial selection of different cultivars (cultivated varieties) in the genus Brassica. The scrubby Eurasian weed colewort (*Brassica oleracea*) is the ancestor of broccoli, brussels sprouts, cabbage, cauliflower, kale, and kohlrabi (rutabaga). All these vegetables are considered to belong to the same species, but since the origin of agriculture, each has been selected for a specific form that is now recognized as a distinct crop.

Marshall D. Sundberg

See also: Adaptations and their mechanisms; Adaptive radiation; Coevolution; Colonization of the land; Dendrochronology; Development and ecological strategies; Evolution: definition and theories; Evolution: history; Evolution of plants and climates; Extinctions and evolutionary explosions; Gene flow; Genetic drift; Genetically modified foods; Isolating mechanisms; Natural selection; Nonrandom mating, genetic drift, and mutation; Paleoecology; Population genetics; Punctuated equilibrium vs. gradualism; Speciation; Species loss.

Sources for Further Study
Bowman, Robert I., Margaret Berson, and Alan E. Leviton. *Patterns of Evolution in Galápagos Organisms*. San Francisco: California Academy of Sciences, 1983.
Darwin, Charles. "Journal and Remarks: 1832-1836." In *Narrative of the Surveying Voyages of His Majesty's Ships Adventure and Beagle Between the Years 1826 and 1836: Describing Their Examination of the Southern Shores of South America, and the Beagle's Circumnavigation of the Globe*, edited by Robert Fitzroy. Vol. 3. Reprint. New York: AMS Press, 1966.
Harris, James G., and Melinda Woolf Harris. *Plant Identification Terminology: An Illustrated Glossary*. Spring Lake, Utah: Spring Lake, 1994.
Uno, Gordon, Richard Storey, and Randy Moore. *Principles of Botany*. New York: McGraw-Hill, 2001.

DEEP ECOLOGY

Type of ecology: Theoretical ecology

Deep ecology is a school of environmental philosophy based on environmental activism and ecological spirituality.

The term "deep ecology" was first used by Norwegian philosopher Arne Naess in 1972 to suggest the need to go beyond the anthropocentric view that nature is merely a resource for human use. The concept has been used in three major ways. First, it refers to a commitment to deep questioning about environmental ethics and the causes of environmental problems. Such questioning leads to critical reflection on the fundamental worldviews that underlie specific environmental ideas and practices. Second, deep ecology refers to a platform of generally agreed upon values that a variety of environmental activists share. These values include an affirmation of the intrinsic value of nature, the recognition of the importance of biodiversity, a call for a reduction of human impact on the natural world, greater concern with quality of life rather than material affluence, and a commitment to changing economic policies and the dominant view of nature. Third, deep ecology refers to particular philosophies of nature that tend to emphasize the value of nature as a whole (ecocentrism), an identification of the self with the natural world, and an intuitive and sensuous communion with the earth.

Because of its emphasis on fundamental worldviews, deep ecology is often associated with non-Western spiritual traditions such as Buddhism and Native American cultures, as well as radical Western philosophers such as Baruch Spinoza and Martin Heidegger. It has also drawn on the nature writing of Henry David Thoreau, John Muir, Robinson Jeffers, and Gary Snyder. Deep ecology's holistic tendencies have led to associations with the Gaia hypothesis, and its emphasis on diversity and intimacy with nature has linked it to bioregionalism. Deep ecological views have also had a strong impact on environmental activism, including the Earth First! movement.

Deep ecologists have sometimes criticized the animal rights perspective for continuing the traditional Western emphasis on individuals while neglecting whole systems, as well as for a revised speciesism that still values certain parts of nature (animals) over others. Some deep ecologists have also been critical of mainstream environmental organizations such as the Sierra Club for not confronting the root causes of environmental degradation.

On the other hand, deep ecology has been criticized by ecofeminists for failing to consider gender differences in the experience of the self and nature, the lack of an analysis of the tie between the oppression of women and nature, and promoting a holism that supposedly disregards the reality and value of individuals and their relationships. Social ecologists have criticized deep ecology for a failure to critique the relationship between environmental destruction on the one hand and social structure and political ideology on the other. In addition, a distrust of human interference with nature has led some thinkers to present the ideal as pristine wilderness with no human presence. In rare and extreme cases, deep ecologists have implied a misanthropic attitude. In some instances, especially early writings by deep ecologists, such criticisms have considerable force. However, these problematic views are not essential to deep ecology, and a number of thinkers have developed a broadened view that overlaps with ecofeminism and social ecology.

David Landis Barnhill

See also: Balance of nature; Biomes: determinants; Biomes: types; Biosphere concept; Ecology: definition; Ecosystems: definition and history; Sustainable development; .

Sources for Further Study

Naess, Arne. *Ecology, Community, and Lifestyle: Outline of an Ecosophy.* Translated and revised by David Rothenberg. New York: Cambridge University Press, 1989.

_____. *Spinoza and the Deep Ecology Movement.* Delft: Eburon, 1993.

DEFENSE MECHANISMS

Types of ecology: Behavioral ecology; Chemical ecology; Physiological ecology

All organisms represent a potential resource for their predators. Several have evolved ingenious ways to prevent themselves from becoming a predator's next meal.

All organisms are composed of fixed carbon, biomolecules, and mineral nutrients and therefore represent energy and nutrient resources for consumers. To be successful in life, animals must avoid, tolerate, or defend themselves against natural enemies such as predators, parasites, and competitors. The term "defense" can be attributed to any trait that reduces the likelihood that an organism, or part of an organism, will be consumed by a predator. Several categories of defenses have evolved in animals, including structural defenses, chemical defenses, associational defenses, behavioral defenses, autotomy, and nutritional defenses. Animals often possess more than one type of defense, thereby having backup plans in case the first line of defense fails. The number of defenses devised by organisms is a reflection of the strong selective pressure exerted by predators.

Structural Defenses: Being Hard and Sharp

Structures that defend animals can act as external shields: sharp spines located externally or internally, skeletal materials that make tissues too hard to bite easily, or weaponry such as horns, teeth, and claws. External structures that protect vulnerable soft tissues include the chitonous exoskeletons of crustaceans, the calcareous shells of corals, mollusks, and barnacles, the tests (skeletal plates) of echinoderms, the tough tunics of ascidians, and the hard plates of armadillos. The pretty shells that tourists collect along beaches were once used to protect a soft, delicate animal that lived inside the shell. Hard, protective shells remain after the animal dies and can be used by other animals for protection. For example, small fish will retreat into empty conch shells when they feel threatened by predators, and hermit crabs live inside empty snail shells to protect their soft, vulnerable abdomens.

Some animals cover their bodies with sharp structures that puncture predators that try to bite them. The porcupine is a good example of a mammal that uses this defensive strategy. Porcupines are covered with tens of thousands of long, pointed spines, or quills, growing from their back and sides. The quills have needle-sharp ends containing hundreds of barbs that make the quills difficult to remove. Sea urchins are also covered with long, sharp spines that deter would-be predators. Urchins can move their spines, and will direct them toward anything that comes in contact with them, such as a predator. While porcupines and urchins are covered with multiple spines, stingrays defend themselves from enemies by inflicting a wound with a single barbed spine. The wound is extremely painful, giving these rays their common name.

Predators have sharp claws and teeth that help them grasp, subdue, and consume their prey. These same structures, used offensively in hunting, can also be used to protect themselves from their own predators. Small predators such as badgers, raccoons, and foxes can fend off larger predators such as wolves and mountain lions with their weaponry. Rather than risk injury, the larger predators will avoid a fight with the smaller predator and seek a less risky meal, such as a rabbit or mouse.

Chemical Defenses: Poor Taste, Bad Smell, or Toxic Chemicals

Both plants and animals defend themselves by using compounds that are distasteful, toxic, or otherwise repulsive to consumers. Most defensive compounds are secondary metabolites of unique structures, but can also include more generic compounds such as sulfuric acid or calcium carbonate. Secondary metabolites get their name because they are not involved in basic metabolic pathways such as respiration or photosynthesis (that is, primary metabolic reactions), not because they are of secondary importance. Indeed, many organisms probably could not survive in their natural environment without the protection of their secondary metabolites.

Stink bugs get their names because of the smelly secondary metabolites they release from pores located on the sides of their thorax. These smelly compounds repel predators, and may even indicate toxicity to the predator. These insects are common garden pests that are usually controlled with chemical pesticides. However, it appears that the eggs of stink bugs are not defended against roly-poly pill bugs, which can control stink bug numbers (and hence, garden damage) by preying on eggs.

Bombardier beetles take chemical defenses a step further, erupting a boiling hot spray of chemicals in the direction of a predator. To accomplish this, the bombardier beetle has a pair of glands that open at the tip of its abdomen. Each gland has two compartments, one that contains a solution of hydroquinone and hydrogen peroxide, and the other that contains a mixture of enzymes. When threatened by a predator, the bombardier beetle squeezes the hydroquinone and hydrogen peroxide mixture into the enzyme compartment, where an exothermic reaction that produces quinone takes place. The large amount of heat generated brings the quinone mixture to its boiling point, and it is forcefully emitted as a vapor toward the threat. An average bombardier beetle can produce about twenty loud discharges of repulsive, hot chemicals in quick succession.

Chemical defenses are common among small, slow animals such as insects, sponges, cnidarians, and sea slugs, which might be limited in their ability to flee from predators. However, chemical defenses are rather rare among large, fast animals. One of the few mammals that uses chemical defenses is the black-and-white-striped skunk. Most people are familiar with the smelly chemical brew emitted from these animals, as it is distinctly detectable along roads when skunks get hit by cars, and can be detected up to a mile from the location where a skunk sprays. These mammals hold their smelly musk in glands located below their tail, and squirt the liquid through ducts that protrude from the anus. When threatened by a predator, the skunk raises its tail and directs its rear end toward the predator. A predator that has had prior experience with a skunk might retreat from this display, but if the predator is persistent at harassing the skunk, the striped mammal will deliver a spray of smelly chemicals that usually sends the predator running. The musk also causes intense pain and temporary blindness if it gets in the eyes of the predator.

Associational Defenses
Associational defenses occur when a species gains protection from a natural enemy by associating with a protective species, such as when humans gain protection from enemies by keeping a guard dog on their property. Types of protection provided to the defended species through this coevolution can be structural, chemical, or aggressive.

Small animals can avoid predators by using a defended species as habitat. For example, small fish defend themselves by associating with sea urchins, gaining protection by hiding among the sharp spines. Some species of shrimp inhabit the cavities and canals of sponges. Sponges are known to be chemically and structurally defended against most predators, with the exception of angel fish and parrot fish. Finally, much of the diverse coral reef fauna seek protection among the cracks and the crevices in the reef. Reefs, slowly built by coral animals, are the largest structures ever made by living organisms, and serve a protective role for thousands of species that inhabit reefs.

Associational defenses can also be chemically mediated. For example, bacteria that grow symbiotically on shrimp eggs produce secondary metabolites that protect the egg from a parasitic fungus. The numerous examples of sequestration of chemical defenses can be categorized as associational defenses, as they involve associating with chemically defended prey.

An organism might even be defended by protective species that aggressively attack would-be predators, especially if the protected species is a resource for the aggressive defender. For example, humans are protected by guard dogs because dogs view people as a resource that provides them with food, water, and shelter. Stop feeding the dog, and it is likely to look elsewhere for somebody to protect. There are several nonhuman examples of aggressive defensive associations, especially among ants. Aphids are insects that feed on the sugary phloem stream of plants. In the process of feeding and processing phloem, the aphids secrete large amounts of honeydew, which the ants harvest and consume; that is, aphids provide ants with a resource. Ants tend to aphids in the same way that dairy farmers tend to their cows. The ants carry aphids to prime feeding locations, defend aphids from predators, and periodically "milk" the aphids of their honeydew by stroking them with their antennae.

Aposematic Coloration and Mimicry

Being chemically defended does not protect an animal from being accidentally eaten. Therefore, chemically defended animals often advertise the fact that they are nasty to avoid such accidents. This advertisement is often in the form of outlandish colors and patterns that flaunt the animal's distastefulness to predators. Using bright warning patterns is called aposematic coloration.

One problem with aposematic coloration is the training of predators: Bright coloration is useful only if the predator understands the warning. Otherwise, the coloration simply makes the animal a conspicuous prey item. One way that different species with aposematic coloration share the cost of training naïve predators is through mimicry. A predator that eats an individual of species A (assume species A is bright red with blue stripes) and vomits shortly thereafter may learn to avoid things that are red with blue stripes, though at the cost of that first individual's life. This educated predator will now avoid other members of species A, and any other organism that looks like species A (the mimic), whether the mimic is toxic or not. If the mimic is toxic, the system is termed Müllerian mimicry. If the mimic is a palatable species that looks like a toxic model, the system is termed Batesian mimicry.

Mimicry is common within groups of closely related organisms (for example, snakes, butterflies, and bees) which are already similar in appearance. However, mimicry can also occur even when the model and mimic are distantly related. For example, there are caterpillars that mimic the head of a snake, moths that mimic the eyes of a cat, and beetles, moths, and flies that mimic stinging bees and wasps.

Autotomy: Throw the Predator a Bone

Sometimes, despite the best defenses, a predator will get hold of a prey. When this happens, some animals are able to sacrifice a portion of their body to the predator, with the hopes that the remaining parts will survive, and perhaps even regrow the lost parts. This ability to lose a body part intentionally is called autotomy.

Many lower animals, such as sponges, cnidarians, and worms, have great regeneration abilities, and can regrow body parts well. In fact, these animals can even use regeneration as a form of asexual reproduction: Break the animal into four parts, and the parts will generate four complete individuals.

Sea cucumbers, in addition to being chemically defended, are able to eviscerate (autotomy of intestines) when harassed by a predator. These are not fast animals, so this action does not allow them to escape, but it might satisfy (or disgust) the predator enough to make it lose interest in the sea cucumber. Losing a large portion of its digestive tract interferes with feeding, but the sea cucumber can regenerate those parts of the gut that were eviscerated, restoring itself to original function. Sea cucumbers also play an important role in a defensive association with the pearlfish. When the pearlfish feels threatened, it locates the anus of a sea cucumber, then backs into its intestine, where it hides until the danger has passed.

The regenerative ability of higher animals is generally less than that of lower animals. However, autonomy does occur even in some vertebrates. Lizards are well known for their ability to release the tips of their tails when grabbed by a predator. The predator is distracted, and perhaps satisfied, by the wiggling piece of flesh, and in the meantime, the remainder of the lizard scampers off to safety. Geckos release skin instead of tails. The part of the skin that is grabbed by the predator is released, enabling the gecko to break free and escape.

Nutritional Defenses

Some animals, such as corals, jellyfish, anemones, and gorgonians (phylum *Cnidaria*), possess a type of combined structural and chemical defense in the form of specialized stinging cells called nematocysts. When nematocysts are stimulated, they rapidly discharge a barb that punctures the skin of a predator, often releasing toxic chemicals at the same time. The stinging sensation that people get when they come into contact with a jellyfish is caused by nematocysts. Some of these jellyfish stings are so potent that they can result in death.

Not only do many predators avoid jellyfish because they posses nematocysts, but predators may avoid jellyfish because they are jellylike, being composed of more than 95 percent water. It takes time and effort for predators to locate, handle, ingest, and digest prey. If the prey item is basically a bag of seawater (as jellyfish are), then predators might not bother eating these nutrient-deficient animals. Thus, these animals are "nutritionally" defended. Nutritional defenses are also used by plants, but they are generally not an available strategy for animals other than jellyfish, as most animal tissue is relatively nutritious.

Greg Cronin

See also: Allelopathy; Animal-plant interactions; Bioluminescence; Co-evolution; Communities: ecosystem interactions; Metabolites; Poisonous animals; Poisonous plants; Predation; Territoriality and aggression.

Sources for Further Study

Cloudsley-Thompson, John L. *Tooth and Claw: Defensive Strategies in the Animal World*. London: J. M. Dent & Sons, 1980.

Edmunds, Malcolm. *Defence in Animals*. Burnt Mill, England: Longman, 1974.

Evans, David L., and Justin O. Schmidt, eds. *Insect Defenses: Adaptive Mechanisms and Strategies of Prey and Predators*. Albany: State University of New York Press, 1990.

Kaner, Etta. *Animal Defenses: How Animals Protect Themselves*. Toronto: Kids Can Press, 1999.

McClintock, James B., and Bill J. Baker, eds. *Marine Chemical Ecology*. Boca Raton, Fla.: CRC Press, 2001.

Owen, Denis. *Survival in the Wild: Camouflage and Mimicry*. Chicago: University of Chicago Press, 1980.

DEFORESTATION

Types of ecology: Ecotoxicology; Restoration and conservation ecology

Deforestation is the loss of forestlands through encroachment by agriculture, industrial development, or nonsustainable commercial forestry, and other human as well as natural activity.

Concerns about deforestation, particularly in tropical regions, have risen as the role that tropical rain forests play in moderating global climate has become better understood. Environmental activists decried the apparent accelerating pace of deforestation in the twentieth century because of the potential loss of wildlife and plant habitat and the negative effects on biodiversity. By the 1990's research by mainstream scientists had confirmed that deforestation was indeed occurring on a global scale and that it posed a serious threat to global ecology.

Deforestation as a result of expansion of agricultural lands or nonsustainable timber harvesting has occurred in many regions of the world at different periods in history. The Bible, for example, refers to the cedars of Lebanon. Lebanon, like many of the countries bordering the Mediterranean Sea, was thickly forested several thousand years ago. A growing human population, overharvesting, and the introduction of grazing animals such as sheep and goats decimated the forests, which never recovered.

Countries in Latin America, Asia, and Africa have also lost woodlands. While some of this deforestation is caused by a demand for tropical hardwoods for lumber or pulp, the leading cause of deforestation in the twentieth century, as it was several hundred years ago, was the expansion of agriculture. The growing demand by the industrialized world for agricultural products such as beef has led to millions of acres of forestland being bulldozed or burned to create pastures for cattle. Researchers in Central America have watched with dismay as large beef-raising operations have expanded into fragile ecosystems in countries such as Costa Rica, Guatemala, and Mexico.

A tragic irony in this expansion of agriculture into tropical rain forests is that the soil underlying the trees is often unsuited for pastureland or raising other crops. Exposed to sunlight, the soil is quickly depleted of nutrients and often hardens. The once-verdant land becomes an arid desert, prone to erosion, that may never return to forest. As the soil becomes less fertile, hardy weeds begin to choke out the desirable forage plants, and the cattle ranchers move on to clear a fresh tract.

131

Slash-and-Burn Agriculture

Beef industry representatives often argue that their ranching practices are simply a form of slash-and-burn agriculture and do no permanent harm. It is true that many indigenous peoples in tropical regions have practiced slash-and-burn agriculture for millennia, with only a minimal impact on the environment. These farmers burn shrubs and trees to clear small plots of land.

Anthropological studies have shown that the small plots these peasant farmers clear can usually be measured in square feet, not hectares as cattle ranches are, and are used for five to ten years. As fertility declines, the farmer clears a plot next to the depleted one. The farmer's family or village will gradually rotate through the forest, clearing small plots and using them for a few years, and then shifting to new ground, until they eventually come back to where their ancestors began one hundred or more years before.

As long as the size of the plots cleared by farmers remains small in proportion to the forest overall, slash-and-burn agriculture does not contribute significantly to deforestation. If the population of farmers grows, however, more land must be cleared with each succeeding generation. In many tropical countries, traditional slash-and-burn agriculture can then be as ecologically devastating as the more mechanized cattle ranching operations.

Logging

Although logging is not the leading cause of deforestation, it is a significant factor. Tropical forests are rarely clear-cut by loggers, as they typically contain hundreds of different species of trees, many of which have no commercial value. Loggers may select trees for harvesting from each stand. Selective harvesting is a standard practice in sustainable forestry. However, just as loggers engaged in the disreputable practice of high-grading across North America in the nineteenth century, so are loggers high-grading in the early twenty-first century in Malaysia, Indonesia, and other nations with tropical forests.

High-grading is a practice in which loggers cut over a tract to remove the most valuable timber while ignoring the damage being done to the residual stand. The assumption is that, having logged over the tract once, the timber company will not be coming back. This practice stopped in North America, not because the timber companies voluntarily recognized the ecological damage they were doing but because they ran out of easily accessible, old-growth timber to cut. Fear of a timber famine caused logging companies to begin forest plantations and to practice sustainable forestry.

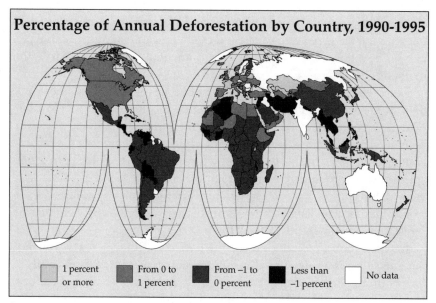

Percentage of Annual Deforestation by Country, 1990-1995

1 percent or more	From 0 to 1 percent	From −1 to 0 percent	Less than −1 percent	No data

Source: United Nations Food and Agriculture Organization

While global satellite photos indicate that significant deforestation has occurred in tropical areas, enough easily harvested old-growth forest remains in some areas that there is no economic incentive for timber companies to switch to sustainable forestry.

Logging may also contribute to deforestation by making it easier for agriculture to encroach on forestlands. The logging company builds roads for use while harvesting trees. Those roads are then used by farmers and ranchers to move into the logged tracts, where they clear whatever trees the loggers have left.

Environmental Impacts

Despite clear evidence that deforestation is accelerating, the extent of the problem remains debatable. The United Nations Food and Agriculture Organization (FAO), which monitors deforestation worldwide, bases its statistics on measurements taken from satellite images. These data indicate that between 1980 and 1990, at least 159 million hectares (393 million acres) of land became deforested. The data also reveal that, in contrast to the intense focus on Latin America by both activists and scientists, the most dramatic loss of forestlands occurred in Asia. The deforestation rate in Latin America was 7.45 percent, while in Asia 11.42 percent of the forests vanished. Environmental activists are particularly concerned about forest losses in Indonesia and Malaysia, two countries where timber companies have been accused of abusing or exploiting native peoples in addition to engaging in environmentally damaging harvesting methods.

Researchers outside the United Nations have challenged the FAO's data. Some scientists claim the numbers are much too high, while others provide convincing evidence that the FAO numbers are too low. Few researchers, however, have tried to claim that deforestation on a global scale is not happening. In the 1990's the reforestation of the Northern Hemisphere, while providing an encouraging example that it is possible to reverse deforestation, was not enough to offset the depletion of forestland in tropical areas. The debate among forestry experts centers on whether deforestation has slowed, and, if so, by how much.

Deforestation affects the environment in a multitude of ways. The most obvious effect is a loss of biodiversity. When an ecosystem is radically altered through deforestation, the trees are not the only thing to disappear. Wildlife species decrease in number and in variety. As forest habitat

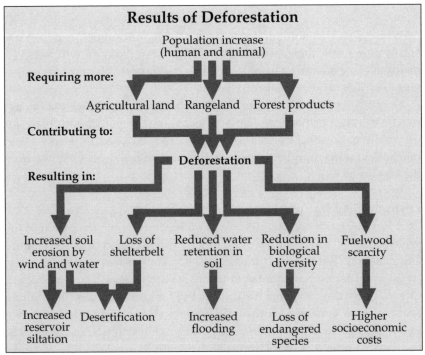

Source: Adapted from A. K. Biswas, "Enviornmental Concerns in Pakistan, with Special Reference to Water and Forests," in *Environmental Conservation*, 1987.

shrinks through deforestation, many plants and animals become vulnerable to extinction. Many biologists believe that numerous animals and plants native to tropical forests will become extinct from deforestation before humans have a chance to even catalog their existence.

Other effects of deforestation may be less obvious. Deforestation can lead to increased flooding during rainy seasons. Rainwater that once would have been slowed or absorbed by trees instead runs off denuded hillsides, pushing rivers over their banks and causing devastating floods downstream. The role of forests in regulating water has long been recognized by engineers and foresters. Flood control was, in fact, one of the motivations behind the creation of the federal forest reserves in the United States during the nineteenth century. More recently, disastrous floods in Bangladesh have been blamed on logging tropical hardwoods in the mountains of Nepal and India.

Conversely, trees can also help mitigate against drought. Like all plants, trees release water into the atmosphere through the process of transpiration. As the world's forests shrink in total acreage, fewer greenhouse gases such as carbon dioxide will be removed from the atmosphere, less oxygen and water will be released into it, and the world will become a hotter, dryer place. Scientists and policy analysts alike agree that deforestation is a major threat to the environment. The question is whether effective policies can be developed to reverse it or if short-term economic greed will win out over long-term global survival.

Nancy Farm Männikkö

See also: Biodiversity; Conservation biology; Endangered plant species; Erosion and erosion control; Forest management; Forests; Grazing and overgrazing; Multiple-use approach; Old-growth forests; Rain forests; Rain forests and the atmosphere; Rangeland; Reforestation; Restoration ecology; Slash-and-burn agriculture; Soil contamination; Sustainable development.

Sources for Further Study

Bevis, William W. *Borneo Log: The Struggle for Sarawak's Forests.* Seattle: University of Washington Press, 1995.

Colchester, Marcus, and Larry Lohmann, eds. *Struggle for Land and the Fate of the Forests.* Atlantic Highlands, N.J.: Zed Books, 1993.

Dean, Warren. *With Broadax and Firebrand: The Destruction of the Brazilian Atlantic Forest.* Berkeley: University of California Press, 1997.

Richards, John F., and Richard P. Tucker, eds. *World Deforestation in the Twentieth Century.* Durham, N.C.: Duke University Press, 1988.

Rudel, Thomas K., and Bruce Horowitz. *Tropical Deforestation: Small Farmers and Land Clearing in the Ecuadorian Amazon*. New York: Columbia University Press, 1993.

Sponsel, Leslie E., Robert Converse Bailey, and Thomas N. Headland, eds. *Tropical Deforestation: The Human Dimension*. New York: Columbia University Press, 1996.

Vajpeyi, Dhirendrea K., ed. *Deforestation, Environment, and Sustainable Development: A Comparative Analysis*. Westport, Conn.: Praeger, 2001.

Wilson, Edward O. *The Future of Life*. New York: Alfred A. Knopf, 2001.

DEMOGRAPHICS

Type of ecology: Population ecology

Demography is the study of the numbers of organisms born in a population within a certain time period, the rate at which they survive to various ages, and the number of offspring that they produce. Many different patterns of birth, survival, and reproduction are found among organisms in nature.

No animal lives forever. Instead, each individual has a generalized life history that begins with fertilization and then goes through embryonic development, a juvenile stage, a period in which it produces offspring, and finally death. There are many variations on this general theme. Still, the life of each organism has two constants: a beginning and an end. Many biologists are fascinated by the births and deaths of individuals in a population and seek to understand the processes that govern the production of new individuals and the deaths of those already present. The branch of biology that deals with such phenomena is called demography.

The word "demography" is derived from Greek; *demos* means "people." For many centuries, demography was applied almost exclusively to humans as a way of keeping written records of new births, marriages, deaths, and other socially relevant information. During the first half of the twentieth century, biologists gradually began to census populations of naturally occurring organisms to understand their ecology more fully. Biologists initially focused on vertebrate animals, particularly game animals and fish. Beginning in the 1960's and 1970's, invertebrate animals, plants, and microbes also became subjects of demographic studies. Studies clearly show that different species of organisms vary greatly in their demographic properties. Often, there is a clear relationship between those demographic properties and the habitat in which these organisms live.

Demographic Parameters

When conducting demographic studies, a demographer must gather certain types of basic information about the population. The first is the number of new organisms that appear in a given amount of time. There are two ways that an organism can enter a population: by being born into it or by immigrating from elsewhere. Demographers generally ignore immigration and concentrate instead on newborns. The number of new individuals born into a population during a specific time interval is termed the natality rate. The natality rate is often based on the number of individuals already in the population. For example, if ten newborns enter a population of a thousand individuals during a given time period, the natality rate is 0.010. A specific time interval must be expressed (days, months, years) for the natality rate to have any meaning.

A second demographic parameter is the mortality rate, which is simply the rate at which individuals are lost from the population by death. Losses that result from emigration to a different population are ignored by most demographers. Like the natality rate, the mortality rate is based on the number of individuals in the population, and it reflects losses during a certain time period. If calculated properly, the natality and mortality rates are directly comparable, and one can subtract the latter from the former to provide an index of the change in population size over time. The population increases whenever natality exceeds mortality and decreases when the reverse is true. The absolute value of the difference denotes the rate of population growth or decline.

When studying mortality, demographers determine the age at which organisms die. Theoretically, each species has a natural life span that no individuals can surpass, even under the most ideal conditions. Normally, however, few organisms reach their natural life span, because conditions are far from ideal in nature. Juveniles, young adults, and old adults can all die. When trying to understand the dynamics of a population, demographers therefore note whether the individuals are dying mainly as adults or mainly as juveniles.

Patterns of Survival
Looking at it another way, demographers want to know the pattern of survival for a given population. This can best be determined by identifying a cohort, which is defined as a group of individuals that are born at about the same time. That cohort is then followed over time, and the number of survivors is counted at set time intervals. The census stops after the last member of the cohort dies. The pattern of survival exhibited by the whole cohort is called its survivorship. Ecologists have examined the survivorship patterns of a wide array of species, including vertebrate animals, invertebrates, plants, fungi, algae, and even microscopic organisms. They have also investigated organisms from a variety of habitats, including oceans, deserts, rain forests, mountain peaks, meadows, and ponds. Survivorship patterns vary tremendously.

Some species have a survivorship pattern in which the young and middle-aged individuals have a high rate of survival, but old individuals die in large numbers. Several species of organisms that live in nature, such as mountain sheep and rotifers (tiny aquatic invertebrates), exhibit this survivorship pattern. At the other extreme, many species exhibit a survivorship pattern in which mortality is heaviest among the young. Those few individuals that are fortunate enough to survive the period of heavy mortality then enjoy a high probability of surviving until the end of their natural life span. Examples of species that have this pattern include marine invertebrates such as sponges and clams, most species of fish, and parasitic worms. An intermediate pattern is also observed, in which the probability of dying stays relatively constant as the cohort gets older. American robins, gray squirrels, and hydras all display this pattern.

These survivorship patterns are usually depicted on a graph that has the age of individuals in the cohort on the x-axis and the number of survivors on the y-axis. Each of the three survivorship patterns gives a different curve when the number of survivors is plotted as a function of age. In the first pattern (high survival among juveniles), the curve is horizontal at first but then swings downward at the right of the graph. In the second pattern (low survival among juveniles), the curve drops at the left of the graph but then levels out to form a horizontal line. The third survivorship pattern (constant mortality throughout the life of the cohort) gives a straight line that runs from the upper-left corner of the graph to the lower right (this is best seen when the y-axis is expressed as the logarithm of the number of survivors).

In the first half of the twentieth century, demographers Raymond Pearl and Edward S. Deevey labeled each survivorship pattern: Type I is high survival among juveniles, type II is constant mortality through the life of the cohort, and type III is low survival among juveniles. That terminology became well entrenched in the biological literature by the 1950's. Few species exhibit a pure type I, II, or III pattern, however; instead, survivorship varies so that the pattern may be one type at one part of the cohort's existence and another type later. Perhaps the most common survivorship pattern, especially among vertebrates, is composed of a type III pattern for juveniles and young adults followed by a type I pattern for older adults. This pattern can be explained biologically. Most species tend to suffer heavy juvenile mortality because of predation, starvation, cannibalism, or the inability to cope with a stressful environment. Juveniles that survive this hazardous period then become strong adults that enjoy relatively low mortality. As time passes, the adults reach old age and ultimately fall victim to disease, predation, and organ-system failure, thus causing a second downward plunge in the survivorship curve.

Patterns of Reproduction

Demographers are not interested only in measuring the survivorship of cohorts. They also want to understand the patterns of reproduction, especially among females. Different species show widely varying patterns of reproduction. For example, some species, such as octopuses and certain salmon, reproduce only once in their life and then die soon afterward. Others, such as humans and most birds, reproduce several or many times in their lives. Species that reproduce only once accumulate energy throughout their lives and essentially put all of it into producing young. Reproduction essentially exhausts them to death. Conversely, those that reproduce several times devote only a small amount of their energy to each reproductive event.

Species also vary in their fecundity, which is the number of offspring that an individual makes when it reproduces. Large mammals have low fecundity, because they produce only one or two progeny at a time. Birds, reptiles, and small mammals have higher fecundity because they typically produce a clutch or litter of several offspring. Fish, frogs, and parasitic worms have very high fecundity, producing hundreds or thousands of offspring.

A species' pattern of reproduction is often related to its survivorship. For example, a species with low fecundity or one that reproduces only once tends to have type I or type II survivorship. Conversely, a species that produces huge numbers of offspring generally shows type III survivorship. Many biologists are fascinated by this interrelationship between survivorship and reproduction. Beginning in the 1950's, some demographers proposed mathematically based explanations as to how the interrelationship might have evolved as well as the ecological conditions in which various life histories would be expected. For example, some demographers predicted that species with low fecundity and type I survival should be found in undisturbed, densely populated areas (such as a tropical rain forest). In contrast, species with high fecundity and type III survival should prevail in places that are either uncrowded or highly disturbed (such as an abandoned farm field). Ecologists have conducted field studies of both plants and animals to determine whether the patterns that actually occur in nature fit the theoretical predictions. In some cases the predictions were upheld, but in others they were found to be wrong and had to be modified.

Age Structures and Sex Ratios

Another feature of a population is its age structure, which is simply the number of individuals of each age. Some populations have an age structure characterized by many juveniles and only a few adults. Two situations could account for such a pattern. First, the population could be rapidly expanding, with the adults successfully reproducing many progeny that are enjoying high survival. Second, the population could be producing many offspring that have type III survival. In this second case, the size of the population can remain constant or even decline. Other populations have a different age structure, in which the number of juveniles only slightly exceeds the number of adults. Those populations tend to remain relatively constant over time. Still other populations have an age structure in which there are relatively few juveniles and many adults. Those populations are probably declining or are about to decline because the adults are not successfully reproducing.

Since most animals are unisexual, an important demographic characteristic of a population is its sex ratio, defined as the ratio of males to females. While the ratio for birds and mammals tends to be 1:1 at conception (the fertilization of an egg), it tends to be weighted toward males at birth, because female embryos are slightly less viable. After birth, the sex ratio for mammals tends to favor females, because young males suffer higher mortality. The posthatching ratio in birds tends to remain skewed toward males, because females devote considerable energy to producing young and suffer higher mortality. As a result, male birds must compete with one another for the opportunity to mate with the scarcer females.

The Age-Specific Approach
To understand the demography of a particular species, one must collect information about its survivorship and reproduction. The best survivorship data are obtained when a demographer follows a group of newly born organisms (this being a cohort) over time, periodically counting the survivors until the last one dies. Although that sounds relatively straightforward, many factors complicate the collection of survivorship data; demographers must be willing to adjust their methods to fit the particular species and environmental conditions.

First, a demographer must decide how many newborns should be included in the cohort. Survivorship is usually based on one thousand newborns, but few studies follow that exact number. Instead, demographers follow a certain number of newborns and multiply or divide their data so that the cohort is expressed as one thousand newborns. For example, one may choose to follow five hundred newborns; the number of survivors is then multiplied by two. Demographers generally consider cohorts composed of fewer than one hundred newborns to be too small. Second, methods of determining survivorship are much more different for highly motile organisms, such as mammals and birds, than for more sedentary ones, such as bivalves (oysters and clams). To determine survivorship of a sedentary species, demographers often find some newborns during an initial visit to a site and then periodically revisit that site to count the number of survivors. Highly motile animals are much more difficult to census because they do not stay in one place waiting to be counted. Vertebrates and large invertebrates can be tagged, and individuals can be followed by subsequently recapturing them. Some biologists use small radio transmitters to follow highly active species. The demography of small invertebrates such as insects is best determined when there is only one generation per year and members of the population are all of the same age-class. For such species, demographers merely count the number present at periodic intervals.

Third, the frequency of the census periods varies from species to species. Short-lived species, such as insects, must be censused every week or two. Longer-lived species need to be counted only once a year. Fourth, the definition of a "newborn" may be troublesome, especially for species with complex life cycles. Demographic studies usually begin with the birth of an infant. Some would argue, however, that the fetus should be included in the analysis because the starting point is really conception. Many sedentary marine invertebrates (sponges, starfish, and barnacles) have highly motile larval stages, and these should be included in the analysis for survivorship to be completely understood. Parasitic roundworms and flatworms that have numerous juvenile stages, each found inside a different host, are particularly challenging to the demographer.

The Time-Specific Approach
The survivorship of long-lived species, such as large mammals, is really impossible to determine by the methods given above. Because of their sheer longevity, one could not expect a scientist to be willing to wait decades or centuries until the last member of a cohort dies. Demographers attempt to overcome this problem by using the age distribution of organisms that are alive at one time to infer cohort survivorship. This is often termed a "horizontal" or "time-specific" approach, as opposed to the "vertical" or "age-specific" approach that requires repeated observations of a single cohort. For example, one might construct a time-specific survivorship curve for a population of fish by live-trapping a sufficiently large sample, counting the rings on the scales on each individual (which for many species is correlated with the age in years), and then determining the number of one-year-olds, two-year-olds, and so on. Typically, demographers who use age distributions to infer age-specific survivorship automatically assume that natality and mortality remain constant from year to year. That is often not

the case, however, because environmental conditions often change over time. Thus, demographers must be cautious when using age distribution data to infer survivorship.

Methods for determining fecundity are relatively straightforward. Typically, fertile individuals are collected, their ages are determined, and the number of progeny (eggs or live young) are counted. Species that reproduce continually (parasitic worms) or those that reproduce several times a year (small mammals and many insects) must be observed over a period of time.

Demographers usually want to determine whether the production of new offspring (natality) balances the losses attributable to mortality. To accomplish this, they construct a life table, which is a chart with several columns and rows. Each row represents a different age of the cohort, from birth to death. The columns show the survival and fecundity of the cohort. By recalculating the survivorship and fecundity information, demographers can compute several interesting aspects of the cohort, including the life expectancy of individuals at different ages, the cohort's reproductive value (which is the number of progeny that an individual can expect to produce in the future), the length of a generation for that species, and the growth rate for the population.

Uses of Demography
Demographic techniques have been applied to nonhuman species, particularly by wildlife managers, foresters, and ecologists. Wildlife managers seek to understand how a population is surviving and reproducing within a certain area, and therefore to determine whether it is increasing or decreasing over time. With that information, a wildlife biologist can then estimate the effect of hunting or other management practice on the population. By extension, fisheries biologists can also make use of demographic techniques to determine the growth rate of the species of interest. If the population is determined to be increasing, it can be harvested without fear of depleting the population. Alternatively, one can conduct demographic analyses to see whether certain species are being overfished.

An often unappreciated benefit of survivorship analyses is that they can help ecologists pinpoint factors that limit population growth in an area. This may be especially important in efforts to prevent rare animals and plants from becoming extinct. Once the factor is identified, the population can be appropriately managed. Increasing amounts of public and private money are allocated each year to biologists who conduct demographic studies on rare species.

Kenneth M. Klemow

See also: Adaptive radiation; Biodiversity; Biogeography; Clines, hybrid zones, and introgression; Convergence and divergence; Extinctions and evolutionary explosions; Gene flow; Genetic diversity; Genetic drift; Human population growth; Insect societies; Nonrandom mating, genetic drift, and mutation; Population analysis; Population fluctuations; Population genetics; Population growth; Reproductive strategies.

Sources for Further Study

Begon, Michael, John L. Harper, and Colin R. Townsend. *Ecology: Individuals, Populations, and Communities*. 3d ed. Boston: Blackwell, 1996.

Begon, Michael, Martin Mortimer, and David J. Thompson. *Population Ecology: A Unified Study of Animals and Plants*. 3d ed. Cambridge, Mass.: Blackwell, 1996.

Brewer, Richard. *The Science of Ecology*. 2d ed. Fort Worth, Tex.: Saunders College Publishing, 1994.

Elseth, Gerald D., and Kandy D. Baumgardner. *Population Biology*. New York: Van Nostrand, 1981.

Gotelli, Nicholas J. *A Primer of Ecology*. 2d ed. Sunderland, Mass.: Sinauer Associates, 1998.

Hutchinson, G. Evelyn. *An Introduction to Population Ecology*. New Haven, Conn.: Yale University Press, 1978.

Smith, Robert Leo. *Elements of Ecology*. 4th ed. San Francisco, Calif.: Benjamin/Cummings, 2000.

Wilson, Edward O., and William Bossert. *A Primer of Population Biology*. Sunderland, Mass.: Sinauer Associates, 1977.

DENDROCHRONOLOGY

Types of ecology: Evolutionary ecology; Paleoecology

Dendrochronology is the science of examining and comparing growth rings in both living and aged woods to draw inferences about past ecosystems and environmental conditions.

In forested regions with seasonal climates, trees produce a growth ring to correspond with each growing season. At the beginning of the growing season, when conditions are optimum, the vascular cambium produces many files of large xylem cells that form wood. As the conditions become less optimal, the size and number of cells produced decreases until growth stops at the end of the growing season. These seasonal differences in size and number of cells produced are usually visible to the unaided eye. The layers produced during rapid early growth appear relatively light-colored because the volume of the large cells is primarily intracellular space. These layers are frequently called springwood because in northern temperate regions spring is the beginning of the growing season. Wood formed later, summerwood, is darker because the cells are smaller and more tightly compacted. The juxtaposition of dark summerwood of one year with the light springwood of the following year marks a distinct line between growth increments. The width of the ring between one line and the next measures the growth increment for a single growing season. If there is a single growing season per year, as in much of the temperate world, then a tree will produce a single annual ring each year.

Tree Rings and Climate
Leonardo da Vinci is credited with counting tree rings in the early 1500's to determine "the nature of past seasons," but it was not until the early 1900's that dendrochronology was established as a science. Andrew Douglass, an astronomer interested in relating sunspot activity to climate patterns on Earth, began to record the sequences of wide and narrow rings in the wood of Douglas firs and ponderosa pines in the American Southwest. Originally, trees were cut down in order to examine the ring patterns, but in the 1920's Douglass began to use a Swedish increment borer to remove core samples from living trees. This instrument works like a hollow drill that is screwed into a tree by hand. When the borer reaches the center of the tree it is unscrewed, and the wood core sample inside is withdrawn with the borer. The small hole quickly fills with sap, and the tree is unharmed.

Borers range in size from 20 centimeters to 100 centimeters or more in length, so with care, samples can be taken from very large, very old living trees.

Counting backward in the rings is counting backward in time. By correlating the size of a ring with the known regional climate of the year the ring was produced, a researcher can calibrate a core sample to indicate the surrounding climate during any year of the tree's growth. By extending his work to sequoias in California, Douglass was able to map a chronology extending back three thousand years.

Tree Rings and History
In order to extend his chronologies so far back in time, Douglass devised the method of cross-dating. By matching distinctive synchronous ring patterns from living and dead trees of the same species in a region, researchers can extend the pattern further into the past than the lifetime of the younger tree. Archaeologists quickly realized that this was a tool that could help to assign the age of prehistoric sites by determining the age of wood artefacts and construction timbers. In this way archaeologists could calculate the age of pre-Columbian southwestern ruins, such as the cliff dwellings at Mesa Verde, Arizona, by cross-dating living trees with dead trees and the latter with timbers from the sites. In 1937 Douglass established the Laboratory of Tree-Ring Research at the University of Arizona, which continues to be a major center of dendrochronological research.

Fine-Tuning
In the mid-1950's Edmund Schulman confirmed the great age of living bristlecone pines in the Inyo National Forest of the White Mountains of California. In 1957 he discovered the Methuselah Tree, which was more than forty-six hundred years old. The section of forest in which he worked is now known as the Ancient Bristlecone Pine Forest. During the next thirty years, Charles Ferguson extended the bristlecone chronology of this area back 8,686 years. This sequence formed the basis for calibrating the technique of radiocarbon dating. In the 1960's, radiocarbon analysis began to be used to determine the age of organic (carbon-based) artefacts from ancient sites. It has the advantage of being applicable to any item made of organic material but the disadvantage of having a built-in uncertainty of 2 percent or more. Tree-ring chronologies provide an absolute date against which radiocarbon analyses of wood samples from a site can be compared.

At about the same time, Valmore LaMarche, a young geologist, began to study root growth of the ancient trees to determine how they could be used to predict the erosional history of a site. By cross-referencing growth ring

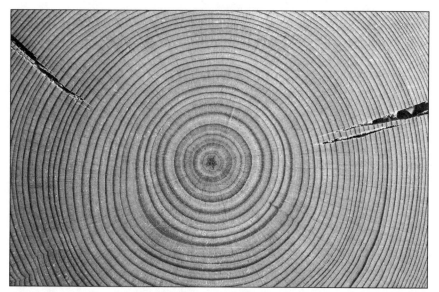

Dendrochronology, or tree-ring counting, can be used to assess the age of a tree because the width of the ring between one line and the next measures the growth increment for a single growing season. (PhotoDisc)

asymmetry to degree of exposure and slope profiles, he was able to estimate rates of soil erosion and rock weathering, which in turn could be cross-referenced to the climatic conditions predicted by growth rings in the stem. LaMarche and his colleagues, particularly Harold Fritts, continued to "fine-tune" the reading of growth rings to be able to take into account factors such as soil characteristics, frost patterns, and daily, weekly, and monthly patterns.

The Oldest Tree
The Methuselah Tree, mentioned above, is the oldest known living tree. Schulman also cored a 4,700-year-old living specimen in the White Mountains, but he did not name it or identify its location. While most of the living specimens older than 4,000 years are found in the White Mountains, the oldest living tree was discovered in the Wheeler Peak area of what is now Great Basin National Park in eastern Nevada. This tree, variously known as WPN-114 and the Prometheus Tree, was estimated to be between 4,900 and 5,100 years old when it was cut down in 1964 as part of a research project. The controversy that followed has left many interesting but unanswered questions.

Marshall D. Sundberg

Dendrochronology

See also: Evolution of plants and climates; Forests; Global warming; Old-growth forests; Paleoecology.

Sources for Further Study

Cohen, Michael P. *Garden of Bristlecones: Tales of Change in the Great Basin.* Reno: University of Nevada Press, 2000.

Cook, E. R., and L. A. Kairiustis, eds. *Methods of Dendrochronology: Applications in the Environmental Sciences.* Boston: Kluwer Academic, 1987.

Harlow, William M. *Inside Wood: Masterpiece of Nature.* Washington, D.C.: The American Forestry Association, 1970.

McGraw, Donald J. *Andrew Ellicott Douglass and the Role of the Giant Sequoia in the Development of Dendrochronology.* Lewiston, N.Y.: Edwin Mellen Press, 2001.

Stokes, Marvin A., and Terah L. Smiley. *An Introduction to Tree-Ring Dating.* Tucson: University of Arizona Press, 1996.

DESERTIFICATION

Type of ecology: Ecosystem ecology

Desertification is the degradation of arid, semiarid, and dry, subhumid lands as a result of human activities or climatic variations, such as a prolonged drought.

Desertification is recognized by scientists and policymakers as a major economic, social, and environmental problem in more than one hundred countries. It impacts about one billion people throughout the world. Deserts are climatic regions that receive fewer than 25 centimeters (10 inches) of precipitation per year. They constitute the most widespread of all climates of the world, occupying 25 percent of the earth's land area. Most deserts are surrounded by semiarid climates referred to as steppes, which occupy 8 percent of the world's lands. Deserts occur in the interior of continents, on the leeward side of mountains, and along the west sides of continents in subtropical regions. All the world's deserts risk further desertification.

Deserts of the World

The largest deserts are in North Africa, Asia, Australia, and North America. Four thousand to six thousand years ago, these desert areas were less extensive and were occupied by prairie or savanna grasslands. Rock paintings found in the Sahara Desert show that humans during this era hunted buffalo and raised cattle on grasslands where giraffes browsed. The region near the Tigris and Euphrates Rivers in the Middle East was also fertile. In the desert of northwest India, cattle and goats were grazed, and people lived in cities that have long since been abandoned. The deserts in the southwestern region of North America appear to have been wetter, according to the study of tree rings (dendrochronology) from that area. Ancient Palestine, which includes the Negev Desert of present-day Israel, was lush and was occupied by three million people.

Scientists use various methods to determine the historical climatic conditions of a region. These methods include studies of the historical distribution of trees and shrubs determined by the deposit patterns in lakes and bogs, patterns of ancient sand dunes, changes in lake levels through time, archaeological records, and tree rings.

The earth's creeping deserts supported approximately 720 million people, or one-sixth of the world's population, in the late 1970's. According to the United Nations, the world's hyperarid or extreme deserts are the Atacama and Peruvian Deserts (located along the west coast of South America), the Sonoran Desert of North America, the Takla Makan Desert of Central Asia, the Arabian Desert of Saudi Arabia, and the Sahara Desert of North Africa, which is the largest desert in the world. The arid zones surround the extreme desert zones, and the semiarid zones surround the arid zones. Areas that surround the semiarid zones have a high risk of becoming desert. By the late 1980's the expanding deserts were claiming about 15 million acres of land per year, or an area approximately the size of the state of West Virginia. The total area threatened by desertification equaled about 37.5 million square kilometers (14.5 million square miles).

Causes of Desertification
Desertification results from a two-prong process: climatic variations and human activities. First, the major deserts of the world are located in areas of high atmospheric pressure, which experience subsiding dry air unfavorable to precipitation. Subtropical deserts have been experiencing prolonged periods of drought since the late 1960's, which causes these areas to be dryer than usual.

The problem of desertification was identified in the late 1960's and early 1970's as a result of severe drought in the Sahel Desert, which extends along the southern margin of the Sahara in West Africa. Rainfall has declined an average of 30 percent in the Sahel. One set of scientific studies of the drought focuses on changes in heat distribution in the ocean. A correlation has been found between sea surface temperatures and the reduction of rainfall in the Sahel. The Atlantic Ocean's higher surface temperatures south of the equator and lower temperatures north of the equator west of Africa are associated with lower precipitation in northern tropical Africa. However, the cause for the change in sea surface temperature patterns has not been determined.

Another set of studies is associated with land-cover changes. Lack of rain causes the ground and soils to get extremely dry. Without vegetative cover to hold it in place, thin soil blows away. As the water table drops from the lack of the natural recharge of the aquifers and the withdrawal of water by desert dwellers, inhabitants are forced to migrate to the grasslands and forests at fringes of the desert. Overgrazing, overcultivation, deforestation, and poor irrigation practices (which can cause salinization of soils) eventually lead to a repetition of the process, and the desert begins to encroach. These causes are influenced by changes in population, climate, and social and economic conditions.

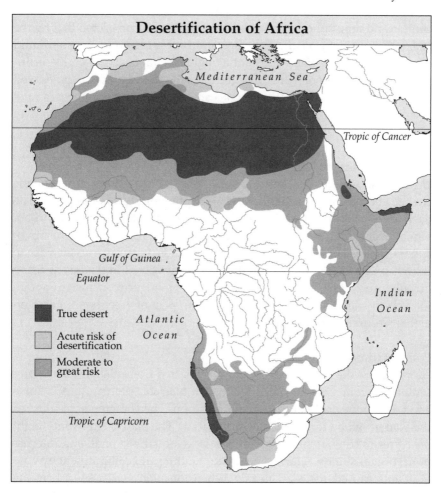

The fundamental cause of desertification, therefore, is human activity. This is especially true when environmental stress occurs because of seasonal dryness, drought, or high winds. Many different forms of social, economic, and political pressure cause the overutilization of these dry lands. People may be pushed onto unsuitable agricultural land because of land shortages, poverty, and other forces, while farmers overcultivate the fields in the few remaining fertile land areas.

Atmospheric Consequences
A reduction in vegetation cover and soil quality may impact the local climate by causing a rise in temperatures and a reduction in moisture. This can, in turn, impact the area beyond the desert by causing changes in the

climate and atmospheric patterns of the region. It is predicted that by the year 2050 substantial changes in vegetation cover in humid and subhumid areas will occur and cause substantial regional climatic changes. Desertification is a global problem because it causes the loss of biodiversity as well as the pollution of rivers, lakes, and oceans. As a result of excessive rainfall and flooding in subhumid areas, fields lacking sufficient vegetation may be eroded by runoff.

Greenhouse Effect
Desertification and even the efforts to combat it may be impacting climatic change because of the emission and absorption of greenhouse gases. The decline in vegetation and soil quality can result in the release of carbon, while revegetation can influence the absorption of carbon from the atmosphere. The use of fertilizer to reclaim dry lands may cause an increase in nitrous oxide emissions. Although scientists involved in studies of rising greenhouse gases have not been able to gather evidence conclusive enough to support such theories, evidence of the impact of greenhouse gases on global warming continues to accumulate.

Policy Actions
As a result of the Sahelian drought, which lasted from 1968 to 1973, representatives from various countries met in Nairobi, Kenya, in 1977 for a United Nations conference on desertification. The conference resulted in the Plan of Action to Combat Desertification. The plan listed twenty-eight measures to combat land degradation by national, regional, and international organizations. A lack of adequate funding and commitment by governments caused the plan to fail. When the plan was assessed by the United Nations Environment Programme (UNEP), it found that little had been accomplished and that desertification had increased.

As a result of the 1977 United Nations conference, several countries developed national plans to combat desertification. One example is Kenya, where local organizations have worked with primary schools to plant five thousand to ten thousand seedlings per year. One U.S.-based organization promotes reforestation by providing materials to establish nurseries, training programs, and extension services. Community efforts to combat desertification have been more successful, and UNEP has recognized that such projects have a greater success rate than top-down projects. The Earth Summit, held in Rio de Janeiro, Brazil, in 1992, supported the concept of sustainable development at the community level to combat the problem of desertification.

Roberto Garza

See also: Biomes: determinants; Biomes: types; Deserts; Ecosystems: definition and history; Erosion and erosion control; Global warming; Greenhouse effect; Hydrologic cycle; Rain forests and the atmosphere; Soil; Soil contamination.

Sources for Further Study

Bryson, Reid A., and Thomas J. Murray. *Climates of Hunger: Mankind and the World's Changing Weather.* Madison: University of Wisconsin Press, 1977.

Glantz, Michael H., ed. *Desertification: Environmental Degradation in and Around Arid Lands.* Boulder, Colo.: Westview Press, 1977.

Grainger, Alan. *Desertification: How People Make Deserts, Why People Can Stop, and Why They Don't.* London: International Institute for Environment and Development, 1982.

Hulme, Mike, and Mick Kelly. "Exploring the Links Between Desertification and Climate Change." *Environment,* July/August, 1993.

Mainguet, Monique. *Aridity: Droughts and Human Development.* New York: Springer, 1999.

_____. *Desertification: Natural Background and Human Mismanagement.* 2d ed. New York: Springer-Verlag, 1994.

Matthews, Samuel W. "What's Happening to Our Climate." *National Geographic,* November, 1976.

Postel, Sandra. "Land's End." *Worldwatch,* May/June, 1989.

DESERTS

Types of ecology: Biomes; Ecosystem ecology

Regions characterized by 10 inches or less of precipitation per year are considered deserts. Desert ecosystems are subject to disruption by human activities.

Deserts are regions, or biomes, too dry to support grasslands or forest vegetation but with enough moisture to allow specially adapted plants to live. In deserts, hot days alternate with cold nights. Ninety percent of incoming solar radiation reaches the ground during the day, and 90 percent of that is radiated back out into space at night, the result of the absence of clouds, low humidity, and sparse vegetation. The surface of the ground in desert areas is devoid of a continuous layer of plant litter and is usually rocky or sandy. Nutrient cycling in deserts is tight, with phosphorus and nitrogen typically in short supply.

How Deserts Form

Most large landmasses have interior desert regions. Air masses blown inland from coastal areas lose their moisture before reaching the interior. Examples include the Gobi Desert in Mongolia and parts of the Sahara Desert in Africa.

Another factor in the formation of deserts is the rain-shadow effect. If moisture-laden air masses bump up against a mountain range, the air mass is deflected upward. As the air mass rises, it cools, and moisture precipitates as rain or snow on the windward side of the mountain range. As the air mass passes over the mountain range, it begins to descend. Because it lost most of its moisture on the windward side, the air mass is dry. As it descends, the air heats up, creating drier conditions on the leeward side of the mountain range. Sometimes these differences in moisture are so pronounced that different plant communities grow on the windward and leeward sides.

Latitude can also influence desert formation. Most deserts lie between 15 and 35 degrees north or south latitude. At the equator, the sun's rays hit the earth straight on. Moist equatorial air, warmed by intense heat from the sun, rises. As this air rises, it cools and loses its moisture, which falls as rain; this is why it usually rains every day in the equatorial rain forests. The Coriolis force causes the air masses to veer off, to the north in the Northern Hemisphere and to the south in the Southern Hemisphere. The now-dry air begins to descend and warm, reaching the ground between 15 and 35 degrees north and south latitude, creating the belt of deserts circling the globe between these latitudes.

Deserts can also form along coastlines next to cold-water ocean currents, which chill the air above them, decreasing their moisture content. Offshore winds blow the air above cold ocean waters back out to sea. In deserts, rain is infrequent, creating great hardships for the native plants and animals. The main source of moisture for the plants and animals of coastal deserts is fog.

Types of Deserts
Depending upon whether the precipitation comes from rain or snow, deserts can be divided into hot (rain) or cold (snow) deserts. The deserts of Arabia, Australia, Chihuahua, Kalahari, Monte, Sonora, and Thar are all considered hot deserts, found in lower latitudes. Cold deserts, found at higher latitudes, include the Atacama, Gobi, Basin, Iranian, Namib, and Turkestan deserts. Regardless of whether the desert is hot or cold, organisms living within desert biomes have to adapt to cope with the scarcity of water and violent swings of temperature. North America contains four different deserts that are usually defined by their characteristic vegetation, which ecologists call indicator species.

Although desert climates are characterized by less than 10 inches of precipitation per year, they can support a broad variety of plant and animal life adapted to survive such arid climates. Cacti, for example, have developed a method of photosynthesis that maximizes water retention, as well as protective spines that inhibit predators that might feed on them. (PhotoDisc)

In Mexico's Chihuahuan Desert, lechuguilla (*Agave lechuguilla*) is the indicator species. Fibers from lechuguilla can be made into nets, baskets, mats, ropes, and sandals. Its stems yield a soap substitute, and its pulp has been used as a spot remover. Certain compounds in lechuguilla are poisonous and were once used to poison the tips of arrows and as fish poisons. Two of the most common plants in the Chihuahuan Desert are creosote bush (*Larrea divaricata*) and soaptree yucca (*Yucca elata*). Cacti in this desert are numerous and diverse, especially the prickly pears and chollas.

The Joshua tree (*Yucca brevifolia*), is the indicator species of the Mojave Desert in Southern California. Nearly one-fourth of all the Mojave Desert plants are endemics, including the Joshua tree, Parry saltbush, Mojave sage, and woolly bur sage.

The Great Basin Desert, situated between the Sierra Nevada and the Rocky Mountains, is a cold desert, with fewer plant species than other North American deserts. Great Basin Desert plants are small to medium-size shrubs, usually sagebrushes or saltbushes. The indicator species of the Great Basin is big sagebrush (*Artemisia tridentata*). Other common plants are littleleaf horsebrush and Mormon tea. The major cactus species is the Plains prickly pear.

In the Sonoran Desert in Mexico, California, and Arizona, plants come in more shapes and sizes than in the other North American deserts, especially in the *Cactaceae*. The indicator species of the Sonoran Desert is the saguaro cactus (*Carnegiea gigantecus*).

The Sahara Desert of northern Africa is the world's largest, at 3.5 million square miles. The Northern Hemisphere also contains the Arabian, Indian, and Iranian deserts and the Eurasian deserts: the Takla Makan, Turkestan, and Gobi. Deserts in the Southern Hemisphere include the Australian, Kalahari, Namib, Atacama-Peruvian (the world's driest), and the Patagonian.

Desert Vegetation

Many typical desert perennial plants, such as members of the *Cactaceae* (the cactus family), have thick, fleshy stems or leaves with heavy cuticles, sunken stomata (pores), and spiny defenses against browsing animals. The spines also trap a layer of air around the plant, retarding moisture loss. Desert plants, many of which photosynthesize using C_4 or CAM (crassulean acid metabolism), live spaced out from other plants. Many desert plants are tall and thin, to minimize the surface area exposed to the strongest light. For example, the entire stem of the Saguaro cactus is exposed to sunlight in the early morning and late afternoon; at noon, only the tops of the stems receive full sun. These traits allow the plants to cope with heat stress and competition for water and avoid damage from herbivores (plant-eating animals).

Where the mixture of heat and water stress is less severe, perennial bushes of the *Chenopodiaceae* (goosefoot family) or *Asteraceae* (sunflower family) form clumps of vegetation surrounded by bare ground. Numerous annuals, called ephemerals, can grow prolifically, if only briefly, following rainfall.

Unrelated plant families from different desert areas of the world show similar adaptations to desert conditions. This has resulted from a process called convergent evolution.

Animal Life in Deserts
Animals living in the desert include mammals, birds and fish, reptiles and amphibians, and insects and spiders. A few examples of mammals include bats, bighorn sheep, bobcats, and coyotes. Desert bats, members of the suborder *Microchioptera*, are often considered to be flying mice, but they are more closely related to primates. Bats are unique among mammals because they can fly. Most bat species also possess a system of acoustic orientation, technically known as echolocation. The bighorn sheep lives in dry, desert mountain ranges and foothills near rocky cliffs. Its body is compact and muscular; the muzzle is narrow and pointed; the ears, short and pointed; the tail, very short. The fur is deerlike and usually a shade of brown, with whitish rump patches. Bighorns are grazers, consuming grasses, sedges, and other low-lying plants. The bobcat (*Felis rufus*) has long legs, large paws, and a short tail (six to seven inches long), with average body weight of fifteen to twenty pounds. However, it is quite fierce and is equipped to kill animals as large as deer. The desert coyote (*Canis latrans*), a member of the dog family, weighs about twenty pounds, less than half the weight of its mountain kin, which can weigh up to fifty pounds. Its body color is light gray or tan, which helps it to reflect heat and blend in in the desert.

A variety of birds reside in the desert due to the abundance of insects and spiders. Golden eagles (*Aquila chrysaetos*) get their name from the golden feathers on the back of their necks. They are birds of the open country, building large stick nests in trees or cliff walls where they have plenty of room to maneuver. Adults weigh 9 to 12.5 pounds, with females usually larger than males. Ravens (*Corvus corax*) are the largest birds of the crow family, averaging twenty-four inches tall, with a wingspan of forty-six to fifty-six inches. Ravens are strong fliers that can soar like a hawk, and they may form large flocks of over several hundred individuals during their autumn migration. The American turkey (*Meleagris gallopavo*), the largest upland game bird in North America, is thirty-six to forty-eight inches long, with a four- to five-foot wingspan. Males average ten inches longer than females, which are paler and of a more buff color. Turkeys inhabit a variety of habitats from open grassland and fields to open woodlands and mature deciduous or coniferous forests.

Many species of reptiles and amphibians live in the desert, including the black-collared lizard (genus *Crotaphytus*), bullfrog (*Rana catesbeiana*), desert dinosaur (orders *Saurischia* and *Ornithischia*), desert iguana (*Dispsosaurus dorsalis*), rattlesnake (genus *Crotalus*), and many others. Insects and spiders include dragonflies (suborder *Anisoptera*), scorpions (order *Scorpionida*), and black widow spiders (*Latrodectus hesperus*). It is nothing short of a miracle that such an abundance of life, both plant and animal, can survive in the extreme conditions of the desert.

Animal Survival in the Desert

Among the thousands of desert animal species, many have remarkable behavioral and structural adaptations for avoiding excess heat. Equally ingenious are the diverse mechanisms various animal species have developed to acquire, conserve, recycle, and actually manufacture water.

Certain species of birds, such as the *Phanopepla*, breed during the relatively cool spring, then leave the desert for cooler areas at higher elevations or along the Pacific coast. The Costa's hummingbird begins breeding in late winter and leaves in late spring when temperatures become extreme. Many birds, as well as other mammals and reptiles, are crepuscular, meaning they are active only at dusk and again at dawn. Many animals, including bats, many snakes, most rodents, and some larger animals such as foxes and skunks, are nocturnal, restricting all their activities to the cooler temperatures of the night and sleeping in a cool den, cave, or burrow by day. A few desert animals, such as the round-tailed ground squirrel, sleep away the hottest part of summer and also hibernate in winter to avoid the cold season. Yet other animals, such as desert toads, remain dormant deep in the ground until the summer rains fill ponds. They then emerge, breed, lay eggs, and replenish their body reserves of food and water for another long period.

Various mechanisms are employed to dissipate heat absorbed by desert animals. Many mammals have long appendages to release body heat into their environment. The enormous ears of jackrabbits, with their many blood vessels, dissipate heat when the animal is resting in a cool, shady location. Their close relatives in cooler regions have much shorter ears. New World vultures, dark in color and thus absorbing considerable heat in the desert, excrete urine on their legs to cool them by evaporation, and circulate the cooled blood back through the body. Many desert animals are paler than their relatives elsewhere, ensuring that they not only suffer less heat absorption, but also are less conspicuous to predators in the bright, pallid surroundings.

The mechanisms by which water is retained by desert animals are even more elaborate. Reptiles and birds excrete metabolic wastes in the form of uric acid, an insoluble white compound, wasting very little water in the process. Other animals retain water by burrowing into moist soil during the dry daylight hours. Some predatory and scavenging animals can obtain their entire water needs from the food they eat. Most mammals, however, need access to a good supply of fresh water at least every few days, if not daily, due to the considerable water loss from excretion of urea, a soluble compound.

Many desert animals obtain water from plants, particularly succulent ones such as cactus and saguaro. Many species of insects thrive in the desert, as they tap plant fluids for water and nectar. The abundance of insect life permits insectivorous birds, bats, and lizards to thrive in the desert. Certain desert animals, such as kangaroo rats, have multiple adaptation mechanisms to acquire and conserve water. First, they live in underground dens that they seal off to block out heat and to recycle the moisture from their own breathing. Second, they have specialized kidneys with extra microscopic projections to extract most of the water from their urine and return it to the bloodstream. Third, and most fascinating of all, they actually manufacture their water metabolically from the digestion of dry seeds. These are just a few examples of the variety of ingenious adaptations animals use to survive in the desert, overcoming the extremes of heat and the paucity of water.

Habitat Loss

Urban and suburban sprawl paves over desert land and destroys habitat for plants and animals, some of which are endemic to specific deserts. Farmers and metropolitan-area builders can tap into critical desert water supplies, changing the hydrology of desert regions. Off-road vehicles can destroy plant and animal life and leave tracks that may last for decades.

Carol S. Radford and Yujia Weng

See also: Biomes: determinants; Biomes: types; Desertification; Ecosystems: definition and history; Erosion and erosion control; Global warming; Greenhouse effect; Hydrologic cycle; Nutrient cycles; Rain forests and the atmosphere; Soil; Soil contamination.

Sources for Further Study

Allaby, Michael. *Deserts*. New York: Facts on File, 2001.

Baylor, Byrd. *The Desert Is Theirs*. Reprint. New York: Aladdin Books, 1987.

Bothma, J. du P. *Carnivore Ecology in Arid Lands*. New York: Springer-Verlag, 1998.

Bowers, Janice. *Shrubs and Trees of the Southwest Deserts*. Tucson, Ariz.: Southwest Parks & Monuments Association, 1993.

Epple, Anne O. *Field Guide to the Plants of Arizona*. Helena, Mont.: Falcon, 1997.

Hare, John. *The Lost Camels of Tartary: A Quest into Forbidden China*. Boston: Little, Brown, 1998.

Larson, Peggy Pickering, Lane Larson, Edward Abbey, and Iy Larson. *The Deserts of the Southwest: A Sierra Club Naturalist's Guide*. 2d ed. San Francisco: Sierra Club Books, 2000.

Raskin, Lawrence, and Debora Pearson. *Fifty-two Days by Camel: My Sahara Adventure*. Toronto: Annick Press, 1998.

DEVELOPMENT AND ECOLOGICAL STRATEGIES

Type of ecology: Evolutionary ecology

The relationship between ontogeny (individual development) and phylogeny (the evolution of species and lineages) is a core concept in the study of ecology and life sciences in general. Changes in developmental timing produce parallels between ontogeny and phylogeny. The subject illuminates the biology of regulation, the evolution of ecological strategies, and the mechanisms for evolutionary change in form.

The idea of a relationship between individual development, or ontogeny, and the evolutionary history of a race, or phylogeny, is an old one. The concept received much attention in the nineteenth century and is often associated with the names of Karl Ernst von Baer and Ernst Haeckel, two prominent German biologists. It was Haeckel who coined the catchphrase and dominant paradigm: Ontogeny recapitulates (or repeats) phylogeny. Since Haeckel's time, however, the relations between ontogeny and phylogeny have been portrayed in a variety of ways, including the reverse notion that phylogeny is the succession of ontogenies. Research in the 1970's and 1980's on the parallels between ontogeny and phylogeny focused on the change of timing in developmental events as a mechanism for recapitulation and on the developmental-genetic basis of evolutionary change.

Early Concepts of Biogenetic Law

During the early nineteenth century, two different concepts of parallels between development and evolution arose. The German J. F. Meckel and the Frenchman E. R. Serres believed that a higher animal in its embryonic development recapitulates the adult structures of animals below it on a scale of being. Baer, on the other hand, argued that no higher animal repeats an earlier adult stage but rather the embryo proceeds from undifferentiated homogeneity to differentiated heterogeneity, from the general to the specific. Baer published his famous and influential four laws in 1828:

(1) The more general characters of a large group of animals appear earlier in their embryos than the more special characters.
(2) From the most general forms, the less general are developed.
(3) Every embryo of a given animal, instead of passing through the other forms, becomes separated from them.
(4) The embryo of a higher form never resembles any other form, only its embryo.

By the late nineteenth century, the notion of recapitulation and Baer's laws of embryonic similarity were recast in evolutionary terms. Haeckel and others established the biogenetic law: that is, ontogeny recapitulates the adult stages of phylogeny. It was, in a sense, an updated version of the Serres-Meckel law but differed in that the notion was valid not only for a chain of being but also for many divergent lines of descent; ancestors had evolved into complex forms and were now considered to be modified by descent. More specifically, Haeckel thought of ontogeny as a short and quick recapitulation of phylogeny caused by the physiological functions of heredity and adaptation. During its individual development, he wrote, the organic individual repeats the most important changes in form through which its forefathers passed during the slow and long course of their paleontological development. The adult stages of ancestors are repeated during the development of descendants but crowded back into earlier stages of ontogeny. Ontogeny is the abbreviated version of phylogeny. These repeated stages reflect the history of the race. Haeckel considered phylogeny to be the mechanical cause of ontogeny.

The classic example of recapitulation is the stage of development in an unhatched bird or unborn mammal when gill slits are present. Haeckel argued that gill slits in this stage represented gill slits of the adult stage of ancestral fish, which in birds and mammals were pressed back into early stages of development. This theory differed from Baer's notion that the gill slit in the human embryo and in the adult fish represented the same stage in development. The gill slits, explained the recapitulationists, got from a large adult ancestor to a small embryo in two ways: first, terminal addition (in which stages are added to the end of an ancestral ontogeny); and second, condensation (in which development is speeded up as ancestral features are pushed back to earlier stages of the embryo). Haeckel also coined another term widely used currently in another sense: "heterochrony." He used the term to denote a displacement in time of the appearance of one organ in ontogeny before another, thus disrupting the recapitulation of phylogeny in ontogeny. Haeckel was not, however, interested primarily in mechanisms or in embryology for its own sake, but rather for the information it could provide for developing evolutionary histories.

Recapitulation in the Twentieth Century

With the rise of mechanistic experimental embryology and with the establishment of Mendelian genetics in the early twentieth century, the biogenetic law was largely repudiated by biologists. Descriptive embryology was out of fashion, and the existence of genes made the two correlate laws to recapitulation—terminal addition and condensation—untenable. One

162

of the most influential modifications for later work on the subject was broached in a paper by Walter Garstang in 1922, in which he reformulated the theory of recapitulation and refurbished the concept of heterochrony. Garstang argued that phylogeny does not control ontogeny but rather makes a record of the former: that is, phylogeny is a result of ontogeny. He suggested that adaptive changes in a larval stage coupled with shifts in the timing of development (heterochrony) could result in radical shifts in adult morphology.

Stephen Jay Gould resurrected the long unpopular concept of recapitulation with his book *Ontogeny and Phylogeny* (1977). In addition to recounting the historical development of the idea of recapitulation, he made an original contribution to defining and explicating the mechanism (heterochrony) involved in producing parallels between ontogeny and phylogeny. He argued that heterochrony—"changes in the relative time of appearance and rate of development for characters already present in ancestors"—was of prime evolutionary importance. He reduced Gavin de Beer's complex eight-mode analysis of heterochrony to two simplified processes: acceleration and retardation. Acceleration occurs if a character appears earlier in the ontogeny of a descendant than it did in an ancestor because of a speeding up of development. Conversely, retardation occurs if a character appears later in the ontogeny of a descendant than it did in an ancestor because of a slowing down of development. To demonstrate these concepts, Gould introduced a "clock model" in order to bring some standardization and quantification to the heterochrony concept.

He considered the primary evolutionary value of ontogeny and phylogeny to be in the immediate ecological advantages for slow or rapid maturation rather than in the long-term changes of form. Neoteny (the opposite of recapitulation) is the most important determinant of human evolution. Humans have evolved by retaining the young characters of their ancestors and have therefore achieved behavioral flexibility and their characteristic form. For example, there is a striking resemblance between some types of juvenile apes and adult humans; this similarity for the ape soon fades in its ontogeny as the jaw begins to protrude and the brain shrinks. Gould also insightfully predicted that an understanding of ontogeny and phylogeny would lead to a rapprochement between molecular and evolutionary biology.

By the 1980's, Rudolf Raff and Thomas Kaufman found this rapprochement by synthesizing embryology with genetics and evolution. Their work focuses on the developmental-genetic mechanisms that generate evolutionary change in morphology. They believe that a genetic program governs ontogeny, that the great decisions in development are made by a small number of genes that function as switches between alternate states or pathways. When these genetic switch systems are modified, evolutionary changes in morphology occur mechanistically. They argue further that regulatory genes—genes that control development by turning structural genes on and off—control the timing of development, make decisions about the fates of cells, and integrate the expression of structural genes to produce differentiated tissue. All this plays a considerable role in evolution.

Description vs. Experimentation

Both embryology and evolution have traditionally been descriptive sciences using methods of observation and comparison. By the end of the nineteenth century, a dichotomy had arisen between the naturalistic (descriptive) and the experimentalist tradition. The naturalists' tradition viewed the organism as a whole, and morphological studies and observations of embryological development were central to their program. Experimentalists, on the other hand, focused on laboratory studies of isolated aspects of function. A mechanistic outlook was compatible with this experimental approach.

Modern embryology uses both descriptive and experimental methods. Descriptive embryology uses topographic, histological (tissue analysis), cytological (cell analysis), and electron microscope techniques supplemented by morphometric (the measurement of form) analysis. Embryos are visualized using either plastic models of developmental stages, schematic drawings, or computer simulations. Cell lineage drawings are also used with the comparative method for phylogenies.

Experimental embryology, on the other hand, uses more invasive methods of manipulating the organism. During this field of study's early period, scientists subjected embryos to various changes to their normal path of development; they were chopped into pieces, transplanted, exposed to chemicals, and spun in centrifuges. Later, fate maps came into usage in order to determine the future development of regions in the embryo. It was found that small patches of cells on the surface of the embryo could be stained, without damaging the cell, by applying small pieces of agar soaked in a vital dye. One could then follow the stained cells to their eventual position in the gastrula. Amphibian eggs are used as material because they are big, easy to procure, and can undergo radical experimental manipulation.

Interdisciplinary Studies

Evolutionary theory primarily uses paleontology (study of the fossil record) to study the evolutionary history of species, yet Gould also used quantification (the clock model, for example), statistics, and ecology to understand the parallels between ontogeny and phylogeny. Most scientists interested in the relationships between ontogeny and phylogeny chiefly use comparative and theoretical methods. For example, they compare structures in different animal groups or compare the adult structures of an animal with the young stage of another. If similarities exist, are the lineages similar? Are the stages in ontogenetic development similar to those of the development of the whole species?

Yet, the study of relationships between ontogeny and phylogeny is an interdisciplinary subject. Not only are methods from embryology and evolutionary theory of help, but also, increasingly, techniques are applied from molecular genetics. Haeckel's method was primarily a descriptive historical one, and he collected myriad descriptive studies of different animals. Although scientists in those days had relatively simple microscopes, they left meticulous and detailed accounts.

A fusion of embryology, evolution, and genetics involves combining different methods from each of the respective disciplines for the study of the relationship between ontogeny and phylogeny. The unifying approach has been causal-analytical, in the sense that biologists have been examining mechanisms that produce parallels between ontogeny and phylogeny as well as the developmental-genetic basis for evolutionary change. The methods are either technical or theoretical. The technical ones include the use of the electron microscope, histological, cytological, and experimental analyses; the theoretical methods include comparison, historical analysis, observations, statistics, and computer simulation.

Ramifications Beyond Science
The relationship between ontogeny and phylogeny is one of the most important ideas in biology and a central theme in evolutionary biology. It illuminates the evolution of ecological strategies, large-scale evolutionary change, and the biology of regulation. This scientific idea has also had far-reaching influences in areas such as anthropology, political theory, literature, child development, education, and psychology.

In the late nineteenth century, embryological development was a major part of evolutionary theory; however, that was not the case for much of the twentieth century. Although there was some interest in embryology and evolution from the 1920's to the 1950's by Garstang, J. S. Huxley, de Beer, and Richard Goldschmidt, during the first three decades of the twentieth century genetics and development were among the most important and active areas in biological thought, yet there were few attempts to integrate the two areas. It is this new synthesis of evolution, embryology, and genetics that has emerged as one of the most exciting frontiers in the life sciences.

Development and ecological strategies

Although knowledge to be gained from a synthesis of development and evolution seems not to have any immediate practical application, it can offer greater insights into mechanisms of evolution, and a knowledge of evolution will give similar insights into mechanisms of development. A study of these relations and interactions also enlarges humankind's understanding of the nature of the development of individuals and their relation to the larger historical panorama of the history of life.

Kristie Macrakis

See also: Evolution: definition and theories; Evolution: history; Extinctions and evolutionary explosions; Gene flow; Genetic drift; Isolating mechanisms; Natural selection; Nonrandom mating, genetic drift, and mutation; Population genetics; Punctuated equilibrium vs. gradualism.

Sources for Further Study

Bonner, J. T. *Morphogenesis: An Essay on Development*. Princeton, N.J.: Princeton University Press, 1952.

De Beer, Gavin. *Embryos and Ancestors*. Oxford, England: Clarendon Press, 1951.

Gould, Stephen Jay. *Ontogeny and Phylogeny*. Cambridge, Mass.: Harvard University Press, 1977.

_____. "Ontogeny and Phylogeny—Revisited and Reunited." *BioEssays* 14, no. 4 (April, 1992): 275-280.

Hall, Brian K. *Evolutionary Developmental Biology*. 2d ed. New York: Chapman & Hall, 1998.

Raff, Rudolf A., and Thomas C. Kaufman. *Embryos, Genes, and Evolution: The Developmental-Genetic Basis of Evolutionary Change*. New York: Macmillan, 1983.

Schwartz, Jeffrey H. *Sudden Origins: Fossils, Genes, and the Emergence of Species*. New York: Wiley, 1999.

DISPLAYS

Type of ecology: Behavioral ecology

Displays are specialized behaviors that act as communication signals within, and occasionally between, species.

Nonlinguistic forms of communication are called displays. Some displays involve, literally and simply, the visual display of a physical feature. Among insects, for example, green or brown coloring is often used for camouflage. Insects that are poisonous do not need camouflage and often advertise themselves with warning colors, such as black and red, or black and orange. This is referred to as aposematic coloration or an aposematic display.

Physical features can also indicate an individual's sex, age, and reproductive status—as do peacock tails, turkey wattles, deer antlers, the canine teeth of male baboons, and the swollen genitals of estrous female chimpanzees. Many such features vary in size, shape, or color in relation to an animal's health, hormones, or social status and are therefore referred to as status badges or signs.

Often, meaningful physical features are highlighted by behavioral displays. A courting peacock or turkey will fan open his tail and shake it back and forth for emphasis; a challenged buck will load plant material onto his antlers so as to exaggerate their size; an angry baboon will curl back his lip to expose his canine teeth; and an estrous chimpanzee will approach a friendly male and assume a posture displaying her fertile state.

A particularly energetic or dramatic behavioral display not only calls attention to a physical feature but also indicates the health and vitality of the performer. The principle of honest signaling refers to the fact that large and healthy individuals tend to have brighter or more contrasting colors, make deeper-pitched and louder sounds, and produce longer, more intense performances than small or weak ones. Such differences in display quality are readily noticed by predators, potential competitors, and potential mates.

Most displays are performed by individuals and are one-way: sender to receiver. Threat displays, however, may involve reciprocal signaling between two challengers or between two groups of challengers. Courtship displays also may occur in groups: In some species, males gather together to perform in what is called a lek or a lekking display. Courtship of monogamous species may include long sequences of frequently repeated, ritual-

Displays

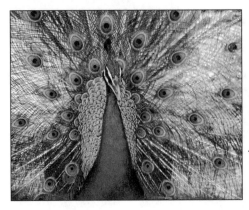

Male peacocks fan their tails and shake them in a courting display to attract females. Such physical features can also signal an individual's health, age, and social status and are therefore called "status badges." (Digital Stock)

ized interactions in which both partners participate. Such pair-bonding displays may continue well into the breeding season and the mateship; initially serving to familiarize the pair with one another and to synchronize their hormones and breeding behavior, they may later serve as greeting displays after separation.

Interpreting Displays

Charles Darwin noted that displays having opposite characteristics often signal opposite meaning. In humans, for example, a face with upturned corners of the mouth (a smile) signals friendliness, whereas a face with downturned corners of the mouth (a frown) signals displeasure. In most animals, loud, deep-pitched sounds (for example, roars and growls) indicate aggression, whereas quiet, high-pitched sounds (for example, mews and peeps) indicate anxiety or fear. Similarly, body postures exaggerating size tend to signal dominance, whereas postures minimizing size tend to signal submission. Darwin called his observation the principle of antithesis.

Although some rules of display can be applied across species, most displays are specialized for intraspecific (within-species) communication—male to female, parent to offspring, or dominant to subordinate—and are therefore species-specific. That is, the ability to perform and interpret a particular display (such as a particular birdsong) is generally characteristic only of individuals of a particular species and is either innate (inborn) or learned from conspecifics (individuals of the same species) during an early critical period of development.

In order that their meaning is easily and quickly conveyed, most displays also tend to be highly ritualized; that is, they are performed only in certain contexts and always in the same way. This consistency in communication prevents errors of interpretation that could be disastrous. It would be a grave mistake, for example, to interpret an aggressive signal as a sexual overture, or an alarm call (predator alert signal) as an offspring's begging call. Mistakes of interpretation are also minimized by signal redundancy; that is, messages are often conveyed simultaneously in more than one sensory modality.

Display Modality

Displays utilize every sensory modality. Visual displays involve the use of bright, contrasting, and sometimes changing colors; changes in body size, shape, and posture; and what ethologists call "intention movements"— brief, suggestive movements which reveal motivational state and likely future actions. Auditory displays include vocal songs and calls, as well as a variety of sounds produced by tapping, rubbing, scraping, or inflating and deflating various parts of the body. Tactile displays include aspects of social grooming, comfort contacts (such as between littermates or parents and offspring), and the seismic signaling of water-striders, elephants, frogs, and spiders which vibrate, respectively, the water, ground, plants, or web beneath them. Olfactory displays include signals from chemicals that have been wafted into the air or water, rubbed onto objects, or deposited in saliva, urine, or feces.

Olfaction (sense of smell) is the most primitive, and therefore the most common and most important, sense in the animal kingdom. Species of almost every taxonomic group use smell to signal their whereabouts and, generally, their sex and reproductive state. (Birds seem to be an exception.) Animals may also use smell to identify particular individuals, to recognize who is related to them and who is not, and to determine the relative dominance status of a conspecific.

Many animals, such as this badger, assume aggressive postures—bared teeth, fluffed fur to make them look larger, growling, and a variety of other signals—to warn off a threatening presence such as a predator or a human being.
(Digital Stock)

Displays

Chemicals used in displays are called pheromones. They may be derived from waste products or hormones, acquired by ingesting certain food items, or obtained directly from plants or other animals. Some pheromones not only communicate information, but also have physical effects on their receivers.

Linda Mealey

See also: Altruism; Communication; Competition; Defense mechanisms; Ethology; Hierarchies; Insect societies; Mammalian social systems; Migration; Mimicry; Pheromones; Poisonous animals; Predation; Reproductive strategies; Territoriality and aggression.

Sources for Further Study

Agosta, William C. *Chemical Communication: The Language of Pheromones.* New York: Scientific American Library, 1992.

Bailey, Winston. *Acoustic Behaviour in Insects.* New York: Chapman and Hall, 1991.

Eibl-Eibesfeldt, I. *Ethology: The Biology of Behavior.* Translated by Erich Klinghammer. 2d ed. New York: Holt, Reinhart and Winston, 1975.

Guthrie, R. Dale. *Body Hot Spots: The Anatomy of Human Social Organs and Behavior.* New York: Van Nostrand Reinhold, 1976.

Johnsgard, Paul A. *Arena Birds: Sexual Selection and Behavior.* Washington, D.C.: Smithsonian Institution Press, 1994.

Morris, Desmond. *Animalwatching.* New York: Crown, 1990.

Owen, Denis. *Camouflage and Mimicry.* Chicago: University of Chicago Press, 1980.

ECOLOGY: DEFINITION

Type of ecology: Theoretical ecology

Ecology is the study of the relationships of organisms to their environments. By examining those relationships in natural ecosystems, ecologists can discover principles that help humankind understand its own role on the planet.

Ecology is the study of how organisms relate to their natural environments. The two principal concerns of ecologists are the distribution and abundance of organisms: Why are animals, plants, and other organisms found where they are, and why are some common and others rare? These questions have their roots in the theory of evolution. In fact, it is difficult (and not often worthwhile) to separate modern ecological matters from the concerns of evolutionary biologists. Ecology can be divided according to several levels of organization: the individual organism, the population, the community, and the ecosystem.

Individual Organisms

An ecologist views organisms as consequences of past natural selection brought about by their environments. That is, each organism represents an array of adaptations that can provide insight into the environmental pressures that resulted in its present form. Adaptations of organisms are also revealed by other features, such as the range of temperature an organism can tolerate, the amount of moisture it requires, or the variety of food it can exploit. Food and space for living are considered resources; factors such as temperature, light, and moisture are conditions that determine the rate of resource utilization. When ecologists have discovered the full range of resources and conditions necessary for an organism's existence, they have discovered its niche.

Many species, such as many insects and plants, have a large reproductive output. This compensates for high mortality imposed by natural selection. Other species, such as large mammals and birds, have fewer offspring. Many of these animals care for their young, thus increasing the chances that their offspring will survive to reproduce. These are two different strategies for success, based upon the principle that organisms have a finite energy budget. Energy acquired from food (animals) or sunlight (plants) must be partitioned among growth, maintenance, and reproduction. The greater the energy allocated to the care of offspring, for example, the fewer the offspring that can be produced.

The concept of an energy budget is a key to understanding evolutionary strategies of organisms, as well as the energetics of ecosystems. The amount of energy fixed and stored by an organism is called net production; this is the energy used for growth and reproduction. Net production is the difference between gross production (the amount of energy assimilated) and respiration (metabolic maintenance cost). The greater the respiration, the less energy will be left over for growth and reproduction. Endothermic animals, which physiologically regulate their body heat (mammals and birds), have a very high respiration rate relative to ectotherms (reptiles, amphibians, fish, and invertebrates), which cannot. Among endotherms, smaller animals have higher respiration rates than larger ones, because the ratio of body surface area (the area over which heat is exchanged with the environment) to volume (the size of the "furnace") decreases with increasing body size.

Populations

Although single organisms can be studied with regard to adaptations, in nature most organisms exist in populations rather than as individuals. Some organisms reproduce asexually (that is, by forming clones), so that a single individual may spawn an entire population of genetically identical individuals. Populations of sexually reproducing organisms, however, have the property of genetic variability, since not all individuals are identical. That is, members of a population have slightly different niches and will therefore not all be equally capable of living in a given environment. This is the property upon which Charles Darwin's theory of natural selection depends: Because not all individuals are identical, some will have greater fitness than others. Those with superior fitness will reproduce in greater numbers and therefore will contribute more genes to successive generations. In nature, many species consist of populations occupying more than a single habitat. This constitutes a buffer against extinction: If one habitat is destroyed, the species will not go extinct, because it exists in other habitats.

Two dynamic features of populations are growth and regulation. Growth is simply the difference between birth and death rates, which can be positive (growing), negative (declining), or zero (in equilibrium). Every species has a genetic capacity for exponential (continuously accelerating) increase, which will express itself to varying degrees depending on environmental conditions: A population in its ideal environment will express this capacity more nearly than one in a less favorable environment. The rate of growth of a population is affected by its age structure—the proportion of individuals of different ages. For example, a population that is growing rapidly will have a higher proportion of juvenile individuals than one that is growing more slowly.

Populations may be regulated (so that they have equal birth and death rates) by a number of factors, all of which are sensitive to changes in population size. A population may be regulated by competition among its members for the resource that is in shortest supply (limiting). The largest population that can be sustained by the available resources is called the carrying capacity of the environment. A population of rodents, for example, might be limited by its food supply such that as the population grows and food runs out, the reproductive rate declines. Thus, the effect of food on population growth depends upon the population size relative to the limiting resource. Similarly, parasites that cause disease spread faster in large, dense populations than in smaller, more diffuse ones. Predators can also regulate populations of their prey by responding to changes in prey availability. Climate and catastrophic events such as storms may severely affect populations, but their effect is not dependent upon density and is thus not considered regulatory.

Communities

Communities of organisms are composed of many populations that may interact with one another in a variety of ways: predation, competition, mutualism, parasitism, and so on. The composition of communities changes over time through the process of succession. In terrestrial communities, bare rock may be weathered and broken down by bacteria and other organisms until it becomes soil. Plants can then invade and colonize this newly formed soil, which in turn provides food and habitat for animals. The developing community goes through a series of stages, the nature of which depends on local climatic conditions, until it reaches a kind of equilibrium. In many cases this equilibrium stage, called climax, is a mature forest. Aquatic succession essentially is a process of becoming a terrestrial community. The basin of a lake, for example, will gradually be filled with silt from terrestrial runoff and accumulated dead organic material from populations of organisms within the lake itself.

Competition occurs between, as well as within, species. Two species are said to be in competition with each other if and only if they share a resource that is in short supply. If, however, they merely share a resource that is plentiful, then they are not really competing for it. Competition is thought to be a major force in determining how many species can coexist in natural communities. There are a number of alternative hypotheses, however, which involve such factors as evolutionary time, productivity (the energy base for a community), heterogeneity of the habitat, and physical harshness of the environment.

Predator-prey interactions are those in which the predator benefits from killing and consuming its prey. These differ from most parasite-host interactions in that parasites usually do not kill their hosts (a form of suicide for creatures that live inside other creatures). Similarly, most plant-eating animals (herbivores) do not kill the plants on which they feed. Many ecologists classify herbivores as parasites for this reason. There are exceptions, such as birds and rodents that eat seeds, and these can be classified as legitimate predator-prey interactions. Predators can influence the number of species in a community by affecting competition among their prey: If populations of competing species are lowered by predators so that they are below their carrying capacities, then there may be enough resources to support colonization by new species.

In many cases, the interaction between two species is mutually beneficial. Mutualism is often thought to arise as a result of closely linked evolutionary histories (coevolution) of different species. Termites harbor protozoans in their guts that produce an enzyme that breaks down cellulose in wood. The protozoans thus are provided with a habitat, and the termites in turn are able to derive nourishment from wood. Some acacia trees in the tropics have hollow thorns which provide a habitat for ants. In return, the ants defend the trees from other insects which would otherwise damage or defoliate them.

Ecosystems
Ecosystems consist of several trophic levels, or levels at which energy is acquired: primary producers, consumers, and decomposers. Primary producers are green plants that capture solar energy and transform it, through the process of photosynthesis, into chemical energy. Organisms that eat plants (herbivores) or animals (carnivores) to obtain their energy are collectively called consumers. Decomposers are those consumers, such as bacteria and fungi, that obtain energy by breaking down dead bodies of plants and animals. These trophic levels are linked together into a structure called a food web, in which energy is transferred from primary producers to consumers and decomposers, until finally all is lost as heat. Each transfer of energy entails a loss (as heat) of at least 90 percent, which means that the total amount of energy available to carnivores in an ecosystem is substantially less than that available to herbivores.

As with individual organisms, ecosystems and their trophic levels have energy budgets. The net production of one trophic level is available to the next-higher trophic level as biomass (mass of biological material). Plants have higher net productivity (rates of production) than animals because their metabolic maintenance cost is lower relative to gross productivity;

herbivores often have higher net productivity than predators for the same reason. For the community as a whole, net productivity is highest during early successional stages, since biomass is being added more rapidly than later on, when the community is closer to climax equilibrium.

In contrast to the unidirectional flow of energy, materials are conserved and recycled from dead organisms by decomposers to support productivity at higher trophic levels. Carbon, water, and mineral nutrients required for plant growth are cycled through various organisms within an ecosystem. Materials and energy are also exchanged among ecosystems: There is no such thing in nature as a "closed" ecosystem that is entirely self-contained.

Methods of Studying Ecology

The science of ecology is necessarily more broadly based than most biological disciplines; consequently, there is more than one approach to it. Ecological studies fall into three categories: descriptive, experimental, and mathematical.

Descriptive ecology is concerned with describing natural history, usually in qualitative terms. The study of adaptations, for example, is descriptive in that one can measure the present "value" of an adaptive feature, but one can only conjecture as to the history of natural selection that was responsible for it. On the other hand, there are some patterns discernible in nature for which hypotheses can be constructed and tested by statistical inference. For example, the spatial distribution (dispersion) of birds on an island may be random, indicating no biological interaction among them. If the birds are more evenly spaced (uniform dispersion) than predicted assuming randomness, however, then it might be inferred that the birds are competing for space; they are exhibiting territorial exclusion of one another. Such "natural experiments," as they are called, depend heavily upon the careful design of statistical tests.

Experimental ecology is no different from any other experimental discipline; hypotheses are constructed from observations of nature, controlled experiments are designed to test them, and conclusions are drawn from the results of the experiments. The basic laboratory for an ecologist is the field. Experiments in the field are difficult because it is hard to isolate and manipulate variable factors one at a time, which is a requisite for any good experiment in science. A common experiment that is performed to test for resource limitation in an organism is enhancement of that resource. If food, for example, is thought to be in short supply (implying competition), one section of the habitat is provided more food than is already present; another section is left alone as a control. If survivorship, growth, or reproductive output is higher in the enhanced portion of the habitat than in the control area, the researcher may infer that the organisms therein were food-limited. Alternatively, an ecologist might have decreased the density of organisms in one portion of the habitat, which might seem equivalent to increasing food supply for the remaining organisms, except that it represents a change in population density as well. Therefore, this second design will not allow the researcher to differentiate between the possibly separate effects of food level and simple population density on organisms in the habitat.

Mathematical ecology relies heavily upon computers to generate models of nature. A model is simply a formalized, quantitative set of hypotheses constructed from sets of assumptions of how things happen in nature. A model of population growth might contain assumptions about the age structure of a population, its genetic capacity for increase, and the average rate of resource utilization by its members. By changing these assumptions, scientists can cause the model population to behave in different ways over time. The utility of such modeling is limited to the accuracy of the assumptions employed.

Modern ecology is concerned with integrating these different approaches, all of which have in common the goal of predicting the way nature will behave in the future, based upon how it behaves in the present. Description of natural history leads to hypotheses that can be tested experimentally, which in turn may allow the construction of realistic mathematical (quantitative) models of how nature works.

Human Impacts and Applications

People historically have viewed nature as an adversary. The "conquest of nature" has traditionally meant human encroachment on natural ecosystems, usually without benefit of predictive knowledge. Such environmental problems as pollution, species extinction, and overpopulation can be viewed as experiments performed on a grand scale without appropriate controls. The problem with such experiments is that the outcomes might be irreversible. A major lesson of ecology is that humans are not separate from nature; we are constrained by the same principles as are other organisms on the earth. One object of ecology, then, is to learn these principles so that they can be applied to our portion of the earth's ecosystem.

Populations that are not regulated by predators, disease, or food limitation grow exponentially. The human population, on a global scale, has grown this way. All the wars and famines in history have scarcely made a dent in this growth pattern. Humankind has yet to identify its carrying capacity on a global scale, although regional famines certainly have provided insights into what happens when local carrying capacity is exceeded. The human carrying capacity needs to be defined in realistic ecological terms, and such constraints as energy, food, and space must be incorporated into the calculations. For example, knowledge of energy flow teaches that there is more energy at the bottom of a food web (producers) than at successively higher trophic levels (consumers), which means that more people could be supported as herbivores than as carnivores.

The study of disease transmission, epidemiology, relies heavily on ecological principles. Population density, rates of migration among epidemic centers, physiological tolerance of the host, and rates of evolution of disease-causing parasites are all the subjects of ecological study.

An obvious application of ecological principles is conservation. Before habitats for endangered species can be set aside, for example, their ecological requirements, such as migratory routes, breeding, and feeding habits, must be known. This also applies to the introduction (intentional or accidental) of exotic species into habitats. History is filled with examples of introduced species that caused the extinction of native species. Application of ecological knowledge in a timely fashion, therefore, might prevent species from becoming endangered in the first place.

One of the greatest challenges humans face is the loss of habitats worldwide. This is especially true of the tropics, which contain most of the earth's species of plants and animals. Species in the tropics have narrow niches, which means that they are more restricted in range and less tolerant of change than are many temperate species. Therefore, destruction of tropical habitats, such as rain forests, leads to rapid species extinction. These species are the potential sources of many pharmaceutically valuable drugs; further, they are a genetic record of millions of years of evolutionary history. Tropical rain forests also are prime sources of oxygen and act as a buffer against carbon dioxide accumulation in the atmosphere. Ecological knowledge of global carbon cycles permits the prediction that destruction of these forests will have a profound impact on the quality of the air.

Lawrence E. Hurd

See also: Balance of nature; Biomes: determinants; Biomes: types; Biosphere concept; Deep ecology; Ecosystems: definition and history; Sustainable development.

Sources for Further Study
Begon, Michael, John L. Harper, and Colin R. Townsend. *Ecology: Individuals, Populations, and Communities.* 3d ed. Cambridge, Mass.: Blackwell Science, 1996.

Bush, M. B. *Ecology of a Changing Planet.* 2d ed. Upper Saddle River, N.J.: Prentice-Hall, 2000.

Carson, Rachel. *Silent Spring.* Boston: Houghton Mifflin, 1962.

Elton, Charles. *Animal Ecology.* New York: Macmillan, 1927.

Hutchinson, G. Evelyn. *The Ecological Theater and the Evolutionary Play.* New Haven, Conn.: Yale University Press, 1969.

Krebs, Charles J. *Ecological Methodology.* 2d ed. Menlo Park, Calif.: Benjamin/Cummings, 1999.

_____. *Ecology: The Experimental Analysis of Distribution and Abundance.* 5th ed. San Francisco: Benjamin/Cummings, 2001.

_____. *The Message of Ecology.* New York: Harper & Row, 1987.

Pianka, Eric R. *Evolutionary Ecology.* 6th ed. San Francisco: Benjamin/Cummings, 2000.

Ricklefs, Robert E. *Ecology.* 4th ed. New York: Chiron Press, 1999.

ECOLOGY: HISTORY

Type of ecology: History of ecology

As a formal discipline, ecology, the science that studies the relationships among organisms and their biotic and abiotic environments, is a relatively new science, which became a focus of study at about the same time evolutionary theories were being proposed.

The study of ecological topics arose in ancient Greece, but these studies were part of a catch-all science called natural history. The earliest attempt to organize an ecological science separate from natural history was made by Carolus Linnaeus in his essay *Oeconomia Naturae* (1749; *The Economy of Nature*), which focused on the balance of nature and the environments in which various natural communities existed. Although the essay was well known, the eighteenth century was dominated by biological exploration of the world, and the science of ecology did not develop.

Early Ecological Studies

The study of fossils led some naturalists to conclude that many species known only as fossils must have become extinct. However, Jean-Baptiste Lamarck argued in his *Philosophie zoologique* (1809; *Zoological Philosophy,* 1914) that fossils represented the early stages of species that evolved into different species that were still living. In order to refute this claim, geologist Charles Lyell mastered the science of biogeography and used it to argue that species do become extinct and that competition from other species seemed to be the main cause. English naturalist Charles Darwin's book *On the Origin of Species by Means of Natural Selection* (1859) blends his own researches with the influence of Linnaeus and Lyell in order to argue that some species do become extinct, but existing species have evolved from earlier ones. Lamarck had underrated and Lyell had overrated the importance of competition in nature.

Although Darwin's book was an important step toward ecological science, he and his colleagues mainly studied evolution rather than ecology. German evolutionist Ernst Haeckel realized the need for an ecological science and coined the name *oecologie* in 1866. It was not until the 1890's, however, that steps were taken to organize this science. Virtually all of the early ecologists were specialists in the study of particular groups of organisms, and it was only in the late 1930's that some efforts were made to write textbooks covering all aspects of ecology. Since the 1890's, many ecologists have viewed themselves as plant ecologists, animal ecologists, marine biologists, or limnologists. Limnology is the study of freshwater aquatic environments.

Nevertheless, general ecological societies were established. The first was the British Ecological Society, which was founded in 1913 and began publishing the *Journal of Ecology* in the same year. Two years later, ecologists in the United States and Canada founded the Ecological Society of America, which began publishing *Ecology* as a quarterly journal in 1920; in 1965 *Ecology* began appearing bimonthly. Other national societies have since been established. More specialized societies and journals also began appearing. For example, the Limnological Society of America was established in 1936 and expanded in 1948 into the American Society of Limnology and Oceanography. It publishes the journal *Limnology and Oceanography*. These and now many other professional organizations, both academic and applied, sponsor not only journals but also Web sites, reports, and symposia.

Although Great Britain and Western Europe were active in establishing the study of ecological sciences, it was difficult for their trained ecologists to obtain full-time employment that utilized their expertise. European universities were mostly venerable institutions with fixed budgets; they already had as many faculty positions as they could afford, and these were all allocated to the older arts and sciences. Governments employed few, if any, ecologists. The situation was more favorable in the United States, Canada, and Australia, where universities were still growing. In the United States, the universities that became important for ecological research and the training of new ecologists were mostly in the Midwest. The reason was that eastern universities were similar to European ones in being well established, with scientists in traditional fields.

Ecology After 1950
Ecological research in the United States was not well funded until after World War II. With the advent of the Cold War, science was suddenly considered important for national welfare. In 1950 the U.S. Congress established the National Science Foundation, and ecologists were able to make the case for their research along with that of the other sciences. The Atomic Energy Commission had already begun to fund ecological research by 1947, and under its patronage the Oak Ridge National Laboratory and the University of Georgia gradually became important centers for radiation ecology research.

Another important source of research funds was the International Biological Program (IBP), which, though international in scope, depended upon national research funds. It got under way in the United States in 1968 and was still producing publications in the 1980's. Even though no new funding sources were created for the IBP, its existence meant that more re-search money flowed to ecologists than otherwise would have.

Ecologists learned to think big. Computers became available for ecolog-ical research shortly before the IBP got under way, and so computers and the IBP became linked in ecologists' imaginations. Earth Day, established in 1970, helped awaken Americans to the environmental crisis. The IBP en-couraged a variety of studies, but in the United States, studies of biomes (large-scale environments) and ecosystems were most prominent. The bio-me studies were grouped under the headings of desert, eastern deciduous forest, western coniferous forest, grassland, and tundra (a proposed tropi-cal forest program was never funded). When the IBP ended, a number of its biome studies continued at a reduced level.

Ecosystem studies are also large-scale, at least in comparison with many previous ecological studies, though smaller in size than a biome. The goal of ecosystem studies was to gain a total understanding of how an eco-system—such as a lake, river valley, or forest—works. IBP funds enabled students to collect data, which computers processed. However, ecologists could not agree on what data to collect, how to compute outcomes, and how to interpret the results. Therefore, thinking big did not always pro-duce impressive results.

Plant Ecology

Because ecology is enormous in scope, it was bound to have growing pains. It arose at the same time as the science of genetics, but because ge-netics is a cohesive science, it reached maturity much sooner than ecology. Ecology can be subdivided in a wide variety of ways, and any collection of ecology textbooks will show how diversely it is organized by different ecologists. Nevertheless, self-identified professional subgroups tend to produce their own coherent findings.

Plant ecology progressed more rapidly than other subgroups and has retained its prominence. In the early nineteenth century, German naturalist Alexander von Humboldt's many publications on plant geography in rela-tion to climate and topography were a powerful stimulus to other bota-nists. By the early twentieth century, however, the idea of plant communi-ties was the main focus for plant ecologists. Henry Chandler Cowles began his studies at the University of Chicago in geology but switched to botany and studied plant communities on the Indiana dunes of Lake Michigan. He received his doctorate in 1898 and stayed at that university as a plant ecologist. He trained others in the study of community succession.

Frederic Edward Clements received his doctorate in botany in 1898 from the University of Nebraska. He carried the concept of plant community succession to an extreme by taking literally the analogy between the growth and maturation of an organism and that of a plant community. His numerous studies were funded by the Carnegie Institute in Washington, D.C., and even ecologists who disagreed with his theoretical extremes found his data useful. Henry Allan Gleason was skeptical; his studies indicated that plant species that have similar environmental needs compete with one another and do not form cohesive communities. Although Gleason first expressed his views in 1917, Clements and his disciples held the day until 1947, when Gleason's individualistic concept received the support of three leading ecologists. Debates over plant succession and the reality of communities helped increase the sophistication of plant ecologists and prepared them for later studies on biomes, ecosystems, and the degradation of vegetation by pollution, logging, and agriculture.

Marine Ecology

Marine ecology is viewed as a branch of either ecology or oceanography. Early studies were made either from the ocean shore or close to shore because of the great expense of committing oceangoing vessels to research. The first important research institute was the Statione Zoologica at Naples, Italy, founded in 1874. Its successes soon inspired the founding of others in Europe, the United States, and other countries. Karl Möbius, a German zoologist who studied oyster beds, was an important pioneer of the community concept in ecology. Great Britain dominated the seas during the nineteenth century and made the first substantial commitment to deep-sea research by equipping the HMS *Challenger* as an oceangoing laboratory that sailed the world's seas from 1872 to 1876. Its scientists collected so many specimens and so much data that they called upon marine scientists in other countries to help them write the fifty large volumes of reports (1885-1895). The development of new technologies and the funding of new institutions and ships in the nineteenth century enabled marine ecologists to monitor the world's marine fisheries and other resources and provide advice on harvesting marine species.

Limnology is the scientific study of bodies of fresh water. The Swiss zoologist François A. Forel coined the term and also published the first textbook on the subject in 1901. He taught zoology at the Académie de Lausanne and devoted his life's researches to understanding Lake Geneva's characteristics and its plants and animals. In the United States in the early twentieth century, the University of Wisconsin became the leading center for limnological research and the training of limnologists, and it has retained that preeminence. Limnology is important for managing freshwater fisheries and water quality.

Frank N. Egerton

See also: Ecology: history; Ecosystems: definition and history; Evolution: history.

Sources for Further Study

Allen, Timothy F. H., and Thomas W. Hoekstra. *Toward a Unified Ecology*. New York: Columbia University Press, 1992.

Hynes, H. Patricia. *The Recurring Silent Spring*. New York: Pergamon Press, 1989.

Leuzzi, Linda. *Life Connections: Pioneers in Ecology*. Danbury, Conn.: Franklin Watts, 2000.

ECOSYSTEMS: DEFINITION AND HISTORY

Types of ecology: Ecosystem ecology; History of ecology; Theoretical ecology

The ecosystem is the fundamental concept in ecology: the basic unit of nature, consisting of the complex of interacting organisms inhabiting a region with all the nonliving physical factors that make up their environment. Ecologists study structural and functional relationships of ecosystem components to be able to predict how the system will respond to natural change and human disturbance.

The ecosystem is essentially an abstract organizing unit superimposed on the landscape to help ecologists study the form and function of the natural world. An ecosystem consists of one or more communities of interacting organisms and their physical environment. Ecosystems have no distinct boundaries; thus the size of any particular ecosystem should be inferred from the context of the discussion. Individual lakes, streams, or strands of trees can be described as distinct ecosystems, as can the entire North American Great Lakes region. Size and boundaries are arbitrary because no ecosystem stands in complete isolation from those that surround it. A lake ecosystem, for example, is greatly affected by the streams that flow into it and by the soils and vegetation through which these streams flow. Energy, organisms, and materials routinely migrate across whatever perimeters the ecologist may define. Thus, investigators are allowed considerable latitude in establishing the scale of the ecosystem they are studying. Whatever the scale, though, the importance of the ecosystem concept is that it forces ecologists to treat organisms not as isolated individuals or species but in the context of the structural and functional conditions of their environment.

Development of the Ecosystem Concept

Antecedents to the ecosystem concept may be traced back to one of America's first ecologists, Stephen Alfred Forbes, an eminent Illinois naturalist who studied food relationships among birds, fish, and insects. During the course of his investigations, Forbes recognized in 1880 that full knowledge of organisms and their response to disturbances would come only from more concentrated research on their interactions with other organisms and with their inorganic (nonliving) physical surroundings. In 1887, Forbes suggested that a lake could be viewed as a discrete system for study: a microcosm. A lake could serve as a scale model of nature that would help biologists understand more general functional relationships among organisms and their environment. Forbes explained how the food supply of a single species, the largemouth bass, was dependent either directly or indirectly upon nearly all the fauna and much of the flora of the lake. Therefore, whenever even one species was subjected to disturbance from outside the microcosm, the effects would probably be felt throughout the community.

In 1927, British ecologist Charles Elton incorporated ideas introduced by Forbes and other fishery biologists into the twin concepts of the food chain and the food web. Elton defined a food chain as a series of linkages connecting basic plants, or food producers, to herbivores and their various carnivorous predators, or consumers. Elton used the term "food cycle" instead of "food web," but his diagrams reveal that his notion of a food cycle—that it is simply a network of interconnecting food chains—is consistent with the modern term.

Elton's diagrams, which traced various pathways of nitrogen through the community, paved the way for understanding the importance of the cycling of inorganic nutrients such as carbon, nitrogen, and phosphorus through ecosystems, a process that is known as biogeochemical cycling. Very simply, Elton illustrated how bacteria could make nitrogen available to algae at the base of the food chain. The nitrogen then could be incorporated into a succession of ever larger consumers until it reached the top of the chain. When the top predators died, decomposer organisms would return the nitrogen to forms that could eventually be taken up again by plants and algae at the base of the food chain, thus completing the cycle.

Elton's other key contribution to the ecosystem concept was his articulation of the pyramid of numbers, the idea that small animals in any given community are far more common than large animals. Organisms at the base of a food chain are numerous, and those at the top are relatively scarce. Each level of the pyramid supplies food for the level immediately above it—a level consisting of various species of predators that generally are larger in size and fewer in number. That level, in turn, serves as prey for a level of larger, more powerful predators, fewer still in number. A graph of this concept results in a pyramidal shape of discrete levels, which today are called trophic (feeding) levels.

Ecosystems in Twentieth Century Thought

Although the basic concept of an ecosystem had been recognized by Forbes as early as 1880, it was not until 1935 that British ecologist Arthur G.

Tansley coined the term. Though he acknowledged that ecologists were primarily interested in organisms, Tansley declared that organisms could not be separated from their physical environments, as organism and environment formed one complete system. As Forbes had pointed out half a century earlier, organisms were inseparably linked to their nonliving environment. Consequently, ecosystems came to be viewed as consisting of two fundamental parts: the biotic, or living components, and the abiotic, or nonliving components.

No one articulated this better than Raymond Lindeman, a limnologist (freshwater biologist) from Minnesota who in 1941 skillfully integrated the ideas of earlier ecological scientists when he published an elegant ecosystem study of Cedar Bog Lake. Lindeman's classic work set the stage for decades of research that centered on the ecosystem as the primary organizing unit of study in ecology. Drawing on the work of his mentor, G. Evelyn Hutchinson, and Charles Elton, Arthur Tansley, and other scientists, Lindeman explained how ecological pyramids were a necessary result of energy transfers from one trophic level to the next.

By analyzing ecosystems in this manner, Lindeman was able to answer a fundamental ecological question that had been posed fourteen years earlier by Elton: Why were the largest and most powerful animals, such as polar bears, sharks, and tigers, so rare? Elton had thought the relative scarcity of top predators was due to their lower rates of reproduction. Lindeman corrected this misconception by explaining that higher trophic levels held fewer animals not because of their reproductive rates but because of a loss of chemical energy with each step up the pyramid. It could be looked upon as a necessary condition of the second law of thermodynamics: Energy transfers yield a loss or degradation of energy. The predators of one food level could never completely extract all the energy from the level below. Some energy would always be lost to the environment through respiration, some energy would not be assimilated by the predators, and some energy simply would be lost to decomposer chains when potential prey died of nonpredatory causes. This meant that each successive trophic level had substantially less chemical energy available to it than was transferred to the one below and, therefore, could not support as many animals.

Ecologists soon expanded the principle of Elton's pyramid of numbers to model other ecosystem processes. They found, for example, that the flow of chemical energy through an ecosystem could be characterized as an energy pyramid; the biomass (the weight of organic material, as in plant or animal tissue) in a community could be plotted in a pyramid of biomass. Collectively, such pyramidal models became known as ecological pyramids.

Energy Production and Transmission

Lindeman subdivided the biotic components of ecosystems into producers, consumers, and decomposers. Producers (also known as autotrophs) produce their own food from compounds in their environment. Green plants are the main producers in terrestrial ecosystems; algae are the most common producers in aquatic ecosystems. Both plants and algae are producers that use sunlight as energy to make food from carbon dioxide and water in the process of photosynthesis. During photosynthesis, plants, algae, and certain bacteria capture the sun's energy in chlorophyll molecules. (Chlorophyll is a pigment that gives plants their green color.) This energy, in turn, is used to synthesize energy-rich compounds such as glucose, which can be used to power activities such as growth, maintenance, and reproduction or can be stored as biomass for later use. These energy-rich compounds can also be passed on in the form of biomass from one organism to another, as when animals (primary consumers) graze on plants or when decomposers break down detritus (dead organic matter).

The energy collected by green plants is called primary production because it forms the first level at the base of the ecological pyramid. Total photosynthesis is represented as gross primary production. This is the amount of the sun's energy actually assimilated by autotrophs. The rate of this production of organic tissue by photosynthesis is called primary productivity. Plants, however, need to utilize some of the energy they produce for their own growth, maintenance, and reproduction. This energy becomes available for such activities through respiration, which essentially is a chemical reversal of the process of photosynthesis. As a result, not all the energy assimilated by autotrophs is available to the consumers in the next trophic level of the pyramid. Consequently, respiration costs generally are subtracted from gross primary production to determine the net primary production, the chemical energy actually available to primary consumers.

Measuring Ecosystem Productivity

The carrying capacity for all the species supported by an ecosystem ultimately depends upon the system's net primary productivity. By knowing the productivity, ecologists can, for example, estimate the number of herbivores that an ecosystem can support. Consequently ecosystem ecologists have developed a variety of methods to measure the net primary productivity of different systems. Productivity is generally expressed in kilocalories per square meter per year when quantifying energy, and in grams per square meter per year when quantifying biomass.

187

Production in aquatic ecosystems may be measured by using the light and dark bottle method. In this technique two bottles containing samples of water and the natural phytoplankton population are suspended for twenty-four hours at a given depth in a body of water. One bottle is dark, permitting respiration but no photosynthesis by the phytoplankton. The other is clear and therefore permits both photosynthesis and respiration. The light bottle provides a measure of net production (photosynthesis minus respiration) if the quantity of oxygen is measured before and after the twenty-four-hour period. (The amount of oxygen produced by photosynthesis is proportional to the amount of organic matter fixed.) Measuring the amount of oxygen in the dark bottle before and after the run provides an estimate of respiration, since no photosynthesis can occur in the dark. Combining net production from the light bottle with total respiration from the dark bottle yields an estimate of gross primary production.

Other studies have concentrated on quantifying the rate of movement of energy and materials through ecosystems. Investigations begun in the 1940's and 1950's by the Atomic Energy Commission to track radioactive fallout were eventually diverted into studies of ecosystems that demonstrated how radionuclides moved through natural environments by means of food-chain transfers. This research confirmed the interlocking nature of all organisms linked by the food relationship and eventually yielded rates at which both organic and inorganic materials could be cycled through ecosystems. As a result of such studies, the Radiation Ecology Section at Oak Ridge National Laboratory in Tennessee became established as a principal center for systems ecology.

Ecologists sometimes extend the temporal boundaries of their studies by utilizing the methods of paleoecology (the use of fossils to study the nature of ecosystems in the past). Research of this type generally centers on the analysis of lake sediments, whose layers often hold centuries of ecosystem history embodied in the character and abundance of pollen grains, diatoms, fragments of zooplankton, and other organic microfossils.

More general trends in the methods of studying ecosystems include a continuing emphasis on quantitative methods, often using increasingly sophisticated computer modeling techniques to simulate ecosystem functions. Equally significant is a trend toward a "big science" approach, modeled on the Manhattan Project, which employed teams of investigators working on different problems related to nuclear fission in different parts of the country. The well-known international biological program, the Hubbard Brook project in New Hampshire, and continuing projects on long-term ecological research all serve as examples of ecosystem studies that involve teams of researchers from a wide range of disciplines.

Responding to Disturbance

One of the practical benefits of studying ecosystems derives from naturalist Stephen Forbes's suggestion, made in 1880, that the knowledge from biological research be used to predict the response of organisms to disturbance. When disturbance is caused by natural events such as droughts, floods, or fires, ecologists can use their knowledge of the structure and function of ecosystems to help resource managers plan for the subsequent recolonization and succession of species.

The broad perspective of the ecosystem approach becomes particularly useful in examining the effects of certain toxic compounds because of the complexity of their interaction within the environment. The synergistic effects that sometimes occur with toxic substances can produce pronounced impacts on ecosystems already stressed by other disturbances. For example, after the atmosphere deposits mercury on the surface of a lake, the pollutant eventually settles in the sediments where bacteria make it available to organisms at the base of the food chain. The contaminant then bioaccumulates as it is passed on to organisms such as fish and fish-eating birds at higher trophic levels. Synergistic effects occur in lakes already affected by acid deposition; researchers have found that acidity somehow stimulates microbes to increase the bioavailability of the mercury. Thus, aquatic ecosystems that have become acidified through atmospheric processes may stress their flora and fauna even further by enhancing the availability of mercury from atmospheric fallout. The complexity of such interactions demands research at the ecosystem level, and ecosystem studies are prerequisite for prudent public policy actions on environmental contaminants.

Robert Lovely

See also: Biomes: determinants; Biomes: types; Communities: ecosystem interactions; Ecology: history; Ecosystems: studies; Food chains and webs; Geochemical cycles; Habitats and biomes; Lakes and limnology; Nutrient cycles; Trophic levels and ecological niches.

Sources for Further Study

Begon, M., J. L. Harper, and C. R. Townsend. *Ecology: Individuals, Populations, and Communities.* Oxford, England: Blackwell Scientific Publications, 1996.

Clark, Tim W., A. Peyton Curlee, Steven C. Minta, and Peter M. Kareiva, eds. *Carnivores in Ecosystems: The Yellowstone Experience.* New Haven, Conn.: Yale University Press, 1999.

Colinvaux, Paul. *Why Big Fierce Animals Are Rare: An Ecologist's Perspective.* Princeton, N.J.: Princeton University Press, 1978.

Golly, Frank Benjamin. *A History of the Ecosystem Concept in Ecology: More than the Sum of the Parts.* New Haven, Conn.: Yale University Press, 1993.

Hagen, Joel B. *An Entangled Bank: The Origins of Ecosystem Ecology.* New Brunswick, N.J.: Rutgers University Press, 1992.

Hunter, Malcolm, Jr. *Maintaining Biodiversity in Forest Ecosystems.* New York: Cambridge University Press, 1999.

Odum, Eugene P. *Ecology and Our Endangered Life-Support Systems.* Sunderland, Mass.: Sinauer Associates, 1989.

Real, Leslie A., and James H. Brown. *Foundations of Ecology: Classic Papers with Commentaries.* Chicago: University of Chicago Press, 1991.

Smith, Robert Leo. *Ecology and Field Biology.* 6th ed. San Francisco: Benjamin/Cummings, 2001.

ECOSYSTEMS: STUDIES

Type of ecology: Ecosystem ecology

The study of ecosystems defines a specific area of the earth and the attendant inter-actions among organisms and the physical-chemical environment present at the site.

Ecosystems are viewed by ecologists as basic units of the biosphere, much as cells are considered by biologists to be the basic units of an organism. Ecosystems are self-organized and self-regulating entities within which energy flows and resources are cycled in a coordinated, interdependent manner to sustain life. Disruptions and perturbations to, or within, the unit's organization or processes may reduce the quality of life there or cause its demise. Ecosystem boundaries are usually defined by the research or management questions being asked. An entire ocean can be viewed as an ecosystem, as can a single tree, a rotting log, or a drop of pond water. Systems with tangible boundaries—such as forests, grasslands, ponds, lakes, watersheds, seas, or oceans—are especially useful to ecosystem research.

Research Principles

The ecosystem concept was first put to use by American limnologist Raymond L. Lindeman in the classic study he conducted on Cedar Bog Lake, Minnesota, which resulted in his article "The Trophic Dynamic Aspect of Ecology" (1942). Lindeman's study, along with the publication of Eugene P. Odum's *Fundamentals of Ecology* (1953), converted the ecosystem notion into a guiding paradigm for ecological studies, thus making it a concept of theoretical and applied significance.

Ecologists study ecosystems as integrated components through which energy flows and resources cycle. Although ecosystems can be divided into many components, the four fundamental ones are abiotic (nonliving) resources, producers, consumers, and decomposers. The ultimate sources of energy come from outside the boundaries of the ecosystem (solar energy or chemothermo energy from deep-ocean hydrothermal vent systems). Because this energy is captured and transformed into chemical energy by producers and translocated through all biological systems via consumers and decomposers, all organisms are considered as potential sources of energy.

Abiotic resources—water, carbon dioxide, nitrogen, oxygen, and other inorganic nutrients and minerals—primarily come from within the boundaries of the ecosystem. From these, producers utilizing energy synthesize biomolecules, which are transformed, upgraded, and degraded as they cycle through the living systems that comprise the various components. The destiny of these bioresources is to be degraded to their original abiotic forms and be recycled.

The ecosystem approach to environmental research is a major endeavor. It requires amassing large amounts of data relevant to the structure and function of each component. These data are then integrated among the components, in an attempt to determine linkages and relationships. This holistic ecosystem approach to research involves the use of systems information theory, predictive models, and computer application and simulations. As ecosystem ecologist Frank B. Golley stated in his book *A History of the Ecosystem Concept in Ecology* (1993), the ecosystem approach to the study of ecosystems is "machine theory applied to nature."

Research Projects

Initially, ecosystem ecologists used the principles of Arthur G. Tansley, Lindeman, and Odum to determine and describe the flow of energy and resources through organisms and their environment. Fundamental academic questions that plagued ecologists concerned controls on ecosystem productivity: What are the connections between animal and plant productivity? How are energy and nutrients transformed and cycled in ecosystems?

Once fundamental insights were obtained, computer-model-driven theories were constructed to provide an understanding of the biochemophysical dynamics that govern ecosystems. Responses of ecosystem components could then be examined by manipulating parameters within the simulation model. Early development of the ecosystem concept culminated, during the 1960's, in defining the approach of ecosystem studies.

Ecosystem projects were primarily funded under the umbrella of the International Biological Program (IBP). Other funding came from the Atomic Energy Commission and the National Science Foundation. The intention of the IBP was to integrate data collected by teams of scientists at research sites that were considered typical of wide regions. Although the IBP was international in scope, studies in the United States received the greatest portion of the funds—approximately $45 million during the life of IBP (1964-1974).

Five major IBP ecosystem studies, involving grasslands, tundra, deserts, coniferous forests, and deciduous forests, were undertaken. The Grasslands Project, directed by George Van Dyne, set the research stage for the other four endeavors. However, because the research effort was so extensive in scope, the objectives of the IBP were not totally realized. Because of the large number of scientists involved, little coherence in results was obtained even within the same project. A more pervasive concern, voiced by environmentalists and scientists alike, was that little of the information obtained from the ecosystem simulation models could be applied to the solution of existing environmental problems.

An unconventional project partially funded by the IBP was called the Hubbard Brook Watershed Ecosystem. Located in New Hampshire and studied by F. Herbert Bormann and Gene E. Likens, the project redirected the research approach for studying ecosystems from the IBP computer-model-driven theory to more conventional scientific methods of study. Under the Hubbard Brook approach, an ecosystem phenomenon is observed and noted. A pattern for the phenomenon's behavior is then established for observation, and questions are posed about the behavior. Hypotheses are developed to allow experimentation in an attempt to explain the observed behavior. This approach requires detailed scrutiny of the ecosystem's subsystems and their linkages. Since each ecosystem functions as a unique entity, this approach has more utility. The end results provide insights specific to the activities observed within particular ecosystems. Explanations for these observed behaviors can then be made in terms of biological, chemical, or physical principles.

Utility of the Concept

Publicity from the massive ecosystem projects and the publication of Rachel Carson's classic *Silent Spring* (1962) helped stimulate the environmental movement of the 1960's. The public began to realize that human activity was destroying the bioecological matrices that sustained life. By the end of the 1960's, the applicability of the IBP approach to ecosystem research was proving to be purely academic and provided few solutions to the problems that plagued the environment. Scientists realized that, because of the lack of fundamental knowledge about many of the systems and their links and because of the technological shortcomings that existed, ecosystems could not be divided into three to five components and analyzed by computer simulation.

The more applied approach taken in the Hubbard Brook project, however, showed that the ecosystem approach to environmental studies could be successful if the principles of the scientific method were used. The Hubbard Brook study area and the protocols used to study it were clearly defined. This ecosystem allowed hypotheses to be generated and experimentally tested. Applying the scientific method to the study of ecosystems had practical utility for the management of natural resources and for testing possible solutions to environmental problems. When perturbations such as diseases, parasites, fire, deforestation, and urban and rural centers disrupt ecosystems from within, this approach helps define potential mitigation and management plans. Similarly, external causative agents within airsheds, drainage flows, or watersheds can be considered.

The principles and research approach of the ecosystem concept are being used to define and attack the impact of environmental changes caused by humans. Such problems as human population growth, apportioning of resources, toxification of biosphere, loss of biodiversity, global warming, acid rain, atmospheric ozone depletion, land-use changes, and eutrophication are being holistically examined. Management programs related to woodlands (the New Forestry program) and urban and rural centers (the Urban to Rural Gradient Ecology, or URGE, program), as well as other governmental agencies that are investigating water and land use, fisheries, endangered species, and exotic species introductions, have found the ecosystem perspective useful.

Ecosystems are also viewed as systems that provide the services necessary to sustain life on earth. Most people either take these services for granted or do not realize that such natural processes exist. Ecosystem research has identified seventeen naturally occurring services, including water purification, regulation, and supply, as well as atmospheric gas regulation and pollination. A 1997 article by Robert Costanza and others, "The Value of the World's Ecosystem Services and Natural Capital," placed a monetary cost to humanity should the service, for some disastrous reason, need to be maintained by human technology. The amount is staggering, averaging $33 trillion per year. Humanity could not afford this; the global gross national product is only about $20 trillion.

Academically, ecosystem science has been shown to be a tool to dissect environmental problems, but this has not been effectively demonstrated to the public and private sectors, especially decision makers and policymakers at governmental levels. The idea that healthy ecosystems provide socioeconomic benefits and services remains controversial. In order to bridge this gap between academia and the public, Scott Collins of the National Science Foundation suggested to the Association of Ecosystem Research Centers that ecosystem scientists be "bilingual"; that is, they should be able speak their scientific language and translate it so that the nonscientist can understand.

Richard F. Modlin

See also: Biodiversity; Biomes: determinants; Biomes: types; Communities: ecosystem interactions; Desertification; Ecosystems: definition and history; Geochemical cycles; Habitats and biomes; Hydrologic cycle; Nutrient cycles; Rain forests and the atmosphere; Trophic levels and ecological niches.

Sources for Further Study

Aber, J. D., and J. M. Melillo. *Terrestrial Ecosystems*. 2d ed. San Diego, Calif.: Harcourt Academic Press, 2001.

Allen, T. F. H., and T. W. Hoekstra. *Toward a Unified Ecology*. New York: Columbia University Press, 1992.

Costanza, Robert, et al. "The Value of the World's Ecosystem Services and Natural Capital." *Nature* 387, no. 6630 (May 15, 1997): 253-260.

Daily, Gretchen C., ed. *Nature's Services: Societal Dependence on Natural Ecosystems*. Washington, D.C.: Island Press, 1997.

Dodson, Stanley I., et al. *Ecology*. New York: Oxford University Press, 1998.

Golley, Frank B. *A History of the Ecosystem Concept in Ecology*. New Haven, Conn.: Yale University Press, 1993.

Likens, Gene E. *The Ecosystem Approach: Its Use and Abuse*. Oldendorf/ Luhe, Germany: Ecology Institute, 1992.

Vogt, Kristiina A., et al. *Ecosystems: Balancing Science with Management*. New York: Springer, 1997.

ENDANGERED ANIMAL SPECIES

Type of ecology: Restoration and conservation ecology

Endangered species are those varieties of plants and animals that are in immediate danger of becoming extinct. Threatened species, by contrast, are those identified as likely to become endangered in the near future.

A t the beginning of 2003, the U.S. Fish and Wildlife Service listed nearly 400 animal species in the United States as "endangered": in immediate danger of extinction. Another 129 species of animals were designated as threatened, meaning likely to become endangered in the near future. Globally, the World Conservation Union (IUCN) has placed thousands of species on its list of endangered and vulnerable species, and more are added every month. The International Council for Bird Preservation has announced that more than one thousand of the nearly ten thousand species of birds are endangered. Fish, especially freshwater fish, are among the most immediately threatened types of animals. About one-fourth of the worldwide number of species is in a state of dangerous decline.

Causes of Species Endangerment

The destruction of animal species is caused in four major ways: Humans have hunted other species out of existence; habitats, the environments in which plants or animals grow and develop, have been destroyed; new species, such as rats, cats, goats, or ground-covering plants, have been introduced into a region and displaced native species; and nonnative plants and animals have introduced diseases into an environment, killing the existing species. For much of history, hunting was the major cause of species extinction. However, hunting has become less of a factor because governments and conservation authorities have imposed strict controls on the practice. In the second half of the twentieth century, habitat destruction and invasion by exotics (nonnative plants and animals) and the diseases they carry caused the most damage. Most biologists agree that whatever the factors involved, the rate of extinction has increased rapidly since the 1950's.

Some people have argued that the destruction of a single species of fish, bird, or flower would make little or no difference to the future of human life or the earth. They also suggest that extinctions have always taken place, even before human beings existed, and therefore are simply part of the natural process of existence. Who, for instance, would really want to have "saved" the dinosaurs? We should be glad they are gone. At least ten million species exist, so who would miss a few dozen or a few hundred of them?

These arguments ignore an important point. Each species inhabits a small part of an entire ecosystem, a community of plants and animals that are closely associated in a chain of survival. For example, plants absorb chemicals and minerals from the soil that are essential to their health. Animals then eat the plants—grasses, fruits, leaves, or flowers—and digest the nutrients they need for energy. Other animals, meat eaters (carnivores), then eat these plant eaters and get their energy from them. If a single species is removed from this chain, the whole ecosystem will experience consequences that are difficult to predict and often negative or disastrous.

The death of an entire species constitutes a loss that cannot always be measured in economic terms. The American biologist William Beebe made the point that any species that is lost diminishes the quality of life for everyone:

> The beauty and genius of a work of art may be reconceived, though its first material expression can be destroyed; a vanished harmony may yet inspire the composer, but when the last individual of a race of living things breathes no more, another heaven and another earth must pass before such a one can be again.

The bald eagle, and other bird populations, became endangered partly as a result of the widespread use of DDT, a pesticide whose toxicity becomes concentrated as it makes its way up the food chain. Agricultural runoff into streams and lakes causes such pesticides to accumulate in the microorganisms on which fish feed, then the fish, and finally the predators that eat the fish, from birds to humans. (PhotoDisc)

How Species Are Lost

The passenger pigeon is one example of species loss, a bird so numerous in the 1820's that John James Audubon, the famous American painter and collector, wrote that the flapping of wings of flocks numbering in the hundreds of millions on the Great Plains sounded like the roar of thunder. More than nine billion of the pigeons were alive in 1850, yet slightly more than sixty years later, exactly one bird, Martha, survived in the Cincinnati Zoo, where she had been taken in 1912. The population fell from nine billion to one in little more than half a century, then to zero when Martha died on September 1, 1917. People had found these pigeons delicious to eat and easy to kill. They formed hundreds of hunting parties, killing more than fifty thousand birds a week. No one dreamed the passenger pigeon could ever be exterminated.

The same fate almost befell the American bison, often called the buffalo. Before the coming of railroads and white settlers in the 1860's and 1870's, the bison numbered more than 100 million. Native Americans hunted the bison, eating their flesh and using the skins for clothing and shelter, but

Poaching, the illegal hunting, trapping, or killing of protected wildlife, remains a worldwide threat to animals whose body parts can be sold on the black market as culinary delicacies, aphrodisiacs, or material for jewelry, clothing, and other artifacts. The demand for the gallbladders and paws of the Asiatic black bear, such as these in Thailand, has brought the species to the brink of extinction. (AP/Wide World Photos)

they killed only what they needed. The settlers, however, saw the bison as a problem that needed to be solved. Huge herds of bison crossed railroad tracks, forcing passenger trains to stop, and the animals interfered with farming, knocking down fences and trampling grain fields. Railroad companies and the U.S. Army sent out hunting parties to get rid of the bison. By 1890 fewer than one thousand bison survived in a herd that had managed to escape far into northern Canada. The extermination ended only after this small herd was given protection by the Canadian government.

Stories of other near extinctions are numerous and frightening but demonstrate that action can be taken to save some if not all of the endangered species. Whales, which had been hunted since the 1600's, faced possible extinction until action was taken to reduce hunting in the 1970's. Whales were easy to kill and provided oil and bone. Whale oil was the major substance burned in lamps until the electric light largely replaced oil-burning lamps in the 1880's. Europeans hunted the Atlantic whale, called the right whale because it was the "right" one to kill, into virtual extinction by the 1860's. When the right whale became too hard to find, hunters turned to the Pacific right whale and then the bowhead whale before action was taken by the world community to save remaining whales.

Greed and Ignorance
Human greed has brought death or near death to many species. The desire for fur coats has nearly killed all jaguars, snow leopards, and various species of fur seals. Pribilof Island fur seals were nearly hunted into extinction in the late 1800's. A treaty between the United States, Canada, and Russia established limits on killing the species in its remote northern Pacific island habitat, but enforcement has proved difficult, and thousands of seals have been slaughtered despite international protection.

The belief that some animals are nuisances has led to the near extinction of wolves, grizzly bears, cougars, and coyotes. These predators have been poisoned and shot by the thousands and have become endangered species as a result. Attempts to kill insects with pesticides in order to control the spread of disease and improve crop yields were successful but had an unfortunate side effect. Chemicals from the pesticides worked their way into ecosystems, killing millions of other forms of life. In the 1950's, dichloro-diphenyl-trichloroethane, or DDT, was used to kill malaria-carrying mosquitoes, but the chemical infested the whole food chain. It entered plants that were eaten by animals and affected birds, fish, and butterflies. Pesticide poisoning also diminished the numbers of the bald eagle and peregrine falcon, which started to come back only after rigid controls on pesticides were established.

Events on the island of Madagascar, a large island in the Indian Ocean off the east coast of Africa, demonstrate most fully the deadly consequences of habitat destruction. About 180 million years ago, the island was attached to the African continent, then was split off after a series of geological catastrophes and ended up 250 miles to the east. The split occurred just at the time mammals were emerging as a class of animals in Africa. One mammal species, the monkeylike lemur, became isolated on Madagascar and increased abundantly. Other animals caught on the island included several kinds of giant birds, one weighing one thousand pounds and standing ten feet tall. The island was isolated for millions of years, allowing hundreds of species found nowhere else in the world to evolve in the diverse island ecosystems. Madagascar had deserts, rain forests, dry forests, and seashores. About 99 percent of its reptiles, 81 percent of its plants, and 99 percent of its frogs were unique and tied specifically to the island's food chain. About nine thousand years ago, the Malagasy people began to arrive on the island. They hunted, fished, began to grow crops, and in the process destroyed more than 90 percent of the forests that covered Madagascar. Dozens of species died as a result, including the giant elephant bird, which was gone by 1700. Ten out of thirty-one species of lemur died out by 1985. The loss of Madagascar's forests caused terrible erosion, which resulted in flooding and the destruction of more trees. Madagascar's entire ecological system is now threatened, and hundreds of unique species are listed as endangered. Only major restrictions on farming and habitat destruction can save these animals.

Endangered Species Acts

The implications of species and habitat destruction were first described in books such as Rachel Carson's *Silent Spring* (1962). Carson, a biologist with the U.S. Department of Interior, wrote about the effects of DDT and insecticides on birds and other animals. Her book inspired Congress to pass the Endangered Species Preservation Act in 1966. This law authorized the secretary of the interior to protect certain fish and wildlife through the creation of a National Wildlife Refuge System. Congress strengthened the law in 1969 by restricting importation of threatened species and adding more domestic species to the list of those deserving protection.

The U.S. Department of the Interior published its first list of endangered species in 1967. The list included seventy-two native species of animals, including grizzly bears, certain butterflies, bats, crocodiles, and trout. No plants were on this list. In 1973, President Richard M. Nixon signed into law an Endangered Species Act that gave the secretaries of the interior and of commerce responsibility for creating a list of endangered animals and

Endangered Animal Species

	U.S.	Foreign	Total
Mammals	65	251	316
Birds	78	175	253
Reptiles	14	64	78
Amphibians	12	8	20
Fishes	71	11	82
Clams	62	2	64
Snails	21	1	22
Insects	35	4	39
Arachnids	12	0	12
Crustaceans	18	0	18
Total	388	516	904

Source: Data are from U.S. Fish & Wildlife Service, Threatened and Endangered Species System (TESS), http://ecos.fws.gov/tess/html/boxscore.html, accessed on March 22, 2003.

plants, or species in immediate danger of extinction. Another list would include species threatened with extinction, or those likely to become endangered in the foreseeable future. Once an animal or plant was on either of the lists, no one could kill, capture, or harm it. Penalties for violators were increased, and international or interstate trade of listed species was prohibited. Fines up to ten thousand dollars could be imposed for knowingly violating the act and one thousand dollars for unwittingly violating it.

A separate provision of the law mandated that federal agencies could not engage in projects that would destroy or modify a habitat critical to the survival of a threatened or endangered plant or animal. This provision became a very important tool in the battle to save species. Friends of wildlife used it to block highway and dam projects, at least until government officials could prove that construction would have no major impact on a fragile ecosystem. The law even called for affirmative measures to aid in the recovery of listed species. The secretaries of the interior and of commerce were required to produce recovery plans detailing steps necessary to bring a species back to a point where it no longer needed protection. Money was appropriated for states to design recovery programs, and most states established plans of their own to deal with local crises.

A major threat to the 1973 law arose in 1978 during the Tellico Dam controversy. The Tennessee Valley Authority, a federal government agency, proposed building a hydroelectric dam on the Tennessee River in Loudon County, Tennessee, in the late 1970's. Shortly after plans were made public, a scientist from the University of Tennessee discovered a three-inch-long fish, the snail darter, that was unique to that area. Building the dam, a $250 million project, would destroy that snail darter's habitat and eliminate the fish. Environmentalists successfully argued in federal court that the dam had to be abandoned. The U.S. Supreme Court supported the ruling of the lower court, arguing that when Congress passed the law, it had intended that endangered species be given the highest priority regardless of the cost or other concerns involved. However, in 1981, Congress enacted a special exemption that excluded "economically important" federal projects from the act. A federal judge then found the Tellico Dam to be without economic importance, and construction was again halted. Congressmen friendly to dam interests then slipped an amendment directing completion of the dam onto an unrelated environmental bill, which passed, and Tellico was constructed. However, the principle of protection remained intact, and the power of the 1973 act remained in force. The snail darter apparently survived, too, as scientists found it living in a river not far from the spot where it was originally discovered.

International Trade Bans
The 1973 act also made the United States a partner in the Convention on International Trade in Endangered Species of Wild Flora and Fauna. This treaty came out of a conference in Washington, D.C., attended by representatives from eighty nations. It created an international system for control of trade in endangered species. Enforced by the World Conservation Union (IUCN) headquartered in Switzerland, the convention has more than one hundred members. The IUCN publishes a series of lists in its *Red Data Book* designating three categories of species. Category one consists of those in immediate danger and therefore absolutely banned from international hunting and trading. Animals and plants in categories two and three are not immediately threatened but require special export permits before they can be bought and sold because their numbers have been seriously reduced.

This convention has a major flaw, a loophole that can be exploited by any member. A nation can make a "reservation" on any listed species, exempting itself from the ban on trade. Japan has been the most frequent user of the reservation, exempting itself from controls on the fin whale, the hawksbill turtle, and the saltwater crocodile, all on the IUCN's most endangered list. Unless this loophole is closed, the IUCN can do little to save extremely threatened species.

Restoring Endangered Species

Several species in the United States—the California condor, the black-footed ferret, whooping cranes, and a bird called the Guam rail—have been saved from extinction because of the 1973 Endangered Species Act. The road to extinction has also been reversed for the brown pelican, found in the southeastern states, the American alligator, which had almost been hunted to extinction in Florida, and the perigrine falcon in the eastern states. Other species, however, such as the dusky seaside sparrow and the Palos Verdes butterfly, have totally disappeared.

The spotted owl found in parts of the rain forest in Oregon and Washington has attracted a good deal of attention because of efforts to save it. The case of the owl points to the most difficult questions raised by the act: Which comes first, the welfare of the plant or animal, or the economic needs of people? Each pair of spotted owls needs six to ten square miles of forest more than 250 years old in which to hunt and breed. The owls also need large hollow trees for nesting, as well as large open fields in which to search for mice and other small animals. A suitable ecosystem that meets all these needs is found only in parts of twelve national forests in the region. At the same time, loggers in the areas need jobs. When the two interests, represented by the lumber industry and environmentalists, collide, it is left to the courts to determine which interest will prevail or whether a compromise can be arranged.

Outside the United States, the future of endangered species appears much grimmer. Scientists at IUCN think several hundred thousand species will disappear by the end of the second decade of the twenty-first century. Many of these species have never even been identified or named. The most endangered habitats in the world are the tropical rain forests, which were reduced by half, nearly 3.5 million square miles, during the twentieth century. About 43,000 square miles are destroyed each year, mainly to provide farms and cattle ranches. The most threatened large animal species in these forests are the large cats, including tigers, jaguars, and leopards; fifteen of the twenty-five species of cats that live in the forests are on the most endangered list.

One solution to forest destruction has been the creation of large wildlife refuges. Several African nations have created one or more of these, but there are limits to the amount of land available for conservation efforts. Another solution is the establishment of more wildlife zoos. Several zoos have successful programs for saving species on the very edge of extinction.

However, capacity is limited, and the very small numbers of animals in a zoo's herd create problems of interbreeding and the handing down of recessive genes.

For many species, it is too late to do very much, so scientists and biologists divide populations into three groups: those that can survive without help, those that would die regardless of whether help is provided, and those species that might survive with help and would certainly die without it. Environmentalisms and restoration ecologists focus their efforts on the plants and animals that fall into the third category. Resources are limited, however, and much work needs to be done, or extinctions will take place at a pace as yet unseen in the history of living things.

Leslie V. Tischauser

See also: Balance of nature; Biodiversity; Communities: ecosystem interactions; Conservation biology; Deforestation; Ecosystems: definition and history; Ecosystems: studies; Endangered plant species; Extinctions and evolutionary explosions; Genetic diversity; Habitats and biomes; Nonrandom mating, genetic drift, and mutation; Old-growth forests; Pollution effects; Reefs; Reforestation; Restoration ecology; Species loss; Sustainable development; Trophic levels and ecological niches; Urban and suburban wildlife; Wildlife management; Zoos.

Sources for Further Study
Beacham, Walton, and Kirk H. Beets, eds. *Beacham's Guide to International Endangered Species*. Osprey, Fla.: Beacham, 1998.
Disilvestro, Roger L. *The Endangered Kingdom: The Struggle to Save America's Wildlife*. New York: John Wiley & Sons, 1989.
Endangered Species of the World. 2 vols. Detroit: Gale Research, 1994.
Endangered Wildlife of the World. 11 vols. New York: Marshall Cavendish, 1993.
Sherry, Clifford J. *Endangered Species: A Reference Handbook*. Santa Barbara, Calif.: ABC-Clio, 1998.
Wilson, Edward O. *The Diversity of Life*. New York: W. W. Norton, 1993.

ENDANGERED PLANT SPECIES

Type of ecology: Restoration and conservation ecology

The World Conservation Union defines "endangered species" as those in immediate danger of extinction and "threatened species" as those at a high risk of extinction but not yet endangered. Vulnerable species are considered ones that are likely to become extinct at some point in the foreseeable future. Rare species are at risk but not yet at the vulnerable, threatened, or endangered levels.

Worldwide, the number of endangered plant species was estimated at more than 33,418 in 1999. This number is much higher than that of all of the endangered or threatened animal species combined. Although extinction is a natural process, and all species will eventually be extinct, human activities threaten the existence of plant and animal life worldwide.

Humans use plants for food; medicine; building materials; energy; to clean water, air, and soil of pollutants; to control erosion; and to convert carbon dioxide to oxygen. The process of extinction increased dramatically during the nineteenth and twentieth centuries because of habitat destruction or loss, deforestation, competition from introduced species, pollution, global warming, and plant hunting, collecting, and harvesting. Over time, pollutants and contaminants accumulate in the soil and remain in the environment, some for many decades. Pollution in the atmosphere also contributes to long-term changes in climate.

Habitat loss

By far the most significant threat to plant species is habitat loss or destruction. Habitat loss can occur because of resource harvesting for food, medicine, and other products; deforestation; and the conversion of wilderness for agricultural, industrial, or urban uses. Wood consumption and tree clearing for agriculture and development threaten the world's forests, especially the tropical forests, which may disappear by the mid-twenty-first century if sufficient preventive action is not taken. Natural disasters, such as climatic changes, meteorites, floods, volcanic eruptions, earthquakes, hurricanes, drought, and tornados, also can be devastating to a habitat.

In Europe and Asia, the plant distribution is complex, with isolated populations of plants spread across a large area. The plants are greatly influenced by the cold climate and by humans. Plant species are disappearing, especially in Europe and the Mediterranean, because of habitat destruction and disturbances including urbanization, road construction,

One of the most endangered plant species in the continental United States is the Catalina mahogany, found only on Catalina Island off the coast of Los Angeles, California. Islands often host unique species that have evolved over thousands of years in isolation from their mainland cousins. (AP/Wide World Photos)

overgrazing, cultivation, forest plantation, fire, pollution, and overexploitation of resources or for use in horticulture.

Mountain plants are threatened by development for industry and tourism, pollution, strip mining, walkers, and skiers. The wetlands are threatened by removal of peat for fuel, water extraction which lowers the water table, and increased drainage for building or agriculture or fear of malaria. Recreational use and susceptibility to pollution such as acid rain or fertilizer run-off present further threats.

In North America, the major causes of endangerment include loss of habitat, overexploitation of resources, introduction of invasive species, and pollution. A massive loss of wilderness has occurred through the clearing of forests, plowing of prairies, and draining of swamplands. For example, in the northeastern United States there are only 13 square miles of alpine habitat, an area in which grow thirty-three at-risk species. This area is heavily used by hikers and mountain bikers. The Florida Everglades are threatened because the water supply is diverted to supply cities, industry, and agriculture.

In California, the Channel Islands are home to seventy-six flowering plants which do not exist on the mainland. Eighteen species are located on just one island. These plants, including the San Clemente broom, bush mallow (*Malacothamnus* Greene), a species of larkspur, and the San Clemente Island Indian paintbrush (*Castilleja grisea*), have been devastated by introduced grazers, browsers, and by invasive other plants. In Hawaii, more than 90 percent of native plants and almost all land birds and invertebrates are found nowhere else in world. The Hawaiian red-flowered geranium *Geranium arboreum* is threatened by introduced feral pigs, agricultural livestock, and competition by nonnative plants.

In developing or highly populated nations in Asia, Africa, Central and South America, the Caribbean, the Pacific Ocean islands, Australia, and New Zealand, habitat loss occurs because of population needs. Land is cleared for agriculture, development, and population resettlement. In Central America and the Caribbean, the Swietenia mahogany is found only in a few protected or remote areas. The Caoba tree (*Persea theobromifolia*) was newly identified as a species as recently as 1977. The lumber is commercially important, and habitat loss has occurred due to the conversion of forests to banana and palm plantations. In Ecuador, only 6 percent of the original rain forest remains standing, because the rest has been converted to farmland. In Asia, including the Philippines, population pressures bring about deforestation and the clearing of land for agriculture.

In southern Africa, land is used for crops, livestock, and firewood production. Overgrazing and the introduction of agriculture have caused the Sahara Desert area to grow rapidly. The island of Madagascar has between ten thousand and twelve thousand plant species, of which 80 percent grow nowhere else in the world. Because of conversion to grassland through farming methods, only about one-fifth of the original species survive. In Australia there are 1,140 rare or threatened plants, and logging, clearing for grazing animals and crops, building developments, and mining have threatened many native species.

Plant Hunting, Collecting, and Harvesting

Habitat damage, the construction of facilities, and the opening of remote areas for human population have made many plants vulnerable to gathering and collecting. Some plants have been overharvested by gardeners, botanists, and horticulturists. One species of lady's slipper orchid (*Cypripedium calceolus*) is rare over much of its natural range except in parts of Scandinavia and the Alps because of collecting. Additionally, many mountain flowers or bulbs such as saxifrages, bellflowers, snowdrops, and cyclamen are endangered. In France and Italy, florulent saxifrage (*Saxifraga florulenta*), an alpine plant, has been overcollected by horticulturists and poachers.

Parts of the southeastern United States have poor soil that is home to the carnivorous or insectivorous plants—those that eat insects. These plants include sundews, bladderworts, Venus's flytrap, and pitcher plants. Collectors or suppliers have stripped many areas of all of these plants. In the Southwest, rare cacti are harvested for sale nationwide and worldwide. Endangered cacti include the Nellie Cory cactus (which has one remaining colony), *Epithelantha micromeres bokei, Ancistrocactus tubuschii*, saguaro cactus (*Carnegiea gigantea*), and *Coryphantha minima*. Near the Sierra Madre, two tree species—Guatemalan fir, or Pinabete, and the Ayuque—are endangered because of harvesting for use as Christmas trees or for the making of hand looms. Additionally, sheep eat the seedlings. In New Mexico, the gypsum wild buckwheat habitat is limited to one limestone hill, and the plants are threatened by cattle, off-road vehicles, and botanists.

In southern Mexico, there are 411 species of epiphytes (air plants or bromeliads in the genus *Tillandsia*), of which several are extremely rare. These plants are threatened by overcollection for the houseplant trade or conservatories. The African violet (*Saintpaulia ionantha*) of Tanzania may soon be extinct in the wild because of the horticultural trade and habitat loss due to encroaching agriculture.

Worldwide, orchids are overcollected for horticulture. Several species have been collected to extinction, are extremely rare, or have been lost because of habitat destruction. Examples include the extremely rare blue vanda (*Vanda caerulea*); *Paphiopedilum druryi*, believed extinct in its native habitat; *Dendrobium pauciflorum*, endangered and possibly extinct—only a single plant was known to exist in the wild in 1970; and the Javan phalaenopsis orchid, *Phalaenopsis javanica*. The latter was believed extinct. When

Endangered Plant Species

	U.S.	Foreign	Total
Flowering plants	570	1	571
Conifers and cycads	2	0	2
Ferns and allies	24	0	24
Lichens	2	0	2
Totals	598	1	599

Source: Data are from U.S. Fish & Wildlife Service, Threatened and Endangered Species System (TESS), http://ecos.fws.gov/tess/html/boxscore.html, accessed on March 22, 2003.

it was rediscovered in 1960's, it was overcollected by commercial orchid dealers and thereby exterminated. There are no other known wild populations.

About 80 percent of the human populations in developing countries rely on traditional medicine, of which 85 percent of ingredients come from plant extracts. In Western medicine, one in four prescription medicines contain one or more plant products. Some at-risk species contain chemicals used in treating medical conditions, such as the African *Prunus africana* tree, whose bark has chemicals used to treat some prostate gland conditions, and the *Strophanthus thollonii*, a root parasite with chemicals used in heart drugs.

The Madagascan periwinkle (*Catharanthus roseus*) is commonly cultivated (its close relative *Catharanthus coriaceus* is rare and its medicinal importance unknown) and produces about seventy chemicals, some of which are useful in the treatment of cancer. The Indian podophyllum (*Podophyllum hexandrum*), a threatened species, is used to treat intestinal worms, constipation, and cancer. Rauwolfia (*Rauvolfia serpentina*), also a threatened species, is used to treat mental disorders, hypertension, and as a sedative. The lily *Amorphopahllus campanulatus* is used to treat stomachaches, and a fig, *Ficus sceptica*, is used to treat fever; both of these species are vulnerable because of habitat destruction.

The Micronesian dragon tree is believed to have magical and medicinal properties. It has been overharvested and is now extinct on several islands. In the United States' Appalachian Mountains, American ginseng is being overcollected because of an escalating demand for this plant's health benefits.

Conservation

The conservation of endangered plant species employs several compelling arguments: Plants enhance the world's beauty, have the right to exist, and are useful to people. The most persuasive argument may be that the survival of the human species depends on a healthy worldwide ecosystem. Three major goals of conservation are recovery, protection, and reintroduction.

Conservation methods depend on increasing public awareness by providing information about endangered or threatened species so that people can take action to reverse damage to the ecosystems. Other important strategies include achieving a widespread commitment to conservation and obtaining funding to protect rare or endangered species. Conservation efforts include setting aside protected areas, such as reserves, wilderness areas, and parks, and recognizing that humans must integrate and protect

Endangered plant species

biodiversity where they live and work. Many countries are actively conserving species through protected areas, endangered-species acts, detailed studies of species and habitat, and information campaigns directed to the public.

Virginia L. Hodges

See also: Balance of nature; Biodiversity; Communities: ecosystem interactions; Conservation biology; Deforestation; Ecosystems: definition and history; Ecosystems: studies; Endangered animal species; Extinctions and evolutionary explosions; Genetic diversity; Habitats and biomes; Nonrandom mating, genetic drift, and mutation; Old-growth forests; Pollution effects; Reefs; Reforestation; Restoration ecology; Species loss; Sustainable development; Trophic levels and ecological niches; Urban and suburban wildlife; Wildlife management; Zoos.

Sources for Further Study

Burton, John A., ed. *Atlas of Endangered Species*. New York: Macmillan, 1999.

Crawford, Mark. *Habitats and Ecosystems: An Encyclopedia of Endangered America*. Santa Barbara, Calif.: ABC-CLIO, 1999.

Freedman, Bill, ed. *Encyclopedia of Endangered Species*. Vol. 2. Detroit: Gale Research, 1999.

Wilson, Edward O. *The Future of Life*. New York: Alfred A. Knopf, 2001.

EROSION AND EROSION CONTROL

Types of ecology: Agricultural ecology; Ecosystem ecology; Restoration and conservation ecology; Soil ecology

Erosion is the loss of topsoil through the action of wind and water. Erosion control is vital because soil loss from agricultural land is a major contributor to nonpoint-source pollution and desertification and represents one of the most serious threats to world food security.

In the United States alone some two billion tons of soil erode from cropland on an annual basis. About 60 percent, or 1.2 billion tons, is lost through water erosion, while the remainder is lost through wind erosion. This is equivalent to losing 0.3 meter (1 foot) of topsoil from two million acres of cropland each year. Although soil is a renewable resource, soil formation occurs at rates of just a few inches per hundred years, which is much too slow to keep up with erosive forces. The loss of soil fertility is incalculable, as are the secondary effects of polluting surrounding waters and increasing sedimentation in rivers and streams.

Erosion removes the topsoil, the most productive soil zone for crop production and the plant nutrients it contains. Erosion thins the soil profile, which decreases a plant's rooting zone in shallow soils, and can disturb the topography of cropland sufficiently to impede farm equipment operation. It carries nitrates, phosphates, herbicides, pesticides, and other agricultural chemicals into surrounding waters, where they contribute to cultural eutrophication. Erosion causes sedimentation in lakes, reservoirs, and streams, which eventually require dredging.

Water Erosion

The common steps in water erosion are detachment, transport, and deposition. Detachment releases soil particles from soil aggregates, transport carries the soil particles away and, in the process, scours new soil particles from aggregates. Finally, the soil particles are deposited when water flow slows. In splash erosion, raindrops impacting the soil can detach soil particles and hurl them considerable distances. In sheet erosion, a thin layer of soil is removed by tiny streams of water moving down gentle slopes. This is one of the most insidious forms of erosion because the effects of soil loss are imperceptible in the short term. Rill erosion is much more obvious because small channels form on a slope. These small channels can be filled in by tillage. In contrast, ephemeral gullies are larger rills that cannot be filled by tillage. Gully erosion is the most dramatic type of water erosion. It leaves channels so deep that even equipment operation is prevented. Gully erosion typically begins at the bottom of slopes where the water flow is fastest and works its way with time to the top of a slope as more erosion occurs.

Wind Erosion

Wind erosion generally accounts for less soil loss than water erosion, but in states such as Arizona, Colorado, Nevada, New Mexico, and Wyoming, it is actually the dominant type of erosion. Wind speeds 0.3 meter (1 foot) above the soil that exceed 16 to 21 kilometers per hour (10 to 13 miles per hour) can detach soil particles. These particles, typically fine- to medium-size sand fewer than 0.5 millimeter (0.02 inch) in diameter, begin rolling and then bouncing along the soil, progressively detaching more and more soil particles by impact. The process, called saltation, is responsible for 50 to 70 percent of all wind erosion. Larger soil particles are too big to become suspended and continue to roll along the soil. Their movement is called surface creep.

Topsoil erosion is one of the most economically devastating forms of erosion, caused not only by wind and water erosion but also by human disturbance resulting from agriculture, overgrazing, deforestation, and soil compaction. (PhotoDisc)

The most obvious display of wind erosion is called suspension, when very fine silt and clay particles detached by saltation are knocked into the air and carried for enormous distances. The Dust Bowl of the 1930's was caused by suspended silt and clay in the Great Plains of the United States. It is also possible to see the effects of wind erosion on the downward side of fences and similar obstacles. Wind passing over these obstacles deposits the soil particles it carries. Other effects of wind erosion are tattering of leaves, filling of road and drainage ditches, wearing of paint, and increasing incidence of respiratory ailments.

Erosion Control
The four most important factors affecting erosion are soil texture and structure, roughness of the soil surface, slope steepness and length, and soil cover. There are several passive and active methods of erosion control that involve these four factors. Wind erosion, for example, is controlled by creating windbreaks, rows of trees or shrubs that shorten a field and reduce the wind velocity by about 50 percent. Tillage perpendicular to the wind direction is also a beneficial practice, as is keeping the soil covered by plant residue as much as possible.

Water erosion is controlled by similar cultural practices. For example, highly erosive, steeply sloped land can be protected by placing it in the U.S.-government-sponsored Conservation Reserve Program. Tillage can be done along the contour of slopes. Long slopes can be shortened by terracing, which also reduces the slope steepness. Permanent grass waterways can be planted in areas of cropland that are prone to water flow. Likewise, grass filter strips can be planted between cropland and adjacent waterways to impede the velocity of surface runoff and cause suspended soil particles to sediment and infiltrate before they can become contaminants.

Conservation tillage practices, such as minimal tillage and no-tillage, are being widely adapted by farmers as a simple means of erosion control. As the names imply, these are tillage practices in which as little disruption of the soil as possible occurs and in which any crop residue remaining after harvest is left on the soil surface to protect the soil from the impact of rain and wind. The surface residue also effectively impedes water flow, which causes less suspension of soil particles. Because the soil is not disturbed, practices such as no-tillage also promote rapid water infiltration, which also reduces surface runoff. No-tillage is rapidly becoming the predominant tillage practice in southeastern states such as Kentucky and Tennessee, where high rainfall and erodible soils occur.

Mark S. Coyne

See also: Desertification; Grazing and overgrazing; Integrated pest management; Multiple-use approach; Rangeland; Reforestation; Slash-and-burn agriculture; Soil; Soil contamination.

Sources for Further Study
Gershuny, Grace, and Joseph Smillie. *The Soul of Soil: A Guide to Ecological Soil Management.* 3d ed. Davis, Calif.: agAccess, 1995.
Morgan, R. P. C. *Soil Erosion and Conservation.* New York: Wiley, 1995.
Plaster, Edward. *Soil Science and Management.* Albany, N.Y.: Delmar, 1997.
Schwab, Glenn O., et al., eds. *Soil and Water Conservation Engineering.* 4th ed. New York: Wiley, 1993.

ETHOLOGY

Type of ecology: Behavioral ecology

Ethology is the study of animal behavior from the perspective of zoology. The information acquired through ethology has helped scientists better understand animals in all their variety.

Ethology is the branch of zoology that investigates the behavior of animals. Behavior may be defined as all the observable responses an animal makes to internal or external stimuli. Responses may be either movements or secretions; however, the study of behavior is much more than a descriptive account of what an animal does in response to particular stimuli. The ethologist is interested in the "how" and "why" of the behaviors they observe. Answering such questions requires an understanding of the physiology and ecology of the species studied. Those who study animal behavior are also interested in the ultimate or evolutionary factors affecting behavior.

The Roots of Ethology

Ethology is a young science, yet it is also a science with a long history. Prior to the late nineteenth century, naturalists had accumulated an abundance of information about the behavior of animals. This knowledge, although interesting, lacked a theoretical framework. In 1859, Charles Darwin published *On the Origin of Species by Means of Natural Selection*, and with it provided a perspective for the scientific study of behavior. Behavior was more central to two of Darwin's later books, *The Descent of Man and Selection in Relation to Sex* (1871) and *Expression of the Emotions of Man and Animals* (1873). By 1973, the science of ethology was sufficiently well developed to be acknowledged by the presentation of the Nobel Prize for Physiology or Medicine to Nikolaas Tinbergen, Konrad Lorenz, and Karl von Frisch for their contributions to the study of behavior. The work of these men was central to the development of modern ethology.

The experimental studies of Frisch revealed the dance language of the honeybee and ways in which the sensory perception of the bees differs from the human sensory world. An awareness of species-specific sensory abilities has provided an important research area and has emphasized a factor that must be considered in the experimental design and interpretation of many types of behavioral research.

Tinbergen studied behavior in a variety of vertebrate and invertebrate organisms. He was good both at observation of animals in their natural habitat and in the design of simple but elegant experiments. His 1951 book *The Study of Instinct* is a classic synthesis of the knowledge that had been gained through the scientific study of animal behavior of that time.

Konrad Lorenz is considered by many to be the founder of ethology, because he discovered and effectively publicized many of the classic phenomena of ethology. Pictures of Lorenz being followed by goslings are almost a standard feature of texts that discuss the specialized form of learning known as imprinting. In natural settings, imprinting allows young animals to identify their parents appropriately. Another contribution of Lorenz was his book *King Solomon's Ring: New Light on Animal Ways*, published in 1952. This extremely readable book raised public awareness of the scientific study of animal behavior and kindled the interest of many who eventually joined the ranks of ethologists.

Ethology and Neurobiology

Many of the features of ethological research characteristic of the work of Lorenz, Tinbergen, and von Frisch have continued to be characteristic of the field. They were concerned that the behavior of animals be understood in the context of the species' natural habitat and that both proximate and ultimate levels of explanation would be examined. Their research strategies have been supplemented by an increase in laboratory-based research and by the introduction of new types of experimental design. These developments have softened the distinctions between ethology and another field of behavioral study, comparative psychology. The focus of comparative psychology is comparative studies of the behavior of nonhuman animals. Initially, questions about learning and development were the major problems investigated in comparative psychology. Although the animals most frequently studied were primates and rodents, those doing the research were interested in gaining insight into the behavior of humans. Comparative psychology was long dominated by behaviorism, a school of thought that assumes that the ultimate basis of behavior is learning. The behaviorists employed rigorous experimental methods. Because such methods require carefully controlled conditions, behavioral research is typically laboratory based, and animals are therefore tested in surroundings remote from their natural environment. Over time, comparative psychology has broadened both the questions it asks and the organisms it studies. The boundaries between comparative psychology and ethology have been further blurred by the rising number of scientists crossing disciplinary lines in their research. Each discipline has learned from the other, and both have also profited from knowledge introduced through neurobiology and behavioral genetics.

Neurobiology investigates the structure and function of the nervous system. One area of ethology that has been directly enriched through neurobiology is the study of sensory perception in animals. The techniques developed in neurobiology allow the investigator to record the response of many individual neurons simultaneously. The neurobiologist examines phenomena such as stimulus filtering at the level of the cell. Stimulus filtering refers to the ability of nerve cells to be selective in their response to stimuli. For example, moths are highly sensitive to sounds in the pitch range of sounds made by the bats that are their chief predators. Neurobiology provides a powerful tool for understanding behavior at the proximate level.

Behavioral Genetics

Another source of information for the ethologist is behavioral genetics. Because of the evolutionary context of ethology, it is important to have an understanding of the genetic basis of behavior. If there were no genetic component in behavior, behavior would not be subject to natural selection. (Natural selection refers to the process by which some genes increase in frequency in a population while alternates decrease because the favored genes have contributed to the reproductive success of those organisms that have them.) While the ethological approach to behavior assumes that behavior patterns are the result of interactions between genes and environment, investigators often ask questions about the genetic programming of behavior.

Early ethologists performed isolation and cross-fostering experiments to discover whether behaviors are learned or instinctive. If a behavior appears in an individual that has been reared in isolation without the opportunity to learn, the behavior is considered instinctive. Observing the behavior of an individual reared by parents of a different species is similarly revealing. When behavior patterns of conspecifics appear in such cross-fostered individuals, such behaviors are regarded as instinctive. Instinctive behaviors typically are innate behaviors that are important for survival. For example, one very common instinctive behavior is the begging call of a newly hatched bird. Isolation and cross-fostering experiments are still a part of the experimental repertoire of ethologists, but behavioral genetics permits the asking of more complex questions. For example, a behavior may accurately be labeled instinctive, but it is more revealing to determine the developmental and physiological processes linking a gene or genes to the instinctive behavior.

Behavioral Ecology

The ethologist is also interested in determining whether behavior is adaptive. It is not sufficient to identify what seems to be a commonsense advantage of the behavior. It is important to show that the behavior does in fact contribute to reproductive success in those that practice the behavior and that the reasons the behavior is adaptive are those that are hypothesized. When behaviors are tested, they frequently do turn out to be adaptive in the ways hypothesized. This type of research, however, has provided many surprises. Research on the adaptive value of behaviors in coping with environmental problems that affect reproductive success is known as behavioral ecology, a major subdiscipline of both ethology and ecology.

Behavioral ecology addresses a variety of questions, in part because the process of evolution is opportunistic. For any environmental problem there are alternate solutions, and the solution a particular species adopts is dependent upon the possibilities inherent in its genes. Questions addressed include such things as whether a species is using the optimum strategy or how a species benefits from living in a group. Because alternate strategies are possible even within a species, behavioral ecologists are interested in evolutionarily stable strategies. An evolutionarily stable strategy is a set of behavioral rules that, when used by a particular proportion of a population, cannot be replaced by any alternative strategy. For example, the sex ratio present in a particular population will determine the optimum sex ratio for the offspring of any individual.

Sociobiology

Sociobiology is another major area of modern ethology. Sociobiology examines animal social behavior within the framework of evolution. Animal species vary in the degree of social behavior they exhibit; other variables include group size and the amount of coordination of activities occurring within the group. The sociobiologist is interested in a number of questions, but prominent among them are the reasons for grouping. Hypotheses such as defense against predators or facilitation of reproduction can be tested. The particular advantage or advantages gained by grouping varies among species. Two important concepts in sociobiology are kin selection and inclusive fitness.

Kin selection refers to the differential reproduction of genes that affect the survival of offspring or closely related kin. Behavior such as the broken-wing display of the killdeer is an example. The behavior carries risk but would be promoted by selection if the offspring of individuals using the display were protected from predators often enough to compensate for the risk. Inclusive fitness is the term used to recognize the concept that fitness includes the total genotype, including those genes that may lower the individual's survival as the price of leaving more genes in surviving kin. The concepts of kin selection and inclusive fitness help to address one problem raised by Darwin, the question of altruistic behavior. Ethology is a young science but a very exciting one, because there are so many questions that can be asked about animal behavior within the context of evolution.

Studying Ethology
The methods and tools of ethology cover the entire spectrum of complexity. One simple, but demanding, method is to collect normative data about a species. In its simplest form, the scientist observes what an animal does and writes it down in a field notebook. Finding and following the animal, coping with field conditions such as bad weather and rugged terrain, and keeping field equipment in operating condition add challenge and variety to this approach. The ethologist uses various techniques to get data as unbiased as possible. One of these is to choose a focal animal at random (or on a rotation) and observe the focal animal for a specific amount of time before switching observation to another member of the population. This prevents bias in which individuals and which behaviors are observed. The sampling of an individual's behavior at timed intervals is an even more effective way of avoiding bias.

When all or most of an animal's behavioral repertoire is known, a list known as an ethogram can be constructed. This catalog can be organized into appropriate categories based on function. Ethograms provide useful baseline information about the behavior of a species. For animals that are difficult or impossible to follow, radio-tracking techniques have been developed. Collars that emit radio signals have been designed for many animals. Miniaturization has made it possible for radio tracking to be used even on relatively small animals.

In field studies, animals are often marked in some way so that observers are able to follow individual animals. A number of techniques have been developed, including banding birds with colored acrylic bands. Color combinations can be varied so that each member of the population has a unique combination. Marking allows the observer to get information such as individual territory boundaries and to determine which animals interact.

Models are frequently used in experiments. For example, a model can be used to determine whether individuals in a species need to learn to identify certain classes of predators. Models were used in many of the classic experiments in ethology. Modern technology has allowed the development of much more sophisticated models. One of the most interesting is a "bee" that can perform a waggle dance (used by bees to indicate location) so effectively that its hivemates can find the food source. Whether a model is simple or sophisticated, it can provide a tool to determine the cues that trigger an animal's response.

Neurobiologists use electrodes and appropriate equipment to stimulate and record the responses of neurons. They can also stimulate specific regions of the nervous system by using tiny tubes to deliver hormones or neurotransmitters. Genetic technology has made it possible to examine the deoxyribonucleic acid (DNA) of individuals in a species. This tool can be used, for example, to determine whether females in monogamous species are completely monogamous or whether some of their offspring are fathered by males other than their mates.

Tape recorders have become very important in studies of animal vocalizations. Recorders are used in two ways. The animal's vocalizations may be recorded and the recording used to make sound spectrographs for analysis. The recordings may also be used to determine whether individuals can discriminate between the vocalizations of neighboring and nonneighboring individuals in their species. Playbacks can also be used to simulate intruders in the territory of an individual and can be applied to many other experimental situations in both the field and laboratory.

The methods used by ethologists are as varied as the problems they investigate. Because the skill of the observer is still a vital link in the investigation of animal behavior, ethology remains one of the more approachable areas of scientific investigation.

Uses of Ethology
The investigation of animal behavior has a number of benefits, both practical and abstract. Some animals are pests, and knowledge of their behavior can be used to manage them. For example, synthetic pheromones have been used to attract members of some insect species. The insects may then be sampled or killed, depending upon the application. To the extent that researchers develop behaviorally based pest management strategies and reduce pesticide use, they will be promoting our own safety as well as that of other species.

It is sometimes important for humans to be able to understand the communication signals of other species and the characteristics of their sensory perception. The knowledgeable individual can recognize the cues that indicate risk that a dog might bite, for example, and can also avoid behavior that the dog will regard as threatening. Understanding the behavior of the wild animals most likely to be encountered in one's neighborhood is an important factor in peaceful coexistence.

The study of animal behavior is providing one of the more fascinating areas of evolutionary biology. Ethology has demonstrated more effectively than most fields of study how diverse the solutions to a given problem can be and has provided insight into human behavior from a biological perspective. Knowledge of animal behavior also enriches human lives simply by satisfying some of our natural curiosity about animals.

Donna Janet Schroeder

See also: Altruism; Communication; Defense mechanisms; Displays; Habituation and sensitization; Herbivores; Hierarchies; Insect societies; Isolating mechanisms; Mammalian social systems; Migration; Mimicry; Omnivores; Pheromones; Poisonous animals; Predation; Reproductive strategies; Territoriality and aggression.

Sources for Further Study

Allen, Colin, and Mark Bekoff. *Species of Mind: The Philosophy and Biology of Cognitive Ethology.* Cambridge, Mass.: MIT Press, 1997.

Barrows, Edward M. *Animal Behavior Desk Reference.* 2d ed. Boca Raton, Fla.: CRC Press, 2000.

Clutton-Brock, T. H., and Paul Harvey, eds. *Readings in Sociobiology.* San Francisco: W. H. Freeman, 1978.

Davies, Nicholas B., and John R. Krebs. *An Introduction to Behavioral Ecology.* 4th ed. Boston, Mass.: Blackwell Scientific Publications, 1997.

Goldsmith, Timothy H. *The Biological Roots of Human Nature: Forging Links Between Evolution and Behavior.* New York: Oxford University Press, 1991.

Krebs, J. R., and N. B. Davies, eds. *Behavioral Ecology: An Evolutionary Approach.* 4th ed. Boston: Blackwell Scientific Publications, 1997.

Lehner, Philip N. *Handbook of Ethological Methods.* 2d ed. New York: Cambridge University Press, 1996.

Wilson, Edward O. *Sociobiology.* Cambridge, Mass.: The Belknap Press of Harvard University Press, 1975.

EUTROPHICATION

Type of ecology: Ecotoxicology

The overenrichment of water by nutrients, eutrophication causes excessive plant growth and stagnation, which leads to the death of fish and other aquatic life.

The word "eutrophic" comes from the Greek *eu*, which means "well," and *trophikos*, which means "food" or "nutrition." Eutrophic waters are well nourished, that is, rich in nutrients; they therefore support abundant life. Eutrophication refers to a condition in aquatic systems (ponds, lakes, and streams) in which nutrients are so abundant that plants and algae grow uncontrollably and become a problem. The plants die and decompose, and the water becomes stagnant. This ultimately causes the death of other aquatic animals, particularly fish, that cannot tolerate such conditions. Eutrophication is a major problem in watersheds and waterways, such as the Great Lakes and Chesapeake Bay, that are surrounded by urban populations.

The stagnation that occurs during eutrophication is attributable to the activity of microorganisms growing on the dead and dying plant material in water. As they decompose the plant material, microbes consume oxygen faster than it can be resupplied by the atmosphere. Fish, which need oxygen in the water to breathe, become starved for oxygen and suffocate. In addition, noxious gases such as hydrogen sulfide (H_2S) can be released during the decay of the plant material. The hallmark of a eutrophic environment is one that is plant-filled, littered with dead aquatic life, and smelly.

Eutrophication is actually a natural process that occurs as lakes age and fill with sediment, as deltas form, and as rivers seek new channels. The main concern with eutrophication in natural resource conservation is that human activity can accelerate the process and can cause it to occur in previously clean but nutrient-poor water. This is sometimes referred to as cultural eutrophication. For example, there is great concern with eutrophication in Lake Tahoe. Much of Lake Tahoe's appeal is its crystal-clear water. Unfortunately, development around Lake Tahoe is causing excess nutrients to flow into the lake and damaging the very thing that attracts people to the lake.

Roles of Nitrogen and Phosphorus
Nitrogen and phosphorus are the key nutrients involved in eutrophication, although silicon, calcium, iron, potassium, and manganese can be im-

portant. Nitrogen and phosphorus are essential in plant and animal growth. Nitrogen compounds are used in the synthesis of amino acids and proteins, whereas phosphate is found in nucleic acids and phospholipids. Nitrogen and phosphorus are usually in limited supply in lakes and rivers. Plants and animals get these nutrients through natural recycling in the water column and sediments and during seasonal variations, as algae and animals decompose, fall to the lower depths, and release their nutrients to be reused by other organisms in the ecosystem. A limited supply of nutrients—as well as variations in optimal temperature and light conditions—prevents any one species of plants or animals from dominating a water ecosystem.

Although nutrient enrichment can have detrimental effects on a water system, an increased supply of nitrogen and phosphorus can have an initial positive effect on water productivity. Much like fertilizers added to a lawn, nutrients added to lakes, rivers, or oceans stimulate plant and animal growth in the entire food chain. Phytoplankton—microscopic algae that grow on the surface of sunlit waters—take up nutrients directly and are able to proliferate. Through photosynthesis, these primary producers synthesize organic molecules that are used by other members of the ecosystem. Increased algal growth thus stimulates the growth of zooplankton—microscopic animals that feed on algae and bacteria—as well as macroinvertebrates, fish, and other animals and plants in the food web. Indeed, many fisheries have benefitted from lakes and oceans that are productive.

Oxygen Depletion

When enough nutrients are added to a lake or river to disrupt the natural balance of nutrient cycling, however, the excess nutrients effectively become pollutants. The major problem is that excess nutrients encourage profuse growth of algae and rooted aquatic weeds, species that can quickly take advantage of favorable growth conditions at the expense of slower-growing species. Algae convert carbon dioxide and water into organic molecules during photosynthesis, a process that produces oxygen. When large blooms of algae and other surface plants die, however, they sink to the bottom of the water to decompose, a process that consumes large amounts of oxygen. The net effect of increased algae production, therefore, is depletion of dissolved oxygen in the water, especially during midsummer.

Reduced oxygen levels (called hypoxia) can have dire consequences for lakes and rivers that support fish and bottom-dwelling animals. Oxygen depletion is greatest in the deep bottom layers of water, because gases from the oxygen-rich surface cannot readily mix with the lower layers. During summer and winter, oxygen depletion in eutrophic waters can cause massive fish kills. In extreme cases of eutrophication, the complete depletion of oxygen (anoxia) occurs, leading to ecosystem crashes and irreversible damage to plant and animal life. Oxygen depletion also favors the growth of anaerobic bacteria, which produce hydrogen sulfide and methane gases, leading to poor water quality and taste.

Excessive algal and plant growth has other negative effects on a water system. Algae and plants at the surface block out sunlight to plants and animals at the lower depths. Loss of aquatic plants can affect fish-spawning areas and encourage soil erosion from shores and banks.

Eutrophication often leads to loss of diversity in a water system, as high nutrient conditions favor plants and animals that are opportunistic and short-lived. Native sea grasses and delicate sea plants often are replaced by hardier weeds and rooted plants. Carp, catfish, and bluegill fish species replace more valuable coldwater species such as trout.

Thick algal growth also increases water turbidity and gives lakes and ponds an unpleasant pea-soup appearance. As algae die and decay, they wash up on shores in stinking, foamy mats.

Algal blooms of unfavorable species can produce toxins that are harmful to fish, animals, and humans. These toxins can accumulate in shellfish and have been known to cause death if eaten by humans. So-called red tides and brown tides are caused by the proliferation of unusual forms of algae, which give water a reddish or tealike appearance and in some cases produce harmful chemicals or neurotoxins.

Assessing Eutrophication
While eutrophication effects are generally caused by nutrient enrichment of a water system, not all cases of nutrient accumulation lead to increased productivity. Overall productivity is based on other factors in the water system, such as grazing pressure on phytoplankton, the presence of other chemicals or pollutants, and the physical features of a body of water. Eutrophication occurs mainly in enclosed areas such as estuaries, bays, lakes, and ponds, where water exchange and mixing are limited. Rivers and coastal areas with abundant flushing generally show less phytoplankton growth from nutrient enrichment because their waters run faster and mix more frequently. On the other hand, activities that stir up nutrient-rich sediments from the bottom, such as development along coastal waters, recreational activities, dredging, and storms, can worsen eutrophication processes.

The nutrient status of a lake or water system is often used as a measure of the extent of eutrophication. For example, lakes are often classified as oligotrophic (nutrient-poor), eutrophic (nutrient-rich), or mesotrophic (moderate in nutrients) based on the concentrations of nutrients and the physical appearance of the lake. Oligotrophic lakes are deep, clear, and unproductive, with little phytoplankton growth, few aquatic rooted plants, and high amounts of dissolved oxygen. In contrast, eutrophic lakes are usually shallow and highly productive, with extensive aquatic plants and sedimentation. These lakes have high nutrient levels, low amounts of dissolved oxygen, and high sediment accumulation on the lake bottom. They often show sudden blooms of green or blue-green algae (or blue-green bacteria, cyanobacteria) and support only warm-water fish species.

Mesotrophic lakes show characteristics in between those of unproductive oligotrophic waters and highly productive eutrophic waters. Mesotrophic lakes have moderate nutrient levels and phytoplankton growth and some sediment accumulation; they support primarily warm-water fish species. As a lake naturally ages over hundreds of years, it usually (but not always) gets progressively more eutrophic, as sediments fill in and eventually convert it to marsh or dryland. Nutrient enrichment from human sources can speed this process greatly.

Limiting Damage

The negative effects of eutrophication can be reduced by limiting the amount of nutrients—in most cases nitrogen and phosphorus—from entering a water system. Nutrients can enter water bodies through streams, rivers, groundwater flow, direct precipitation, and dumping and as particulate fallout from the atmosphere. While natural processes of eutrophication are virtually impossible to control, eutrophication from human activity can be reduced or reversed.

Phosphorus enrichment into water systems occurs primarily as the result of wastewater drainage into a lake, river, or ocean. Phosphate is common in industrial and domestic detergents and cleaning agents. Mining along water systems is also a major source of phosphorus. When phosphorus enters a water system, it generally accumulates in the sediments. Storms and upwelling can stir up sediments, releasing phosphorus. Treatment of wastewater to remove phosphates and the reduction of phosphates in detergents have helped to reduce phosphorus enrichment of water systems.

Nitrogen enrichment is harder to control; it is present in many forms, as ammonium, nitrates, nitrites, and nitrogen gas. The major sources of nitrogen eutrophication are synthetic fertilizers, animal wastes, and agricultural runoff. Some algae species can also fix atmospheric nitrogen directly, converting it to biologically usable forms of nitrogen. Since the atmosphere contains about 78 percent nitrogen, this can be a major source of nitrogen enrichment in waters that already have significant algal populations.

Efforts to control nitrogen and phosphorus levels have examined both point and nonpoint sources of nutrient loading. Point sources are concentrated, identifiable sites of nutrients that include municipal sewage-treatment plants, feed lots, food-processing plants, pulp mills, laundry detergents, and domestic cleaning agents. Nonpoint, or diffuse, sources of nutrients include surface runoff from rainwater, fertilizer from agricultural land and lawns, eroded soil, and roadways.

Linda Hart and Mark S. Coyne

See also: Acid deposition; Biomagnification; Erosion and erosion control; Food chains and webs; Invasive plants; Lakes and limnology; Nutrient cycles; Ocean pollution and oil spills; Pesticides; Phytoplankton; Pollution effects; Slash-and-burn agriculture; Waste management.

Sources for Further Study

Brönmark, Christer. *The Biology of Lakes and Ponds*. New York: Oxford University Press, 1998.

Cole, Gerald A. *Textbook of Limnology*. 4th ed. Prospect Heights, Ill.: Waveland Press, 1994.

Harper, David. *Eutrophication of Freshwaters: Principles, Problems, and Restoration*. London: Chapman & Hall, 1992.

Hinga, Kenneth, Heeseon Jeon, and Noelle F. Lewis. *Marine Eutrophication Review*. Silver Springs, Md.: U.S. Department of Commerce, National Oceanic and Atmospheric Administration, Coastal Ocean Office, 1995.

Horne, Alexander J., and Charles R. Goldman. *Limnology*. 2d ed. New York: McGraw-Hill, 1994.

Schramm, Winfrid, and Pieter H. Nienhuis, eds. *Marine Benthic Vegetation: Recent Changes and the Effects of Eutrophication*. Berlin: Springer, 1996.

Wetzel, Robert G. *Limnological Analyses*. 3d ed. New York: Springer, 2000.

EVOLUTION: DEFINITION AND THEORIES

Types of ecology: Evolutionary ecology; Paleoecology; Speciation

Evolution is change in species through time. From a one-celled ancestor, many billions of species have evolved into a grand diversity of life-forms that populate earth and interact in ways that will affect the evolution of future life-forms.

Inheritance and Natural Selection

"Evolution" comes from the Latin word meaning "to unroll." In a general sense, it refers to any change through time, but it is often restricted to biological change. For most of human history, the universe was thought to be unchanging. Then, in the eighteenth century, expeditions to new continents and the discovery of extinct fossil animals convinced many people that the biological world was not as unchanging as had been thought. There could be no proof for this hypothesis, however, until an explanation of how such change occurred could be found. In 1809, Jean-Baptiste Lamarck became the first to propose an explanation; his theory was based on the inheritance of acquired traits. According to this theory, giraffes, for example, obtained long necks because individual giraffes stretched their neck muscles more and more to reach ever higher leaves, and the longer necks were passed on to the offspring. This idea was quickly shown to be false by experiment. Traits acquired during an individual's lifetime (such as larger muscles acquired through weightlifting) are not passed on to offspring.

It was 1859 when Charles Darwin proposed what is now known to be the actual process by which evolution occurs: natural selection. This process can be divided into three steps: Individuals in a species vary in their traits; some individuals will have more offspring than others, depending on how advantageous their particular traits are; advantageous traits will increase in a species, and disadvantageous traits will be lost through time and as the environment changes. For example, climatic change may cause a forested environment to become a snowy tundra. Creatures with dark coats are best off in a forest because of the concealment of the dark, shadowy environment, but as the lighter, snowy environment becomes dominant, lighter individuals will have an advantage because they are more easily concealed. Over a long period, enough changes will accumulate in a group that an observer might say that a new species had been created. This process has often been called "survival of the fittest," but the "fittest" organisms are not always the fastest, fiercest, or even most competitive. For example, animals that cooperate with other animals or are the most timid and conceal themselves readily may survive more often and produce more offspring.

When the whole species changes at once, "nonbranching" evolution occurs. Many species, however, have wide ranges and occur in many different geographic areas, so often only some populations of a species are subjected to environmental changes. That is an important point because it explains how so many species can be created from one ancestral species. "Branching" evolution is especially common when one of the populations becomes cut off from the others by a barrier of some kind. For example, a new river may form. This river prevents interbreeding and allows each population to form its own pool of traits. In time, differences between the two environments will cause the two populations to become so distinct that they form two different species.

Species have hitherto been described as groups that are visibly distinct enough to be distinguishable from one another; however, there is a much more objective definition of species, based on the criterion of interbreeding. To a biologist, members of a species can produce fertile offspring only when they breed with other members. Therefore, a new species has evolved not when it "looks" sufficiently different from its ancestors or neighboring populations but when it can no longer successfully interbreed with them. This definition of reproductive isolation is important because many closely related species look quite similar yet cannot successfully interbreed.

Evolution by natural selection explains not only how species have changed but also why they are so well adapted to their surroundings: The best-adapted individuals have the most offspring. Further, it explains some crucial aspects of basic anatomy, such as why vestigial, or "remnant," organs exist: They are in the process of being lost. For example, the now-useless human appendix was once an important part of the human digestive tract. Also, it explains why many organisms have similar organs that are used for different purposes, such as five-fingered hands on humans and five-digit organs on bat wings. Such "homologous" organs have been modified from a common ancestor. This modification also explains why many organisms pass through similar embryonic stages; human embryos, for example, have tails and gills like a fish.

Laws of Heredity
After Darwin proposed natural selection as the process of evolution, it was readily accepted, and most scientists have accepted it ever since. The ex-

planation was incomplete, however, in one major area: Darwin could not explain how variation was produced or passed on. The laws of heredity were discovered by Gregor Mendel in 1866 while Darwin was wrestling with this problem. Mendel's work lay unnoticed until the early twentieth century, when other scientists independently discovered the gene as the basic unit of heredity. Genes are now known to be molecular "blueprints" that are repeatedly copied within each cell. They contain instructions on how to build the organism and how to maintain it.

Genes are passed on to the offspring when a sperm and an egg cell unite. The resulting fertilized egg consists of one cell that contains all the genes on strands, or chromosomes, in the cell nucleus. The chromosomes occur in pairs such that one member of each pair is from the father and one is from the mother. As growth occurs, certain genes in each cell will be bio-chemically "read" and will give instructions on what happens next. Genes are composed of the molecule deoxyribonucleic acid (DNA), which is shaped like a twisted ladder and is copied when the "ladder" splits in half at the middle of the "rungs." Once the instructions are copied, they are carried outside the cell nucleus by messenger molecules, which proceed to build proteins (such as enzymes and muscle tissue) using the rungs as a blueprint.

With this added knowledge of genes as the units of heredity, evolutionists could see that natural selection acting on individuals selects not only traits but also the genes that serve as blueprints for those traits. Therefore, as well as being a change in a species' traits through time, evolution is also often defined as a change in the "gene pool" of a species. The gene pool is the total of all the genes contained in a species. Individual variation in a gene pool originally arises via mutations, errors made in the DNA copying process. Usually, mutation involves a change in the DNA sequence that causes a change in the genetic instructions.

Most mutations have little effect, which is fortunate because those that are expressed generally kill or handicap the offspring. That occurs since any organism is a highly integrated, complex system, and any major alterations to it are therefore likely to disrupt it. Nevertheless, rare improvements do occur, and it is these that are passed on and become part of the breeding gene pool. Although mutations provide the ultimate source of variation, the sexual recombination of genes provides the more immediate source. Each organism has a unique combination of genes, and it is the fitness of this combination that determines how well those genes survive and are passed on. Although brothers and sisters have the same parents, they are not alike because genes are constantly shuffled and reshuffled in the production of each sperm and egg cell.

The Geologic Time Scale

MYA	Eon	Era	Period	Epoch	Developments
0.01			Quaternary (1.8 mya-today)	Holocene (11,000 ya-today)	Ice Age ends. Human activities begin to impact biosphere.
1.8				Pleistocene (1.8 mya-11,000 ya)	Ice Age in northwest Europe, Siberia, and North America. Plants migrate in response. New speciations lead to modern plants. Modern humans evolve.
5		Cenozoic (65 mya-today)	Tertiary (65-1.8 mya)	Pliocene (5-1.8 mya)	Cooling period leads to Ice Age.
23				Miocene (23-5 mya)	Erect-walking human ancestors
38				Oligocene (38-23 mya)	Primate ancestors of humans
54				Eocene (54-38 mya)	Intense mountain building occurs (Alps, Himalaya, rockies); modern mammals appear (rodents, hoofed animals); diverse conferous forests
65	Phanerozoic Eon (544 mya-today)			Paleocene (65-54 mya)	Cretaceous-Tertiary even leads to dinosarus' extinction c. 65 mya; seed ferns; bennettites and caytonias dief off; ginkgoes decline; decidous plants rise.
146		Mesozoic (245-208 mya)	Cretaceous (146-65 mya)		Breakup of supercontinents into tosay's form; birds appear. Cycads, other gymnosperms still widespread, but angiosperms dominate by 90 mya; animal-aided pollination begins to evolve.
208			Jurassic (208-146 mya)		Earliest mammals; ginkgoes thrive in moister areas; drier climates in North and South America, parts of Africa, central Asia; rise of modern gymnosperms, such as junipers, pine trees. Earliest angiosperm fossil (from China) dates from end of this period.
245			Triassic (245-208 mya)		Diminished land vegetation, with lack of variation reflecting global frost-free climate; gymnosperms dominate, bennettites and gnetophytes appear; dinosaurs develop.
286		Paleozoic (544-245 mya)	Permian (286-245 mya)		Permian extinction even initiates drier, colder period; supercontinent Pangaea has formed; tree-sized lycophytes, club mosses, cordaites die off; horsetails, peltasperms, cycads, conifers dominate.
325			Carboniferous (360-286 mya)	Pennsylvanian (325-286 mya)	Gymnosperms, ferns, calamites, lycopods thrive (first seed plants); reptiles appear. Plant life diverse, from small creeping forms to tall forest trees. coal beds form in swamp forests from the dominant seedless vascular plants.
360				Mississippian (360-325 mya)	

MYA	Eon	Era	Period	Epoch	Developments
410			Devonian (410-360 mya)		Club mosses, early ferns, lycophytes, progymnosperms; amphibians, diverse insects; horsetails, gymnosperms present by end of period.
440	Phanerozoic Eon (244 mya-today)	Paleozoic (544-245 mya)	Silurian (440-410 mya)		Early land plants: nonvascular bryophytes (mosses, hornworts), followed by seedless vascular plants in now-extinct phyla *Rhyniophyta, Zosterophyllophyta, Trimerophytophyta.*
505			Ordovician (505-440 mya)		Life colonizes land; earliest vertebrates appear in fossil record.
544			Cambrian (544-505 mya)	Tommotian (530-527 mya)	Cambrian diversification of life
900	Precambrian Time (4,500-544 mya)	Proterozoic (2,500-544 mya)	Neoproterozoic (900-544 mya)	Vendian (650-544 mya)	Age of algae, earliest invertebrates
1600			Mesoproterozoic (1,600-900 mya)		Eukaryotic life established.
2500			Paleoproterozoic (2,500-1,600 mya)		Transition from prokaryotic to eukaryotic life leads to multicellular organisms
3800		Archaean (3,800-2,500 mya)			Microbial life (anaerobic and cyanobacteria) as early as 3.5 bya.
4500		Hadean (4,500-3,800 mya)			Earth forms 4.5 bya.

Source: Data on time periods in this version of the geologic time scale are based on new findings in the last decade of the twentieth century as presented by the Geologic Society of America, which notably moves the transition between the Precambrian and Cambrian times from 570 mya to 544 mya.

Notes: bya=billions of years ago; mya = millions of years ago; ya = years ago.

Rates and Patterns of Evolution

A major area of debate is how fast evolution occurs. Some scientists believe that most evolution occurs rapidly. This view has been called "punctuated" evolution. Another group argues that evolution is more often gradual, as Darwin originally proposed. To some extent, this disagreement is a matter of different perspectives. A geneticist working with flies in a laboratory would see the evolution of a new species in ten thousand years as very slow. To a paleontologist, however, who often deals with fossil species lasting millions of years, ten thousand years is brief indeed. Nevertheless, there is more to the debate than perspective alone. Punctuationists argue not only that evolution is rapid but also that species have such tightly integrated gene pools that virtually no change at all occurs during most of a species' existence.

In contrast, gradualists view species as being much less integrated, so that change can be a continuous process. The fossil record at first glance seems to support the punctuated view. The majority of species show very little change for long spans of time and then either disappear or rapidly give rise to another species. The fossil record is very incomplete, however, being full of gaps where no fossils were deposited. As a result, it is often impossible to tell whether the "rapid" change in species is real or only follows a gap in what was actually a gradual sequence. Also, fossils represent only part of the original organism—usually only the hard parts, such as shells, bones, or teeth. Therefore, any changes in soft anatomy, such as tissues or biochemistry, are lost, making it impossible to say with certainty that no change occurred. Whatever the outcome of the debate, all scientists agree that evolutionary rates vary.

In addition to the rate of evolution, much has also been written about the patterns produced by evolution since life arose about 3.5 million years ago. Evolutionary trends are directional changes seen in a group. The most common trend, found in many groups, is an increase in size. Another trend, seen mainly in mammals, is an increase in brain size. Life as a whole has shown an increase in total diversity and complexity. These trends, however, are only statistical tendencies. They are not inevitable "laws," as many have misinterpreted them in the past. Often, groups do not show them, and in those that do, the change is not constant and may reverse itself at times. Finally, trends are often interrupted by mass extinctions. At least five times in the past 600 million years, more than 50 percent of all the species on the earth have been wiped out by catastrophes of different kinds, from temperature changes to impacts of huge meteorites.

Study of Fossils
Fossils, the remains of former life, provide the only record of most evolution, because more than 99 percent of all species that have ever existed are now extinct. Paleontology is the study of fossils. Such study begins with identification of the remains—usually hard parts, such as bones—and ends with measurement of fossil size, shape, and abundance. The extreme incompleteness of the fossil record is a major obstacle to this method, since only some parts are preserved, and these are from strictly limited periods of the evolutionary past. Nevertheless, many evolutionary lineages can be traced through time. Indeed, refined measurements of rate and direction of anatomical change are often possible when used in conjunction with dating techniques.

The study of living organisms permits observation of the complete organism. Comparative anatomy reveals similarities among related species and shows how evolution has modified them since they separated from their common ancestor. For example, humans and chimpanzees are extremely similar in their organ and muscle anatomy. This method is not limited to comparison of adults but includes earlier stages of development as well. Comparative embryology often shows anatomical similarities, such as those between humans and other vertebrates. For example, the human embryo goes through a stage with gills and a tail, resembling stages of an amphibian embryo. Comparative biochemistry is also very useful, revealing similarities in proteins and many other molecules. Such comparisons are based on differences in molecular sequences, such as amino acids. Molecular "clocks" are sometimes calculated in this manner. More distantly related species are thought to have more differences. The accuracy of such clocks, however, is hotly debated.

A major technique is DNA sequencing, whereby the exact genetic information is read directly from the gene. This method will greatly add to knowledge of evolutionary relationships, although it is expensive and time-consuming. Biogeography, or the distribution of organisms in nature, is a method that Darwin used and that is still important today. This technique often reveals populations (races) within a species' overall range that differ from one another because they inhabit slightly different geographic areas. These populations give the scientist a "snapshot" of evolution in progress. Given more time, many of these races would eventually become different species.

Artificial breeding is a method of directly manipulating evolution. The most widely used experimental organism for this purpose is the fruit fly, which is used in part because of its exceptionally large chromosomes; they make the genes easy to identify. A common experiment is to subject the flies to radiation or chemicals that cause mutations and then to analyze the effects. The gene pool is then subjected to extreme artificial selection as the experimenter allows only certain individuals to breed. For example, only those with a gene for a certain kind of wing may reproduce. Although such experiments have often altered the organisms' gene pools and created new varieties within the species, no truly new species has ever been created in the laboratory. Apparently, more time is needed to produce a new species. Outside the laboratory, artificial breeding has been done for thousands of years. Food plants and domesticated animals have had much of their evolution controlled by humans. Analysis of the effect of this breeding on the organisms' gene pools is the most complete and direct method of studying evolution.

Significance

The study of fossils has been a major tool in understanding Earth's history. This understanding has allowed more efficient exploitation of the earth's resources. For example, petroleum and coal provide the major energy resources today and were both formed by organisms of the past. Petroleum comes from the biochemicals of marine organisms, and coal comes from fossilized plants. Most paleontologists are employed in the costly search for these "fossil fuels," and knowing the evolutionary history of these groups helps to determine the most productive places to search. Fossils also form nonenergy resources. Limestone is used in many processes, from making cement to making steel. Most limestone is composed of the fossilized remains of seashells and other marine skeletons. Phosphate minerals, essential for fertilizers in almost all forms of agriculture, come from marine fossil deposits as well.

Darwin's theory of natural selection caused a violent reaction throughout much of the world when it was applied in social contexts (to which Darwin himself disagreed) as "social Darwinism." The notion that humans evolved from lower life-forms such as the ape was truly revolutionary. Instead of creatures of a divine plan, humans were now seen as products of natural, sometimes "random," processes. The impact of this realization on ethics, the arts, and society in general is still being felt. Evolution, however, does not necessarily conflict with religion, as is often thought. Science seeks to find out only how things happen, not the ultimate reasons why they happen. Therefore, most major religions have reconciled their tenets with the fact of evolution by viewing natural selection as simply a mechanism employed by God to meet his ends.

Michael L. McKinney

See also: Adaptations and their mechanisms; Adaptive radiation; Coevolution; Colonization of the land; Convergence and divergence; Dendrochronology; Development and ecological strategies; Evolution: history; Evolution of plants and climates; Extinctions and evolutionary explosions; Gene flow; Genetic drift; Genetically modified foods; Isolating mechanisms; Natural selection; Nonrandom mating, genetic drift, and mutation; Paleoecology; Population genetics; Punctuated equilibrium vs. gradualism; Speciation; Species loss.

Sources for Further Study

Dawkins, Richard. *The Blind Watchmaker: Why the Evidence of Evolution Reveals a Universe Without Design.* New York: W. W. Norton, 1988.

Futuyma, Douglas J. *Evolutionary Biology.* Sunderland, Mass.: Sinauer Associates, 1986.

McNamara, K. J. *Shapes of Time: The Evolution of Growth and Development.* Baltimore: Johns Hopkins University Press, 1997.

Schopf, J. William, ed. *Major Events in the History of Life*. Boston: Jones and Bartlett, 1992.

Stanley, Steven M. *The New Evolutionary Timetable: Fossils, Genes, and the Origin of Species*. New York: Basic Books, 1981.

Stebbins, G. Ledyard. *Darwin to DNA, Molecules to Humanity*. New York: W. H. Freeman, 1982.

EVOLUTION: HISTORY

Types of ecology: Evolutionary ecology; History of ecology

Evolution is the theory that biological species undergo sufficient change with time to give rise to new species. The development of the theory of evolution has contributed much to two later scientific disciplines, genetics and ecology, by providing explanations for the adaptations and interrelationships of species from early times to the present.

The concept of evolution has ancient roots. Anaximander suggested in the sixth century B.C.E. that life had originated in the seas and that humans had evolved from fish. Empedocles (c. 450 B.C.E.) and Lucretius (c. 96-55 B.C.E.), in a sense, grasped the concepts of adaptation and natural selection. They taught that bodies had originally formed from the random combination of parts, but that only harmoniously functioning combinations could survive and reproduce. Lucretius even said that the mythical centaur, half horse and half human, could never have existed because the human teeth and stomach would be incapable of chewing and digesting the kind of grassy food needed to nourish the horse's body.

Early Biological Theory
For two thousand years, however, evolution was considered an impossibility. The theory of forms (also called his theory of ideas) proposed by Plato (c. 428-348 B.C.E.) gave rise to the notion that each species had an unchanging "essence" incapable of evolutionary change. As a result, most scientists from Aristotle (384-322 B.C.E.) to Carolus Linnaeus (1707-1778) insisted upon the immutability of species.

Many of these scientists tried to arrange all species in a single linear sequence known as the scale of being (also called the great chain of being or *scala naturae*), a concept supported well into the nineteenth century by many philosophers and theologians. The sequence in this scale of being was usually interpreted as a static "ladder of perfection" in God's creation, arranged from higher to lower forms. The scale had to be continuous, for any gap would detract from the perfection of God's creation. Much exploration was devoted to searching for missing links in the chain, but it was generally agreed that the entire system was static and incapable of evolutionary change. Pierre-Louis Moreau de Maupertuis and Jean-Baptiste Lamarck (1744-1829) were among the scientists who tried to reinterpret the scale of being as an evolutionary sequence, but this single-sequence idea was later replaced by the concept of branching evolution proposed by Charles Darwin (1809-1882). Georges Cuvier (1769-1832) finally showed that the major groups of animals had such strikingly different anatomical structures that no possible scale of being could connect them all; the idea of a scale of being lost most of its scientific support as a result.

The theory that new biological species could arise from changes in existing species was not readily accepted at first. Linnaeus and other classical biologists emphasized the immutability of species under the Platonic-Aristotelian concept of essentialism. Those who believed in the concept of evolution realized that no such idea could gain acceptance until a suitable mechanism of evolution could be found. Many possible mechanisms were therefore proposed. Étienne Geoffroy Saint-Hilaire (1805-1861) proposed that the environment directly induced physiological changes, which he thought would be inherited, a theory now known as Geoffroyism. Lamarck proposed that there was an overall linear ascent of the scale of being but that organisms could also adapt to local environments by voluntary exercise, which would strengthen the organs used; unused organs would deteriorate. He thought that the characteristics acquired by use and disuse would be passed on to later generations, but the inheritance of acquired characteristics was later disproved. Central to both these explanations was the concept of adaptation, or the possession by organisms of characteristics that suit them to their environments or to their ways of life. In eighteenth century England, the Reverend William Paley (1743-1805) and his numerous scientific supporters believed that such adaptations could be explained only by the action of an omnipotent, benevolent God. In criticizing Lamarck, the supporters of Paley pointed out that birds migrated toward warmer climates before winter set in and that the heart of the human fetus had features that anticipated the changes of function that take place at birth. No amount of use and disuse could explain these cases of anticipation, they claimed; only an omniscient God who could foretell future events could have designed things with their future utility in mind.

Darwin's Theory

The nineteenth century witnessed a number of books asserting that living species had evolved from earlier ones. Before 1859, these works were often more geological than biological in content. Most successful among them was the anonymously published *Vestiges of the Natural History of Creation* (1844), written by Robert Chambers (1802-1871). Books of this genre sold well but contained many flaws. They proposed no mechanism to account for evolutionary change. They supported the outmoded concept of a scale of being, often as a single sequence of evolutionary "progress." In geology, they supported the outmoded theory of catastrophism, an idea that the history of the earth had been characterized by great cataclysmic upheavals. From 1830 on, however, that theory was being replaced by the modern theory of uniformitarianism, championed by Charles Lyell (1797-1875). Darwin read these books and knew their faults, especially their lack of a mechanism that was compatible with Lyell's geology. In his own work, Darwin carefully tried to avoid the shortcomings of these theories. Eventually, he brought about the greatest revolution in biological thought by proposing both a theory of branching evolution and a mechanism of natural selection to explain how it occurred.

Much of Darwin's evidence was gathered during his voyage around the world aboard HMS *Beagle* between 1831 and 1836. Darwin's stop in the Galápagos Islands and his study of tortoises and finchlike birds on these islands is usually credited with convincing him that evolution was a branching process and that adaptation to local environments was an essential part of the evolutionary process. Adaptation, he later concluded, came about through natural selection, a process that led to the deaths of maladapted variations and allowed only the well-adapted ones to survive and pass on their hereditary traits. After returning to England from his voyage, Darwin raised pigeons, consulted with various animal breeders about changes in domestic breeds, and investigated other phenomena that later enabled him to demonstrate natural selection and its power to produce evolutionary change.

Darwin delayed the publication of his book for seventeen years after he wrote his first manuscript version. He might have waited even longer, except that his hand was forced. From the East Indies, another British scientist, Alfred Russel Wallace (1823-1913), had written a description of an identical theory and submitted it to Darwin for his comments. Darwin showed Wallace's letter to Lyell, who urged that both Darwin's and Wallace's contributions be published, along with documented evidence showing that both had arrived at the same ideas independently. Darwin's great book, *On the Origin of Species by Means of Natural Selection*, was published in 1859, and it quickly won most of the scientific community's support of the concept of branching evolution. In his later years, Darwin also published *The Descent of Man and Selection in Relation to Sex* (1871), in which he outlined his theory of sexual selection. According to this theory, the agent that determines the composition of the next generation may often be the opposite sex. An organism may be well adapted to live, but unless it can mate and leave offspring, it will not contribute to the next or to future generations.

After Darwin

In the early 1900's, the rise of Mendelian genetics (named for botanist Gregor Mendel, 1822-1884) initially resulted in challenges to Darwinism. Hugo de Vries (1848-1935) proposed that evolution occurred by random mutations, which were not necessarily adaptive. This idea was subsequently rejected, and Mendelian genetics was reconciled with Darwinism during the period from 1930 to 1942. According to this modern synthetic theory of evolution, mutations initially occur at random, but natural selection eliminates most of them and alters the proportions among those that survive. Over many generations, the accumulation of heritable traits produces the kind of adaptive change that Darwin and others had described. The process of branching evolution through speciation is also an important part of the modern synthesis.

The branching of the evolutionary tree has resulted in the proliferation of species from the common ancestor of each group, a process called adaptive radiation. Ultimately, all species are believed to have descended from a single common ancestor. Because of the branching nature of the evolutionary process, no one evolutionary sequence can be singled out as representing any overall trend; rather, there have been different trends in different groups. Evolution is also an opportunistic process, in the sense that it follows the path of least resistance in each case. Instead of moving in straight lines toward a predetermined goal, evolving lineages often trace meandering or circuitous paths in which each change represents a momentary increase in adaptation. Species that cannot adapt to changing conditions die out and become extinct.

Evolutionary biology is itself the context into which all the other biological sciences fit. Other biologists, including physiologists and molecular biologists, study how certain processes work, but it is evolutionists who study the reasons that these processes came to work in one way and not another. Organisms and their cells are built one way and not another because their structures have evolved in a particular direction and can only be explained as the result of an evolutionary process. Not only does each biological system need to function properly, but it also must have been able to achieve its present method of functioning as the result of a long, historical, evolutionary process in which a previous method of functioning changed into the present one. If there were two or more ways of accomplishing the same result, a particular species used one of them because its ancestors were more easily capable of evolving this one method rather than another.

Eli C. Minkoff

See also: Adaptations and their mechanisms; Adaptive radiation; Coevolution; Colonization of the land; Convergence and divergence; Dendrochronology; Development and ecological strategies; Ecology: history; Ecosystems: definition and history; Evolution: definition and theories; Evolution of plants and climates; Extinctions and evolutionary explosions; Gene flow; Genetic drift; Genetically modified foods; Isolating mechanisms; Natural selection; Nonrandom mating, genetic drift, and mutation; Paleoecology; Population genetics; Punctuated equilibrium vs. gradualism; Speciation; Species loss.

Sources for Further Study

Bowler, Peter J. *Evolution: The History of an Idea.* Rev. ed. Berkeley: University of California Press, 1989.

_____. *Life's Splendid Drama: Evolutionary Biology and the Reconstruction of Life's Ancestry, 1860-1940.* Chicago: University of Chicago Press, 1996.

Brandon, Robert N. *Concepts and Methods in Evolutionary Biology.* New York: Cambridge University Press, 1996.

Grant, Verne. *The Evolutionary Process: A Critical Study of Evolutionary Theory.* 2d ed. New York: Columbia University Press, 1991.

Minkoff, Eli C. *Evolutionary Biology.* Reading, Mass.: Addison-Wesley, 1983.

Zimmer, Carl. *Evolution: The Triumph of an Idea.* New York: HarperCollins, 2001.

EVOLUTION OF PLANTS AND CLIMATES

Types of ecology: Evolutionary ecology; Paleoecology; Speciation

As a result of prehistoric events such as the Permian-Triassic extinction event and the Cretaceous-Tertiary mass extinction event, many plant families and some ancestors of extant plant were extinct before the beginning of recorded history.

The general trend of earth's plant diversification involves four major plant groups that rose to dominance from about the Middle Silurian period to present time. The first major group providing land vegetation comprised the seedless vascular plants, represented by the phyla *Rhyniophyta*, *Zosterophyllophyta*, and *Trimerophytophyta*. The second major group appearing in the late Devonian period was made up of the ferns (*Pterophyta*). The third group, the seed plants (sometimes called the Coal Age plants), appeared at least 380 million years ago (mya). This third group includes the gymnosperms (*Gymnospermophyta*), which dominated land flora for most of the Mesozoic era until 100 mya. The last group, the flowering angiosperms (*Anthophyta*), appeared in the fossil record 130 mya. The fossil record also shows that this group of plants was abundant in most parts of the world within 30 million to 40 million years. Thus, the angiosperms have dominated land vegetation for close to 100 million years.

The Paleozoic Era

The Proterozoic and Archean eons have restricted fossil records and predate the appearance of land plants. Seedless, vascular land plants appeared in the middle of the Silurian period (437-407 mya) and are represented by the rhyniophytes or rhyniophytoids and possibly the *Lycophyta* (lycophytes or club mosses). From the primitive rhyniophytes and lycophytes, land vegetation rapidly diversified during the Devonian period (407-360 mya). Pre-fern ancestors and maybe true ferns (*Pterophyta*) were developed by the mid-Devonian. By the Late Devonian the horsetails (*Sphenophyta*) and gymnosperms (*Gymnospermophyta*) were present. By the end of the period, all major divisions of vascular plants had appeared except the angiosperms.

Development of vascular plant structures throughout the Devonian allowed for greater geographical diversity of plants. One such structure was flattened, planated leaves, which increased photosynthetic efficiency. Another was the development of secondary wood, allowing plants to increase significantly in structure and size, thus resulting in trees and probably forests. A gradual process was the reproductive development of the seed; the earliest structures are found in Upper Devonian deposits.

Ancestors of the conifers and cycads appeared in the Carboniferous period (360-287 mya), but their documentation is poor in the fossil record. During the early Carboniferous in the high and middle latitudes, vegetation shows a dominance of club mosses and progymnosperms (*Progymnospermophyta*). In the lower latitudes of North America and Europe, a greater diversity of club mosses and progymnosperms are found, along with a greater diversity of vegetation. Seed ferns (lagenostomaleans, calamopityaleans) are present, along with true ferns and horsetails (Archaeocalamites).

Late Carboniferous vegetation in the high latitudes was greatly affected by the start of the Permo-Carboniferous Ice Age. In the northern middle latitudes, the fossil record reveals a dominance of horsetails and primitive seed ferns (pteridosperms) but few other plants.

In northern low latitudes, landmasses of North America, Europe, and China were covered by shallow seas or swamps and, because they were close to the equator, experienced tropical to subtropical climatic conditions. The first tropical rain forests appeared there, known as the Coal Measure Forests or the Age of Coal. Vast amounts of peat were laid down as a result of favorable conditions of year-round growth and the giant club mosses' adaptation to the wetland tropical environments. In drier areas surrounding the lowlands, forests of horsetails (calamites, sphenophylls), seed ferns (medullosans, callistophytes, lagenostomaleans), cordaites, and diverse ferns (including marattialean tree ferns) existed in great abundance.

The Permian period (287-250 mya) marks a major transition of the conifers, cycads, glossopterids, gigantopterids, and the peltasperms from a poor fossil record in the Carboniferous to significantly abundant land vegetation. The two most prevalent plant assemblages of the Permian were the horsetails, peltasperms, cycadophytes, and conifers. The second most prevalent were the gigantopterids, peltasperms, and conifers. These two plant assemblages are considered the typical paleo-equatorial lowland vegetation of the Permian. Other plants, such as the tree ferns and giant club mosses, were present in the Permian but not abundant. As a result of the Permian-Triassic extinction event, tropical swamp forests disappeared, with the extinction of the club mosses; the cordaites and glossopterids disappeared from higher latitudes; and 96 percent of all plant and animal species became extinct.

The Mesozoic Era

At the beginning of the Triassic period (248-208 mya), a meager fossil record reveals diminished land vegetation (that is, no coal formed). By the middle to late Triassic, the modern family of ferns, conifers, and a now-extinct group of plants, the bennettites (cycadeoids), inhabited most land surfaces. After the mass extinction, the bennettites moved into vacant lowland niches. They may be significant because of the similarity of their reproductive organs to the reproductive organs of the angiosperms.

Late Triassic flora in the equatorial latitudes are represented by a wide range of ferns, horsetails, pteriosperms, cycads, bennettites, leptostrobaleans, ginkgos, and conifers. The plant assemblages in the middle latitudes are similar but not as species-rich. This lack of plant variation in low and middle latitudes reflects a global frost-free climate.

In the Jurassic period (208-144 mya), land vegetation similar to modern vegetation began to appear, and the ferns of this age can be assigned to modern families: *Dipteridaceae, Matoniaceae, Gleicheniaceae,* and *Cyatheaceae.* Conifers of this age can also be assigned to modern families: *Podocarpaceae, Araucariaceae* (Norfolk pines), *Pinaceae* (pines), and *Taxaceae* (yews). These conifers created substantial coal deposits in the Mesozoic.

During the Early to Middle Jurassic, diverse vegetation grew in the equatorial latitudes of western North America, Europe, Central Asia, and the Far East and comprised the horsetails, pteridosperms, cycads, bennettites, leptostrobaleans, ginkgos, ferns, and conifers. Warm, moist conditions also existed in the northern middle latitudes (Siberia and northwest Canada), supporting Ginkgoalean forests and leptostrobaleans. Desert conditions existed in central and eastern North America and North Africa, and the presence of bennettites, cycads, peltasperms, and cheirolepidiacean conifers there are plant indicators of drier conditions. The southern latitudes had vegetation similar to that of the equatorial latitudes, but owing to drier conditions, cheirolepidiacean conifers were abundant, ginkgos scarce. This southern vegetation spread into very high latitudes, including Antarctica, because of the lack of polar ice.

In the Cretaceous period (144-66.4 mya), arid, subdesert conditions existed in South America, Central and North Africa, and central Asia. Thus, the land vegetation was dominated by cheirolepidiacean conifers and matoniacean ferns. The northern middle latitudes of Europe and North America had a more diverse vegetation comprising bennettites, cycads, ferns, peltasperms, and cheirolepidiacean conifers with the southern middle latitudes dominated by bennettites and cheirolepidiaceans.

A major change in land vegetation took place in the late Cretaceous with the appearance and proliferation of flowering seed plants, the angiosperms. The presence of the angiosperms marked the end of the typical gymnosperm-dominated Mesozoic flora and a definite decline in the leptostrobaleans, bennettites, ginkgos, and cycads.

During the late Cretaceous in South America, central Africa, and India, arid conditions prevailed, resulting in tropical vegetation dominated by palms. The southern middle latitudes were also affected by desert conditions, and the plants that fringed these desert areas were horsetails, ferns, conifers (araucarias, podocarps), and angiosperms, specifically Nothofagus (southern beech). The high-latitude areas were devoid of polar ice; owing to the warmer conditions, angiosperms were able to thrive. The most diverse flora was found in North America, with the presence of evergreens, angiosperms, and conifers, especially the redwood, Sequoia.

The Cretaceous-Tertiary (K/T) mass extinction event occurred at about 66.4 mya. This event has been hypothesized to be a meteoritic impact; whatever the cause, at this time an event took place that suddenly induced global climatic change and initiated the extinction of many species, notably the dinosaurs. The K/T had a greater effect on plants with many families than it did on plants with very few families. Those that did become extinct, such as the bennettites and caytonias, had been in decline. The greatest shock to land vegetation occurred in the middle latitudes of North America. The pollen and spore record just above the K/T boundary in the fossil record shows a dominance of ferns and evergreens. Subsequent plant colonization in North America shows a dominance of deciduous plants.

The Cenozoic Era
Increased rainfall at the beginning of the Paleogene-Neogene period (66.4-1.8 mya) supported the widespread development of rain forests in southerly areas. Rain forests are documented by larger leaf size and drip tips at leaf edge, typical characteristics of modern rain-forest floras.

Notable in this period was the polar Arcto-Tertiary forest flora found in northwest Canada at paleolatitudes of 75-80 degrees north. Mild, moist summers alternated with continuous winter darkness, with temperatures ranging from 0 to 25 degrees Celsius. These climatic conditions supported deciduous vegetation that included *Platanaceae* (sycamore), *Judlandaceae* (walnut), *Betulaceae* (birch), *Menispermaceae*, *Cercidophyllaceae*, *Ulmaceae* (elm), *Fagaceae* (beech), *Magnoliaceae*; and gymnosperms such as *Taxodiaceae* (redwood), *Cypressaceae* (cypress), *Pinaceae* (pine), and *Ginkgoaceae* (gingko). This flora spread across North America to Europe via a land bridge between the continents.

About eleven million years ago, during the Miocene epoch, a marked change in vegetation occurred, with the appearance of grasses and their subsequent spread to grassy plains and prairies. The appearance of this widespread flora supported the development and evolution of herbivorous mammals.

The Quaternary period (1.8 mya to present) began with continental glaciation in northwest Europe, Siberia, and North America. This glaciation affected land vegetation, with plants migrating north and south as a response to glacial and interglacial fluctuations. Pollen grains and spores document the presence of Aceraceae (maple), hazel, and Fraxinus (ash) during interglacial periods.

Final migrations of plant species at the close of the last ice age (about eleven thousand years ago), formed the modern geographical distribution of land plants. Some areas, such as mountain slopes or islands, have unusual distribution of plant species as a result of their isolation from the global plant migrations.

Mariana Louise Rhoades

See also: Adaptations and their mechanisms; Adaptive radiation; Biomes: determinants; Biomes: types; Coevolution; Colonization of the land; Convergence and divergence; Dendrochronology; Evolution: definition and theories; Evolution: history; Extinctions and evolutionary explosions; Isolating mechanisms; Natural selection; Nonrandom mating, genetic drift, and mutation; Paleoecology; Punctuated equilibrium vs. gradualism; Speciation.

Sources for Further Study

Cleal, Christopher J., and Barry A. Thomas. *Plant Fossils: The History of Land Vegetation*. Suffolk, England: Boydell Press, 1999.

Stewart, Wilson N., and Gar A. Rothwell. *Paleobotany and the Evolution of Plants*. New York: Cambridge University Press, 1993.

EXTINCTIONS AND
EVOLUTIONARY EXPLOSIONS

Types of ecology: Evolutionary ecology; Paleoecology; Population
ecology; Speciation

*The history of life has been punctuated by episodes of great change, some marked by
the loss of large numbers of organisms, others by explosive development. Explanations proposed for these fluctuations have a bearing on current extinction levels
and the extent to which they can be controlled.*

Extinction of species is a continuous process, and evidence of its occurrence abounds in the fossil record. It has been estimated that marine
species persist for about four million years, which translates into an overall
loss of about two or three species each year. This is the "background" extinction rate, and it is balanced by speciation events that result in the development of new species. Mass extinctions are events during which the rate
of extinction rises dramatically above this background rate, and a number
of these have been recognized in the Phanerozoic era. In each of these
events, at least 40 percent of the genera of shallow marine organisms were
eliminated. Using statistical methods, it has been estimated that at least 65
percent of species became extinct at each of these events, with 77 percent
being eliminated at the event at the end of the Cretaceous period and 95
percent at the event at the end of the Permian period. These mass extinctions were balanced by periods of explosive development that often followed, as organisms moved into vacant adaptive zones during periods of
adaptive radiation. The most important of these was at the base of the
Cambrian period, 544 million years ago, when all the major groups in existence originated, but other radiations occurred in the Early Triassic period
and at the start of the Tertiary period.

Causes of Mass Extinctions
Attempts to explain the causes of mass extinctions have centered on terrestrial phenomena such as sea-level changes, climatic changes, or volcanic
activity. The sea level has shown regular fluctuations on a global level during the Phanerozoic era, and these appear to be related to the melting or
formation of polar ice caps or to major tectonic events such as continental
splits or the collision and uplift or subsidence of ocean ridges. Extinction
events appear to be correlated mostly with periods of marine regression.

During such a regression, the withdrawal of the ocean leaves a much smaller habitat for shallow marine organisms. This leads to increased crowding and competition and ultimately to an increased extinction rate. Reduction of large terrestrial vertebrates during these regressions, as happened during the events at the end of the Permian and Cretaceous periods, may be related to increased seasonality caused by the loss of the ameliorating influence of the shallow epicontinental seas.

It has also been shown that some extinctions are related to transgressive events (the spread of the sea over land areas), possibly resulting from the spread of anoxic (oxygen-poor) waters across epicontinental areas. Climatic changes seem to be correlated with eustatic events (worldwide changes in sea level), and the evidence implicating temperature as the main cause of extinctions seems weak. For example, the most important extinction event at the end of the Permian period occurred at a time of climatic amelioration marked by the disappearance of the Gondwanaland ice sheet. Volcanic activity has been presented as a possible cause of the extinctions that occurred at the end of the Cretaceous period. The Deccan Traps of northern India were erupting at that time and would have produced large quantities of volatile emissions that could have resulted in global cooling, ozone-layer depletion, and changes in ocean chemistry. However, no evidence exists as yet for the involvement of volcanic activity in other extinction events.

Although various extraterrestrial causes for mass extinction events have been suggested in the past, these ideas have gained greater credence since the publication in the early 1980's of work by Luis and Walter Alvarez, who ascribe the end-Cretaceous extinction event to the effects of the impact of a large bolide, or extraterrestrial object, perhaps ten kilometers in diameter. The impact of such a large object would have resulted in some months of darkness because of the global dust clouds generated, and this would have halted photosynthesis and resulted in the collapse of both terrestrial and marine food chains. Although cold would initially have accompanied the darkness, greenhouse effects and global warming would follow as atmospheric gases and water vapor trapped infrared energy radiating from the earth. Physical evidence for an impact rests on the presence in the period boundary layers of high concentrations of iridium and other elements generally rare at the earth's surface but abundant in asteroids. In addition, these layers often contain shocked quartz grains, otherwise found only in impact craters and at nuclear test sites, and microtectites, glassy droplets formed by impact. Although the evidence for extraterrestrial impacts having caused the other major extinction events is slight, this causal factor has been linked with the apparently regular 26-million-year periodicity exhibited by extinctions. Scientists suggest that the regular passage of an unidentified planetary body by the Oort Cloud of comets and the subsequent perturbation could result in increased asteroid impacts and extinction events on earth.

Historical Mass Extinctions

The first mass extinction event that can be recognized in the fossil record occurred in the Middle Vendian period, about 650 million years ago, when microorganisms underwent a severe decline. This event has been linked to climatic cooling related to glaciation. The extinction in the Late Ordovician period was a major event in which 22 percent of marine families became extinct. As there were two main pulses of extinction and no iridium anomaly was found, an extraterrestrial cause seems unlikely. However, sea-level and temperature changes have been cited as likely causes. In addition, biologically toxic bottom waters might have been brought to the surface during periods of climatic change. The event at the end of the Devonian period had a devastating effect on brachiopods, which lost about 86 percent of genera, and on reef-building organisms such as corals. Shallow-water faunas were most severely affected; only 4 percent of species survived, although 40 percent of deeper-water species did, and cool-water faunas also survived better. This event has been linked to a significant drop in global temperatures of unknown cause.

The mass extinction event at the end of the Permian period was the most severe of the Phanerozoic era and resulted in the extinction of up to 95 percent of all marine invertebrate species. On land, amphibians and mammal-like reptiles were both badly affected, and plant diversity fell by 50 percent. No iridium anomaly was found, and the most likely explanation is climatic instability caused by continental amalgamation and the simultaneous occurrence of marine regressions. These occurrences would have disrupted food webs on a major scale. The event that occurred at the end of the Triassic period was much less severe but still involved extensive reductions in marine invertebrates and reptiles. On land, a major faunal turnover took place. Primitive amphibians, early reptile groups, and mammal-like reptiles died out and were replaced by advanced reptiles and mammals. No evidence of an impact event has been found, and the extinctions are generally correlated with widespread marine regressions.

The extinction that took place at the end of the Cretaceous period has become the most hotly debated, in large part because of the bolide impact hypothesis. Although the broad pattern of extinction among marine organisms is known, the detailed picture only encompasses microorganisms such as planktonic foraminifera and calcareous nannoplankton. Study of the ranges of these microorganisms shows that the extinctions occurred over an extended period, starting well before and finishing well after the boundary. Although much has been made of the extinction of ammonites at the end of the Cretaceous period, there are too few ammonite-bearing sections to show if it was gradual or abrupt. On land, evidence of an increase in the population of ferns just above the boundary suggests the presence of wildfires, as ferns are usually the first plants to recolonize an area devastated by fire. However, in many sections, a return of the Cretaceous vegetation is seen above the fern increase, indicating little extinction.

Post-Extinction Recoveries

Among the vertebrates, a picture of gradual change is seen for mammals, with drastic reductions occurring only in the marsupials. The boundary also does not seem to have been a barrier for turtles, crocodiles, lizards, and snakes, all of which came through virtually unscathed. The dinosaurs did become extinct, and much argument has centered on whether this was abrupt or occurred after a slow decline. In this context, it must be noted that there is only one area where a dinosaur-bearing sedimentary transition across the boundary can be seen, and that is in Alberta, Canada, and the northwestern United States. Records of dinosaurs in this area during the upper part of the Cretaceous period show a gradual decline in diversity, with a drop from thirty to seven genera over the last eight million years. Although explanations of the extinction of dinosaurs have ranged from mammals eating their eggs to terminal allergies caused by the rise of flowering plants to the current ideas about bolide impacts, the answer may be climate related. A major regression of the oceans occurred at this point, resulting in a drop in mean annual temperatures and an increase in seasonality. The bolide impact may have served as the death blow to taxa (animals in classification groups) that were already declining.

The main period of evolutionary expansion in the Phanerozoic era is at the base of the Cambrian period, 544 million years ago. Termed the "Cambrian explosion," it marks the development of all the modern phyla of organisms, and as many as one hundred phyla may have existed during the Cambrian period. This period seems to have lasted only about 5 million years, and the subsequent history of animal life consists mainly of variations on the anatomical themes developed during this short period of intense creativity. This period is represented in the fossil record by the remarkably well-preserved Burgess Shale fauna of British Columbia, which has been extensively described, and faunas of similar age from China and Greenland. Why the Cambrian explosion could establish all major anatomical designs so quickly is not clear. Some scientists believe that the lack of complex organisms before the explosion had left large areas of ecological space open, and when experimentation took place, particularly with the advent of hard skeletons, any novelty could find a niche. Also, the earliest multicellular organisms may have maintained a genetic flexibility that became greatly reduced as organisms became locked into stable and successful designs. Why some of the innovations were successful in the long term and others were not is unknown, as no recognized traits unite the successful taxa. It has even been suggested that success may be due to no more than the luck of the draw.

In contrast, the recoveries after the major extinctions at the end of the Permian and Cretaceous periods did not result in the development of new phyla. The earliest Triassic ecosystems were more vacant than at any time since the Cambrian period, yet no new phyla or classes appear in the Triassic period. This suggests that despite the overwhelming nature of the extinctions, the pattern was insufficient to permit major morphological innovations, in part probably because no adaptive zone was entirely vacant. Hence, despite the fact that the mass extinction at the end of the Permian period triggered an explosion in marine diversity described as the Mesozoic marine revolution, persisting species may have limited the success of broad evolutionary jumps.

Reading the Fossil Records

All understanding of extinction events or of evolutionary explosions depends on the fossil record. The study of the diversity of organisms through time—the number of different types of organisms that occur at a particular time and place—is therefore very important. The basic data consists of compilations of extinctions of taxa plotted against similar compilations of origination of taxa. Periods when either extinction or origination were unusually high show as peaks or troughs on a graph. Unfortunately, biases in the preservation, collection, and study of fossils have conspired to obscure patterns of change in diversity.

The geological history of patterns of diversity are obscured by a variety of filters, many of which are sampling biases that cause the observed fossil record to differ from the actual history of the biosphere. The most severe bias is the loss of sedimentary rock volume and area as the age of the record increases because the volume and area correlate strongly with the diversity of organisms described from a stratigraphic interval. The quality of the record also tends to fall with increasing age because the rocks are exposed to changes that may destroy the fossils they contain. The differences in levels of representation among the paleoenvironments in the stratigraphic record also influences the composition of the fossil record; for example, shallow-marine faunas are much better represented than are terrestrial faunas.

Diversity patterns are studied at a variety of levels, from the species upward, that vary in their quality and inclusiveness. A basic problem is that many of the processes that are of interest occur at the species level or even below it, but the biases of the fossil record mean that data is best at higher levels. Diversity of shallow-marine organisms for the Phanerozoic era cannot be read directly at the species level because the record is too fragmentary. The record at the family level is much more complete because the preservation of one species in a family allows the family to be recorded. For this reason, paleodiversity studies are often conducted at the family level. However, higher taxon diversity is a poor predictor of species diversity. For example, an analysis of the mass extinction at the end of the Permian period indicates that the 17 percent reduction in marine orders and 52 percent reduction in marine families probably represents a 95 percent reduction in the number of species. Another problem with the study of fossils is that soft-bodied and poorly skeletonized groups may leave little or no record. It has generally been assumed that the ratio of heavily skeletonized to non-skeletonized species has remained approximately constant through the Phanerozoic era; however, there is no data to support this and some evidence that skeletons have become more robust through time in response to newly evolving predators. The net result of these biases is severe. Only 10 percent of the skeletonized marine species of the geologic past and far fewer of the soft-bodied species are known.

Despite these problems, it has been possible to show that diversity of organisms has varied in a number of ways during the Phanerozoic era. Tabulations of classes, orders, and families have been used to show that there were significant periods of increased extinction or increased evolutionary rates. One of the most important uses of this data has been the tabulation at the family level that appears to show a regular periodicity of about 26 million years for extinction events and that has been used to support ideas about periodic extraterrestrial events. However, although fluctuations occurred, it has also been possible to show that the number of marine orders increased rapidly to the Late Ordovician period and has remained approximately constant since then.

The Ebb and Flow of Life on Earth

Mass extinctions and evolutionary explosions are the opposite faces of the pattern of diversity of organisms through time. During periods of mass extinction, the diversity of organisms on earth has dropped drastically and in some cases entire lineages have been wiped out. Evolutionary explosions, on the other hand, resulted in enormous innovation, particularly at the beginning of the Cambrian period, and the development of new variations on established morphotypes (animal and plant forms and structures) later in the geologic record. Understanding the processes that caused these events is of major importance because people have reached the point where they are capable of influencing their environment in drastic ways.

Studies of extinction events have shown that they have a variety of causes, some of which appear to be environmental changes brought about by natural processes while others may be the result of extraterrestrial forces. The most severe of these extinction events occurred at the end of the Permian period, 245 million years ago, and resulted in the loss of up to 95 percent of marine invertebrate species. The cause of this extinction appears primarily to be that continents were amalgamating and oceans were retreating, which resulted in a major reduction in the habitat of shallow-marine organisms. Terrestrial habitats were also affected as the increase in continental area and loss of the ameliorating effect of extensive areas of shallow ocean brought about climatic changes. Although climatic changes are thought to be the main culprit in the majority of extinction events, some scientists believe that large bolides, or extraterrestrial bodies, struck the earth with such force as to create major changes in the environment that significantly reduced diversity. This theory has enjoyed the most popularity as the explanation for the event at the end of the Cretaceous period, during which the dinosaurs became extinct, but evidence for an extraterrestrial body's involvement in any of the other events is slight.

Whatever the cause, environmental change that results in habitat reduction is the main reason for species decline. As humans have risen to dominance over other species, the extinction rate has accelerated, and in the last half-century, this rate has climbed considerably above natural attrition as populations have increased and habitats have been altered. Although the levels of extinction have not yet reached those recorded during major extinction events of the past, some scientists believe people may be facing an ecological disaster. A better understanding of the processes surrounding past extinction events and the rebounds that followed them will help people prepare for and deal with the future.

David K. Elliott

See also: Adaptations and their mechanisms; Adaptive radiation; Coevolution; Colonization of the land; Convergence and divergence; Dendrochronology; Development and ecological strategies; Evolution: definition and theories; Evolution: history; Evolution of plants and climates; Gene flow; Genetic drift; Genetically modified foods; Isolating mechanisms; Natural selection; Nonrandom mating, genetic drift, and mutation; Paleoecology; Population genetics; Punctuated equilibrium vs. gradualism; Speciation; Species loss.

Sources for Further Study

Allen, Keith C., and Derek E. Briggs, eds. *Evolution and the Fossil Record.* Washington, D.C.: Smithsonian Press, 1989.

Alvarez, W. T. *Rex and the Crater of Doom.* Princeton, N.J.: Princeton University Press, 1997.

Archibald, J. D. *Dinosaur Extinction and the End of an Era.* New York: Columbia University Press, 1996.

Bakker, Robert T. *The Dinosaur Heresies.* New York: William Morrow, 1986.

Briggs, Derek E., and Peter R. Crowther, eds. *Palaeobiology: A Synthesis.* Oxford, England: Blackwell Scientific Publications, 1990.

Donovan, Stephen K., ed. *Mass Extinctions: Processes and Evidence.* London: Belhaven Press, 1989.

Drury, Stephen. *Stepping Stones: Evolving the Earth and Its Life.* New York: Oxford University Press, 1999.

Erwin, Douglas H. *The Great Paleozoic Crisis.* New York: Columbia University Press, 1993.

Frankel, Charles. *The End of the Dinosaurs: Chicxulub Crater and Mass Extinctions.* New York: Cambridge University Press, 1999.

Gould, Stephen J. *Wonderful Life: The Burgess Shale and the Nature of History.* New York: W. W. Norton, 1989.

Leakey, Richard E., and Roger Lewin. *The Sixth Extinction: Patterns of Life and the Future of Humankind.* New York: Doubleday, 1995.

McGhee, George R. *The Late Devonian Mass Extinction.* New York: Columbia University Press, 1996.

McMenamin, Mark A., and Dianna L. McMenamin. *The Emergence of Animals: The Cambrian Breakthrough.* New York: Columbia University Press, 1989.

Martin, Paul S., and Richard G. Klein, eds. *Quaternary Extinctions: A Prehistoric Revolution.* Tucson: University of Arizona Press, 1984.

Muller, Richard. *Nemesis: The Death Star.* New York: Weidenfeld & Nicolson, 1988.

Officer, Charles, and Jake Page. *The Great Dinosaur Extinction Controversy.* Boston: Addison-Wesley, 1996.

Raup, David M. *Extinction: Bad Genes or Bad Luck?* New York: W. W. Norton, 1991.

Runnegar, Bruce, and James W. Schopf, eds. *Major Events in the History of Life*. Boston: Jones and Bartlett, 1992.
Stanley, Steven M. *Extinction*. New York: Scientific American Library, 1987.
Stearns, Beverly Petersen, and Stephen C. Stearns. *Watching, from the Edge of Extinction*. New Haven, Conn.: Yale University Press, 1999.
Ward, Peter D., and Don Brownlee. *Rare Earth: Why Complex Life Is Uncommon in the Universe*. New York: Copernicus, 2000.

FOOD CHAINS AND WEBS

Types of ecology: Community ecology; Ecoenergetics

The concept of food chains and webs allows ecologists to interconnect the organisms living in an ecosystem and to trace mathematically the flow of energy from plants through animals to decomposers.

The food chain concept provides the basic framework for production biology and has major implications for agriculture, wildlife biology, and calculating the maximum amount of life that can be supported on the earth. As early as 1789, naturalists such as Gilbert White described the many sequences of animals eating plants and themselves being eaten by other animals. However, the use of the term "food chain" dates from 1927, when Charles Elton described the implications of the food chain and food web concept in a clear manner. His solid exposition advanced the study of two important biological concepts: the complex organization and interrelatedness of nature, and energy flow through ecosystems.

Food Chains in Ecosystem Description

Stephen Alfred Forbes, founder of the Illinois Natural History Survey, contended in 1887 that a lake comprises a system in which no organism or process can be understood unless its relationship to all the parts is understood. Forty years later, Elton's food chains provided an accurate way to diagram these relationships. Because most organisms feed on several food items, food chains were cross-linked into complex webs with predictive power. For instance, algae in a lake might support an insect that in turn was food for bluegill. If unfavorable conditions eliminated this algae, the insect might also disappear. However, the bluegill, which fed on a wider range of insects, would survive because the loss of this algae merely increases the pressure on the other food sources. This detailed linkage of food chains advanced agriculture and wildlife management and gave scientists a solid overview of living systems. When Arthur George Tansley penned the term ecosystem in the 1930's, it was food-chain relationships that described much of the equilibrium of the ecosystem.

Today most people still think of food chains as the basis for the "balance of nature." This phrase dates from the controversial 1960 work of Nelson G. Hairston, Frederick E. Smith, and Lawrence B. Slobodkin. They proposed that if only grazers and plants are present, grazing limits the plants. With predators present, however, grazers are limited by predation, and the plants are free to grow to the limits of the nutrients available. Such explanations of the "balance of nature" were commonly taught in biology books throughout the 1960's and 1970's.

Food Chains in Production Biology

Elton's explanation of food chains came just one year after Nelson Transeau of Ohio State University presented his calculations on the efficiency with which corn plants converted sunlight into plant tissue. Ecologists traced this flow of stored chemical energy up the food chain to herbivores that ate plants and on to carnivores that ate herbivores. Food chains therefore undergirded the new "production biology" that placed all organisms at various trophic levels and calculated the extent to which energy was lost or preserved as it passed up the food chain.

With data accumulating from many ecologists, Elton extended food chains into a pyramid of numbers. The food pyramid, in which much plant tissue supports some herbivores that are in turn eaten by fewer carnivores, is still referred to as an Eltonian pyramid. In 1939 August Thienemann added decomposers to reduce unconsumed tissues and return the nutrients of all levels back to the plants. Early pyramids were based on the amount of living tissues, or biomass.

Calculations based on the amount of chemical energy at each level, as measured by the heat released when food is burned (calories), provided even more accurate budgets. Because so much energy is lost at each stage in a food chain, it became obvious that this inefficiency was the reason food chains are rarely more than five or six links long and why large, fierce animals are uncommon. It also became evident that because the earth intercepts a limited amount of sunlight energy per year, there is a limit on the amount of plant life—and ultimately upon the amount of animal life and decomposers—that can be fed. Food chains are also important in the accounting of carbon, nitrogen, and water cycling.

Value of Food Chains in Environmental Science

Unlike calories, which are dramatically reduced at each step in a food chain, some toxic substances become more concentrated as the molecules are passed along. The concentration of molecules along the food chain was first noticed by the Atomic Energy Commission, which found that radioactive iodine and strontium released in the Columbia River were concentrated in tissue of birds and fish. However, the pesticide DDT provided the most notorious example of biological magnification: DDT was found to be deposited in animal body fat in ever-increasing concentrations as it moved up the food chain to ospreys, pelicans, and peregrine falcons. High levels of DDT in these birds broke down steroid hormones and interfered with eggshell formation.

Because humans are omnivores, able to feed at several levels on the food chain (that is, both plants and other animals), it has been suggested that a higher world population could be supported by humans moving down the food chain and becoming vegetarians. A problem with this argument is that much grazing land worldwide is unfit for cultivation, and therefore the complete cessation of pig or cattle farming does not necessarily free up substantial land to grow crops.

While the food chain and food web concepts are convenient theoretical ways to summarize feeding interactions among organisms, real field situations have proved far more complex and difficult to measure. Animals often switch diet between larval and adult stages, and they are often able to shift food sources widely. It is often difficult to draw the boundaries of food chains and food webs.

John Richard Schrock

See also: Balance of nature; Biomass related to energy; Geochemical cycles; Herbivores; Hydrologic cycle; Nutrient cycles; Omnivores; Phytoplankton; Rain forests and the atmosphere; Trophic levels and ecological niches.

Sources for Further Study

Colinvaux, Paul A. *Why Big Fierce Animals Are Rare*. New York: Penguin Books, 1990.

Elton, Charles. *Animal Ecology*. New York: October House, 1966.

Golley, Frank B. *A History of the Ecosystem Concept in Ecology*. New Haven, Conn.: Yale University Press, 1993.

Quamman, David. *The Song of the Dodo: Island Biogeography in an Age of Extinction*. New York: Simon & Schuster, 1997.

"Something's Fishy." *Science News* 146, no. 1 (July 2, 1994): 8.

Svitil, Kathy A. "Collapse of a Food Chain." *Discover* 16, no. 7 (July, 1995): 36.

FOREST FIRES

Type of ecology: Ecosystem ecology

Whether natural or caused by humans, fires destroy life and property in forest-lands but are also vital to the health of forests.

Evidence of forest fires is routinely found in soil samples and tree borings. The first major North American fires in the historical record were the Miramichi and Piscataquis fires of 1825. Together, they burned 3 million acres in Maine and New Brunswick. Other U.S. fires of significance were the Peshtigo fire in 1871, which raged over 1.28 million acres and took fourteen hundred human lives in Wisconsin; the fire that devastated northern Idaho and northwestern Montana in 1910 and killed at least seventy-nine firefighters; and a series of fires that joined forces to sweep across one-third of Yellowstone National Park in 1988.

Fire Behavior

Fires need heat, fuel, and oxygen. They spread horizontally by igniting particles at their edge. At first, flames burn at one point, then move outward, accumulating enough heat to keep burning on their own. Topography and weather affect fire behavior. Fires go uphill faster than downhill because warm air rises and preheats the uphill fuels. Vegetation on south- and west-facing slopes receives maximal sunlight and so is drier and burns more easily. Heat is pulled up steep, narrow canyons, as it is up a chimney, increasing heat intensity. For several reasons, only one-third of the vegetation within a large fire usually burns. This mosaic effect may be caused by varied tree species that burn differently, old burns that stop fire, strong winds that blow the fire to the leeward side of trees, and varied fuel moisture.

Ecological Benefits

Some plant species require very high temperatures for their seed casings to split for germination. After fire periodically sweeps through the forest, seeds will germinate. Other species, such as the fire-resistant ponderosa pine, require a shallow layer of decaying vegetable matter in which to root. Fires burn excess debris and small trees of competing species and leave an open environment suitable for germination. Dead material on the forest floor is processed into nutrients more quickly by fire than by decay, and in

a layer of rich soil, plants will sprout within days to replace those destroyed in the fire.

Ecological Destruction

Erosion is one of the devastating effects of a fire. If the fuels burn hot, tree
oils and resins can be baked into the soil, creating a hard shell that will not
absorb water. When it rains, the water runoff gathers mud and debris, creating flash floods and extreme stream sedimentation. Culverts and storm
drains fill with silt, and streams flood and change course. Fish habitat is destroyed, vegetation sheltering stream banks is ripped away, and property
many miles downstream from the forest is affected.

When a fire passes through timber it generally leaves pockets of green,
although weakened, stands. Forest pests, such as the bark beetle, are attracted to the burned trees and soon move to the surviving trees, weakening them further. Healthy trees outside the burn area may also fall to pest
infestation unless the burned trees are salvaged before pests can take hold.

*During the twentieth century,
low fires that otherwise would
have burned through the forest
at ground level every five to
twenty-five years were
suppressed. As a result, the
natural cycle of frequent fires
moving through an area was
broken, and when fire did erupt,
accumulated kindling burned
hot and fast, exploding into
devastating crown fires and
completely destroying the local
community's habitat.*
(PhotoDisc)

The ash and smoke from hot, fast-burning forest fires can be transported for miles, affecting air quality many miles from the actual fire.

Ecological Impact of Fire Suppression

One of the early criteria of forest management was fire protection. In the second quarter of the twentieth century, lookout towers, firebreaks, and trails were built to locate fires as quickly as possible. Low fires that otherwise would have burned through the forest at ground level and cleared out brush every five to twenty-five years were suppressed. As a result, the natural cycle of frequent fires moving through an area was broken. Fallen trees, needles, cones, and other debris collected as kindling on the forest floor, rather than being incinerated every few years.

It took foresters and ecologists fifty years to realize that too much fire suppression was as bad as too little. Infrequent fires cause accumulated kindling to burn hot and fast and explode into treetops. The result is a devastating crown fire, a large fire that advances as a single front. Burning embers of seed cones and sparks borne by hot, strong winds created within the fire are tossed into unburned areas to start more fires.

Prescribed Burns

In the 1970's prescribed burning was added to forest management techniques used to keep forests healthy. Fires set by lightning are allowed to burn when the weather is cool, the area isolated, and the risk of the fire exploding into a major fire low. More than 70 percent of prescribed burning takes place in the southeastern states, where natural fires burn through an area more frequently than in the West.

In some areas, prescribed fires are set in an attempt to re-create the natural sequence of fire. In Florida, prescribed burns provide wildlife habitat by opening up groves to encourage healthy growth. Other fires start accidentally but are allowed to burn until they reach a predetermined size.

Although a prescribed fire is an attempt to duplicate natural fire, it is not as efficient, because private and commercial property within the fire path must be protected. Once a fire has occurred, burned timber deteriorates quickly, either through insect infestation or blueing—a mold that stains the wood. Private landowners can move quickly to salvage fire-damaged trees and plant new seedlings to harness erosion. On federal land, regulations governing the salvage of trees can delay logging of the burned snags until deterioration makes it uneconomical to harvest them.

Causes

Forest fires may be caused by natural events or human activity. Most natural fires are started by lightning strikes. Dozens of strikes can be recorded from one lightning storm. When a strike seems likely, fire spotters watch for columns of smoke, and small spotter planes will fly over the area, looking for smoke. Many of the small fires simply smoulder and go out, but if the forest is dry, multiple fires can erupt from a single lightning storm.

The majority of forest fires are human-caused, and most are the result of carelessness rather than arson. Careless campers may leave a campsite without squelching their campfire completely, and winds may then whip the glowing embers into flames. A smoker may toss a cigarette butt from a car window. Sparks from a flat tire riding on the rim may set fire to vegetation alongside the highway. The sun shining through a piece of broken glass left by litterers may ignite dry leaves.

Fire Fighting
In fire fighting, bulldozers are used to cut fire lines ahead of the approaching fire, and fuels between fire lines and the fire are backburned. Helicopters and tanker planes drop water with a fire-retardant additive, or bentonite, a clay, at the head of the fire to smother fuels. Firefighters are equipped with fire shelters in the form of aluminized pup tents, which they can pull over themselves if a fire outruns them. Despite technological advances, one of the best tools for fighting fires—along with the shovel—remains the pulaski, a combination ax and hoe, first produced commercially in 1920. This tool, in the hands of on-the-ground firefighters, is used to cut fire breaks and to throw dirt on smoldering debris.

Public Policy and Public Awareness
Since the early twentieth century, forest fires have engendered public policy in the United States. In the aftermath of major fires in 1903 and 1908 in Maine and New York, state fire organizations and private timber protective associations were formed to provide fire protection. These, in turn, contributed to the Weeks Act of 1911, which permitted cooperative fire protection between federal and state governments.

People who make their homes in woodland settings in or near forests face the danger of forest fire, and government agencies provide information to help people safeguard themselves and their property. Homes near forests should be designed and landscaped with fire safety in mind, using fire-resistant or noncombustible materials on the roof and exterior. Landscaping should include a clear safety zone around the house. Hardwood trees, less flammable than conifers, and other fire-resistant vegetation should be planted.

J. A. Cooper

See also: Communities: structure; Erosion and erosion control; Forest management; Forests; Mediterranean scrub; Mountain ecosystems; Old-growth forests; Paleoecology; Rain forests; Savannas and deciduous tropical forests; Taiga; Trophic levels and ecological niches.

Sources for Further Study

Fuller, Margaret. *Forest Fires: An Introduction to Wildland Fire Behavior, Management, Firefighting, and Prevention.* New York: Wiley, 1991.

Johnson, Edward A. *Fire and Vegetation Dynamics: Studies from the North American Boreal Forest.* New York: Cambridge University Press, 1992.

Kasischke, Eric S., and Brian J. Stocks, eds. *Fire, Climate Change, and Carbon Cycling in the Boreal Forest.* New York: Springer, 2000.

Pringle, Laurence P. *Fire in the Forest: A Cycle of Growth and Renewal.* New York: Simon and Schuster, 1995.

Pyne, Stephen J. *Fire in America: A Cultural History of Wildland and Rural Fire.* 2d ed. Seattle: University of Washington Press, 1997.

Whelan, Robert J. *The Ecology of Fire.* New York: Cambridge University Press, 1995.

FOREST MANAGEMENT

Type of ecology: Restoration and conservation ecology

Forest management includes reforestation programs as well as techniques to manage logging practices, provide grazing lands, support mining operations, maintain infrastructure networks, or slow the destruction of rain forests.

Forests provide lumber for buildings, wood fuel for cooking and heating, and raw materials for making paper, latex rubber, resin, dyes, and essential oils. Forests are also home to millions of plants and animal species and are vital in regulating climate, purifying the air, and controlling water runoff. A 1993 global assessment by the United Nations Food and Agriculture Organization (FAO) found that three-fourths of the forests in the world still have some tree cover, but less than one-half of these have intact forest ecosystems. Deforestation is occurring at an alarming rate, and management practices are being sought to try to halt this destruction.

Thousands of years ago, forests and woodlands covered almost 15 billion acres of the earth. Approximately 16 percent of the forests have been cleared and converted to pasture, agricultural land, cities, and nonproductive land. The remaining 11.4 billion acres of forests cover about 30 percent of the earth's land surface. Clearing forests has severe environmental consequences. It reduces the overall productivity of the land, and nutrients and biomass stored in trees and leaf litter are lost. Soil once covered with plants, leaves, and snags becomes prone to erosion and drying. When forests are cleared, habitats are destroyed and biodiversity is greatly diminished. Destruction of forests causes water to drain off the land instead of being released into the atmosphere by transpiration or percolation into groundwater. This can cause major changes in the hydrologic cycle and ultimately in the earth's climate. Because forests remove a large amount of carbon dioxide from the air; the clearing of forests causes more carbon dioxide to remain, thus upsetting the delicate balance of atmospheric gases.

Rain Forests
Rain forests provide habitats for at least 50 percent (some estimates are as high as 90 percent) of the total stock of plant, insect, and other animal species on earth. They supply one-half of the world's annual harvest of hardwood and hundreds of food products, such as chocolate, spices, nuts, coffee, and tropical fruits. Tropical rain forests also provide the main ingredients in 25 percent of prescription and nonprescription drugs, as well as 75 per-

cent of the three thousand plants identified as containing chemicals that fight cancer. Industrial materials, such as natural latex rubber, resins, dyes, and essential oils, are also harvested from tropical forests.

Tropical forests in Asia, Africa, and Latin America are rapidly being cleared to produce pastureland for large cattle ranches, establish logging operations, construct large plantations, grow narcotic plants, develop mining operations, or build dams to provide power for mining and smelting operations. In 1985 the FAO's Committee on Forest Development in the Tropics developed the Tropical Forestry Action Plan to combat these practices, develop sustainable forest methods, and protect precious ecosystems. Fifty nations in Asia, Africa, and Latin America have adopted the plan.

Management Techniques

Several management techniques have been successfully applied to slow the destruction of tropical forests. Sustainable logging practices and reforestation programs have been established on lands that allow timber cutting, with complete bans of logging on virgin lands. Certain regions have set up extractive reserves to protect land for the native people who live in the forest and gather latex rubber and nuts from mature trees. Sections of some tropical forests have been preserved as national reserves, which attract tourists while preserving trees and biodiversity.

Developing countries have been encouraged to protect their tropical forests by using a combination of debt-for-nature swaps and conservation easements. In debt-for-nature swaps, tropical countries act as custodians of the tropical forest in exchange for foreign aid or relief from debt. Conservation easement involves having another country, private organization, or consortium of countries compensate a tropical country for protecting a specific habitat.

Another management technique involves putting large areas of the forest under the control of indigenous people who use slash-and-burn agriculture (also known as swidden or milpa agriculture). This traditional, productive form of agriculture follows a multiple-year cycle. Each year farmers clear a forest plot of several acres in size to allow the sun to penetrate to the ground. Leaf litter, branches, and fallen trunks are burned, leaving a rich layer of ashes. Fast-growing crops, such as bananas and papayas, are planted and provide shade for root crops, which are planted to anchor the soil. Finally, crops such as corn and rice are planted. Crops mature in a staggered sequence, thus providing a continuous supply of food. Use of mixed perennial polyculture helps prevent insect infestations, which can destroy monoculture crops. After one or two years, the forest begins to take over the agricultural plot. The farmers continue to pick the perennial crops but essentially allow the forest to reclaim the plot for the next ten to fifteen years before clearing and planting the area again. Slash-and-burn agriculture can, however, post hazards: A drought in Southeast Asia in 1997 caused such fires to burn for months when monsoon rains did not materialize, polluting the air and threatening the health of millions of Indonesians. In 1998, previous abuse of the technique resulted in flooding and mudslides in Honduras after the onset of Hurricane Mitch.

U.S. Forest Management
Forests cover approximately one-third of the land area of the continental United States and comprise 10 percent of the forests in the world. Only about 22 percent of the commercial forest area in the United States lies within national forests. The rest is primarily managed by private companies that grow trees for commercial logging. Land managed by the U.S. Forest Service provides inexpensive grazing lands for more than three million cattle and sheep every year, supports multimillion-dollar mining operations, and consists of a network of roads eight times longer than the U.S. interstate highway system. Almost 50 percent of national forest land is open for commercial logging. Nearly 14 percent of the timber harvested in the United States each year comes from national forest lands. Total wood production in the United States has caused the loss of more than 95 percent of the old-growth forests in the lower forty-eight states. This loss includes not only high-quality wood but also a rich diversity of species not found in early-growth forests.

National forests in the United States are required by law to be managed in accordance with principles of sustainable yield. Congress has mandated that forests be managed for a combination of uses, including grazing, logging, mining, recreation, and protection of watersheds and wildlife. Healthy forests also require protection from pathogens and insects. Sustainable forestry, which emphasizes biological diversity, provides the best management. Other management techniques include removing only infected trees and vegetation, cutting infected areas and removing debris, treating trees with antibiotics, developing disease-resistant species of trees, using insecticides and fungicides, and developing integrated pest management plans.

Two basic systems are used to manage trees: even-aged and uneven-aged. Even-aged management involves maintaining trees in a given stand that are about the same age and size. Trees are harvested, then seeds are replanted to provide for a new even-aged stand. This method, which tends toward the cultivation of a single species or monoculture of trees, empha-

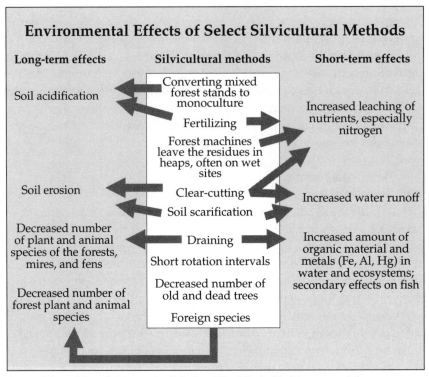

Source: Adapted from I. Stjernquist, "Modern Wood Fuels," in *Bioenergy and the Environment*, edited by Pasztor and Kristoferson, 1990.

sizes the mass production of fast-growing, low-quality wood (such as pine) to give a faster economic return on investment. Even-aged management requires close supervision and the application of both fertilizer and pesticides to protect the monoculture species from disease and insects.

Uneven-aged management maintains trees at many ages and sizes to permit a natural regeneration process. This method helps sustain biological diversity, provides for long-term production of high-quality timber, allows for an adequate economic return, and promotes a multiple-use approach to forest management. Uneven-aged management also relies on selective cutting of mature trees and reserves clear-cutting for small patches of tree species that respond favorably to such logging methods.

Harvesting Methods

The use of a particular tree-harvesting method depends on the tree species involved, the site, and whether even-aged or uneven-aged management is being applied. Selective cutting is used on intermediate-aged or mature trees in uneven-aged forests. Carefully selected trees are cut in a prescribed stand to provide for a continuous and attractive forest cover that preserves the forest ecosystem.

Shelterwood cutting involves removing all the mature trees in an area over a period of ten years. The first harvest removes dying, defective, or diseased trees. This allows more sunlight to reach the healthiest trees in the forest, which will then cast seeds and shelter new seedlings. When the seedlings have turned into young trees, a second cutting removes many of the mature trees. Enough mature trees are left to provide protection for the younger trees. When the young trees become well established, a third cutting harvests the remaining mature trees, leaving an even-aged stand of young trees from the best seed trees to mature. When done correctly, this method leaves a natural-looking forest and helps reduce soil erosion and preserve wildlife habitat.

Seed-tree cutting harvests almost every tree at one site, with the exception of a few high-quality seed-producing and wind-resistant trees, which will function as a seed source to generate new crops. This method allows a variety of species to grow at one time and aids in erosion control and wildlife conservation.

Clear-cutting removes all the trees in a single cutting. The clear-cut may involve a strip, an entire stand, or patches of trees. The area is then replanted with seeds to grow even-aged or tree-farm varieties. More than two-thirds of the timber produced in the United States, and almost one-third of the timber in national forests, is harvested by clear-cutting. A clear-cut reduces biological diversity by destroying habitat. It can make trees in bordering areas more vulnerable to winds and may take decades to regenerate.

Forest Fires

Forest fires can be divided into three types: surface, crown, and ground fires. Surface fires tend to burn only the undergrowth and leaf litter on the forest floor. Most mature trees easily survive, as does wildlife. These fires occur every five years or so in forests with an abundance of ground litter and help prevent more destructive crown and ground fires. Such fires can release and recycle valuable mineral nutrients, stimulate certain plant seeds, and help eliminate insects and pathogens.

Crown fires are very hot fires that burn both ground cover and tree tops. They normally occur in forests that have not experienced fires for several decades. Strong winds allow these fires to spread from deadwood and ground litter to treetops. They are capable of killing all vegetation and wildlife, leaving the land prone to erosion.

Ground fires are more common in northern bogs. They can begin as surface fires but burn peat or partially decayed leaves below the ground surface. They can smolder for days or weeks before anyone notices them, and they are difficult to douse.

Natural forest fires can be beneficial to some plant species, including the giant sequoia and the jack pine trees, which release seeds for germination only after being exposed to intense heat. Grassland and coniferous forest ecosystems that depend on fires to regenerate are called fire climax ecosystems. They are managed for optimum productivity with prescribed fires.

The Society of American Foresters has begun advocating a concept called new forestry, in which ecological health and biodiversity, rather than timber production, are the main objectives of forestry. Advocates of new forestry propose that any given site should be logged only every 350 years, wider buffer zones should be left beside streams to reduce erosion and protect habitat, and logs and snags should be left in forests to help replenish soil fertility. Proponents also wish to involve private landowners in the cooperative management of lands.

Toby R. Stewart and Dion Stewart

See also: Communities: structure; Conservation biology; Deforestation; Erosion and erosion control; Forest fires; Forests; Grazing and overgrazing; Integrated pest management; Multiple-use approach; Old-growth forests; Reforestation; Restoration ecology; Species loss; Sustainable development; Urban and suburban wildlife; Wildlife management.

Sources for Further Study
Davis, Kenneth P. *Forest Management.* New York: McGraw-Hill, 1996.
Davis, Lawrence S., and K. Norman Johnson. *Forest Management.* 3d ed. New York: McGraw-Hill, 1987.
Hunter, Malcolm L., Jr. *Wildlife, Forests, and Forestry.* Englewood Cliffs, N.J.: Prentice-Hall, 1990.
McNeely, Jeffrey A. *Conserving the World's Biological Diversity.* Washington, D.C.: WRI, 1990.
Nyland, Ralph D. *Silviculture: Concepts and Applications.* 2d ed. Boston: McGraw-Hill, 2002.
Robbins, William G. *American Forestry.* Lincoln: University of Nebraska Press, 1985.
Smith, David M. *The Practice of Silviculture.* 9th ed. New York: Wiley, 1997.
Spurr, Stephen H., and Burton V. Barnes. *Forest Ecology.* 4th ed. New York: Wiley, 1998.

FORESTS

Types of ecology: Biomes; Ecosystem ecology

Forests are complex ecosystems in which trees are the dominant type of plant. There are three main forest biomes: tropical, temperate, and boreal.

Both humans and animals depend on forests for food, shelter, and other resources. Forests once covered much of the world and are still found from the equator to the Arctic regions. A forest may vary in size from only a few acres to thousands of square miles, but generally any natural area in which trees are the dominant type of plant can be considered a forest. For a plant to be called a tree, the standard definition requires that the plant must attain a mature height of at least 8 feet (about 3 meters), have a woody stem, and possess a distinct crown. Thus, size makes roses shrubs and apples trees, even though apples and roses are otherwise close botanical relatives. Foresters generally divide the forests of the world into three general categories: tropical, temperate, and boreal.

Tropical Rain Forest

The tropical rain forest is a forest consisting of a dizzying variety of trees, shrubs, and other plants that remain green year-round. The growth is lush and usually includes both a dense canopy formed by the crowns of the largest trees and a thick understory of smaller trees and shrubs. Growth is often continuous, rather than broken into periods of dormancy and active growth, so that fruiting trees are occasionally seen bearing blossoms and mature fruit simultaneously.

Temperate Forest

The temperate forest lies between the tropical forest and the boreal, or northern, forest. The forests of the Mediterranean region of Europe as well as the forests of the southern United States are temperate forests. Trees in temperate forests can be either deciduous or coniferous. Although coniferous trees are generally thought of as evergreen, the distinction between types is actually based on seed production and leaf shape. Coniferous trees, such as spruces, pines, and hemlocks, produce seeds in cones and have needle-like leaves. Deciduous trees, such as maples, poplars, and oaks, have broad leaves and bear seeds in other ways. Some conifers, such as tamarack, do change color and drop their needles in the autumn, while some deciduous trees, particularly in the southerly regions of the temperate forest, are evergreen.

Deciduous trees are also referred to as hardwoods, while conifers are softwoods, a classification that refers more to the typical density of the wood than to how difficult it is to nail into it. Softwoods are lower in density and will generally float in water while still green. Hardwoods are higher in density on average and will sink.

Like tropical forests, temperate forests can be quite lush. While the dominant species vary from area to area, depending on factors such as soil types and available rainfall, a dense understory of shade-tolerant species often thrives beneath the canopy. Thus, a mature temperate forest may have thick stands of rhododendrons 20 to 30 feet (6 to 9 meters) high thriving in the shade of 80-foot (24-meter) oaks and tulip poplar. As the temperate forest approaches the edges of its range and the forest makes the transition to boreal, the understory thins out, disappearing almost completely or consisting only of low shrubs. Even in temperate forests, the dominant species may prevent an understory from forming. Stands of southern loblolly pine, for example, often have a parklike feel, as the thick mulch created by fallen needles chokes out growth of other species.

Boreal Forest

The boreal forest, which lies in a band across the northern United States, Canada, northern Europe, and northern Asia, is primarily a coniferous forest. The dominant species are trees such as white spruce, hemlock, and white pine. Mixed stands of northern hardwoods, such as birch, sugar maple, and red oak, may be found along the southern reaches of the boreal forest. As the forest approaches the Arctic, trees are fewer in type, becoming primarily spruce, birch, and willows, and smaller in size. The understory is generally thin or nonexistent, consisting of seedlings of shade-tolerant species, such as maple, and low shrubs. Patches of boreal-type forest can be found quite far south in higher elevations in the United States, such as the mountains of West Virginia. The edge of the temperate forest has crept steadily northward following the retreat of the glaciers at the end of the Ice Age twenty thousand years ago.

Forest Ecology and Resources

In all three types of forest a complex system of interrelationships governs the ecological well-being of the forest and its inhabitants. Trees and animals have evolved to fit into particular environmental niches. Some wildlife may need one resource provided by one species of tree in the forest during one season and a resource provided by another during a different

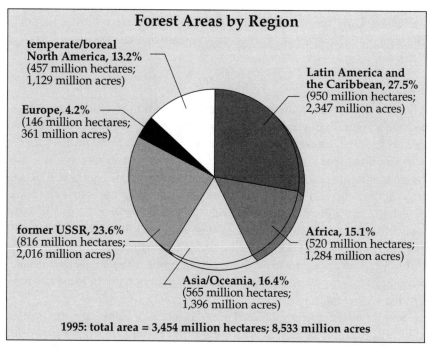

Forest Areas by Region

temperate/boreal
North America, 13.2%
(457 million hectares;
1,129 million acres)

**Latin America and
the Caribbean, 27.5%**
(950 million hectares;
2,347 million acres)

Europe, 4.2%
(146 million hectares;
361 million acres)

former USSR, 23.6%
(816 million hectares;
2,016 million acres)

Africa, 15.1%
(520 million hectares;
1,284 million acres)

Asia/Oceania, 16.4%
(565 million hectares;
1,396 million acres)

1995: total area = 3,454 million hectares; 8,533 million acres

Source: Data are from United Nations Food and Agriculture Organization (FAOSTAT Database, 2000)

time of year, while other animals become totally dependent on one specific tree. Whitetail deer, for example, browse on maple leaves in the summer, build reserves of fat by eating acorns in the fall, and survive the winter by eating evergreens. Deer are highly adaptable in contrast with other species, such as the Australian koala, which depends entirely on eucalyptus leaves for its nutritional needs. Just as the animals depend on the forest, the forest depends on the animals to disperse seeds and thin new growth. Certain plant seeds, in fact, will not sprout until being abraded as they pass through the digestive tracts of birds.

Humans also rely on the forest for food, fuel, shelter, and other products. Forests provide wood for fuel and construction, fibers for paper, and chemicals for thousands of products often not immediately recognized as deriving from the forest, such as plastics and textiles. In addition, through the process of transpiration, forests regulate the climate by releasing water vapor into the atmosphere while removing harmful carbon compounds. Forests play an important role in the hydrology of a watershed. Rain that falls on a forest will be slowed in its passage downhill and is often absorbed into the soil rather than running off into rivers

271

and lakes. Thus, forests can moderate the effects of severe storms, reducing the dangers of flooding and preventing soil erosion along stream and river banks.

Threats to the Forest

The primary threat to the health of forests around the world comes from humans. As human populations grow, three types of pressure are placed on forests. First, forests are cleared to provide land for agriculture or for the construction of new homes. This process has occurred almost continuously in the temperate regions for thousands of years, but it did not become common in tropical regions until the twentieth century. Often settlers level the forest and burn the fallen trees to clear land for farming (slash-and-burn agriculture) without the wood itself being utilized in any way. Tragically, the land thus exposed can become infertile for farming within a few years. After a few years of steadily diminishing crops, the land is abandoned. With the protective forest cover removed, it may quickly become a barren, eroded wasteland.

Second, rising or marginalized populations in developing nations often depend on wood or charcoal as their primary fuel for cooking and for home heat. Forests are destroyed as mature trees are removed for fuel wood faster than natural growth can replace them. As the mature trees disappear, younger and younger growth is also removed, and eventually the forest is gone completely.

Finally, growing populations naturally demand more products derived from wood, which can include everything from lumber for construction to chemicals used in cancer research. Market forces can drive forest products companies to harvest more trees than is ecologically sound as stockholders focus on short-term individual profits rather than long-term environmental costs. The challenge to foresters, ecologists, and other scientists is to devise methods that allow humanity to continue to utilize the forest resources needed to survive without destroying the forests as complete and healthy ecosystems.

Nancy Farm Männikkö

See also: Biomes: types; Deforestation; Erosion and erosion control; Forest fires; Forest management; Grazing and overgrazing; Mountain ecosystems; Multiple-use approach; Old-growth forests; Paleoecology; Rain forests; Rain forests and the atmosphere; Reforestation; Restoration ecology; Savannas and deciduous tropical forests; Species loss; Sustainable development; Taiga; Trophic levels and ecological niches; Tundra and high-altitude biomes; Urban and suburban wildlife; Wildlife management.

Sources for Further Study

Holland, Israel I. *Forests and Forestry*. 5th ed. Danville, Ill.: Interstate, 1997.

Kimmins, J. P. *Balancing Act: Environmental Issues in Forestry*. 2d ed. Vancouver: UBC Press, 1997.

Page, Jake. *Forest*. Rev. ed. Alexandria, Va.: Time-Life Books, 1987.

Walker, Laurence C. *The Southern Forest*. Austin: University of Texas Press, 1991.

GENE FLOW

Types of ecology: Community ecology; Evolutionary ecology;
Population ecology; Speciation

Gene flow represents a recurrent exchange of genes between populations. This exchange results when immigrants from one population interbreed with members of another.

Charles Darwin published *On the Origin of Species by Means of Natural Selection* in 1859. Since then, scientists have modified and added new concepts to the theory of evolution by natural selection. One of those concepts, which was only dimly understood in Darwin's lifetime, is the importance of genetics in evolution, especially the concepts of migration and gene flow.

Genes
Genes are elements within the cells of a living organism that control the transmission of hereditary characteristics by specifying the structure of a particular protein or by controlling the function of other genetic material. Within any species, the exchange of genes via reproduction is constant among its members, ensuring genetic similarity. If a new gene or combination of genes appears in the population, it is rapidly dispersed among all members of the population through inbreeding. New alleles (forms of a gene) may be introduced into the gene pool of a breeding population (thus contributing to the evolution of that species) in two ways: mutation and migration. Gene flow is integral to both processes.

A mutation occurs when the DNA code of a gene becomes modified so that the product of the gene will also be changed. Mutations occur constantly in every generation of every species. Most of them, however, are either minor or detrimental to the survival of the individual and thus are of little consequence. A very few mutations may prove valuable to the survival of a species and are spread to all of its members by migration and gene flow.

Separation and Migration
In nature, gene flow occurs on a more or less regular basis between demes, geographically isolated populations, races, and even closely related species. Gene flow is more common among the adjacent demes of one species. The amount of migration between such demes is high, thus ensuring that

their gene pools will be similar. This sort of gene flow contributes little to the evolutionary process, as it does little to alter gene frequencies or to contribute to variation within the species.

Much more significant for the evolutionary process is gene flow between two populations of a species that have not interbred for a prolonged period of time. Populations of a species separated by geographical barriers (as a result, for example, of seed dispersal to a distant locale) often develop very dissimilar gene combinations through the process of natural selection. In isolated populations, dissimilar alleles become fixed or are present in much different frequencies. When circumstances do permit gene flow to occur between populations, it results in the breakdown of gene complexes and the alteration of allele frequencies, thereby reducing genetic differences in both. The degree of this homogenization process depends on the continuation of interbreeding among members of the two populations over extended periods of time.

Hybridization
The migration of a few individuals from one breeding population to another may, in some instances, also be a significant source of genetic variation in the host population. Such migration becomes more important in the evolutionary process in direct proportion to the differences in gene frequencies—for example, the differences between distinct species. Biologists call interbreeding between members of separate species hybridization. Hybridization usually does not lead to gene exchange or gene flow, because hybrids are not often well adapted for survival and because most are sterile. Nevertheless, hybrids are occasionally able to breed (and produce fertile offspring) with members of one or sometimes both the parent species, resulting in the exchange of a few genes or blocks of genes between two distinct species. Biologists refer to this process as introgressive hybridization. Usually, few genes are exchanged between species in this process, and it might be more properly referred to as "gene trickle" rather than gene flow.

Introgressive hybridization may, however, add new genes and new gene combinations, or even whole chromosomes, to the genetic architecture of some species. It may thus play a role in the evolutionary process, especially in plants. Introgression requires the production of hybrids, a rare occurrence among highly differentiated animal species but quite common among closely related plant species. Areas where hybridization takes place are known as contact zones or hybrid zones. These zones exist where populations overlap. In some cases of hybridization, the line between what constitutes different species and what constitutes different populations of

Gene flow

the same species becomes difficult to draw. The significance of introgression and hybrid zones in the evolutionary process remains an area of some contention among life scientists.

Speciation
Biologists often explain, at least in part, the poorly understood phenomenon of speciation through migration and gene flow—or rather, by a lack thereof. If some members of a species become geographically isolated from the rest of the species, migration and gene flow cease. Such geographic isolation can occur, for example, when populations are separated by water (as occurs on different islands or other landmasses) or valleys (different hillsides). The isolated population will not share in any mutations, favorable or unfavorable, nor will any mutations that occur among its own members be transmitted to the general population of the species. Over long periods of time, this genetic isolation will result in the isolated population becoming so genetically different from the parent species that its members can no longer produce fertile progeny should one of them breed with a member of the parent population. The isolated members will have become a new species, and the differences between them and the parent species will continue to grow as time passes. Scientists, beginning with Darwin, have demonstrated that this sort of speciation has occurred on the various islands of the world's oceans and seas.

Paul Madden

See also: Adaptive radiation; Clines, hybrid zones, and introgression; Convergence and divergence; Evolution: definition and theories; Evolution of plants and climates; Extinctions and evolutionary explosions; Genetic diversity; Genetic drift; Isolating mechanisms; Natural selection; Nonrandom mating, genetic drift, and mutation; Punctuated equilibrium vs. gradualism; Speciation.

Sources for Further Study
Christiansen, Freddy B. *Population Genetics of Multiple Loci.* New York: Wiley, 2000.
Endler, John A. *Geographic Variation, Speciation, and Clines.* Princeton, N.J.: Princeton University Press, 1977.
Foster, Susan A., and John A. Endler, eds. *Geographic Variation in Behavior: Perspectives on Evolutionary Mechanisms.* New York: Oxford University Press, 1999.
Leapman, Michael. *The Ingenious Mr. Fairchild: The Forgotten Father of the Flower Garden.* New York: St. Martin's Press, 2001.

Mousseau, Timothy A., Barry Sinervo, and John A. Endler, eds. *Adaptive Genetic Variation in the Wild*. New York: Oxford University Press, 2000.

Real, Leslie A., ed. *Ecological Genetics*. Princeton, N.J.: Princeton University Press, 1994.

Stuessy, Tod F., and Mikio Ono, eds. *Evolution and Speciation of Island Plants*. New York: Cambridge University Press, 1998.

GENETIC DIVERSITY

Types of ecology: Community ecology; Ecosystem ecology; Population
ecology; Restoration and conservation ecology

Genetic diversity includes the inherited traits encoded in the DNA of all living or-
ganisms and can be examined on four levels: among species, among populations
(in communities), within populations, and within individuals. Populations with
higher levels of diversity are better able to adapt to changes in the environment and
are more resistant to the deleterious effects of inbreeding.

Genetic diversity is the most fundamental level of biological diversity
because genetic material is responsible for the variety of life. For new
species to form, genetic material must change. Changes in the inherited
properties of populations occur deterministically through gene flow (mating between individual organisms representing formerly separated populations) and through natural or artificial selection (which occurs when
some types of individuals breed more successfully than others). Change
can also occur randomly through mutations or genetic drift (when the relative proportions of genes change by chance in small populations). Populations with higher levels of diversity tend to do better—to have more survival options—as surroundings change than do populations (particularly
smaller ones) with lower levels of genetic diversity.

Preservation Efforts

Conservation efforts directed at maintaining genetic diversity involve
both germ plasm preservation (germ plasm kept in a steady state for periods of time) and germ plasm conservation (germ plasm kept in a natural,
evolving state). The former usually involves ex situ laboratory techniques
in which genetic resources are removed from their natural habitats. They
include seminatural strategies such as botanical gardens, arboreta, nurseries, zoos, farms, aquaria, and captive fisheries, as well as completely artificial methods such as seed reserves or "banks," microbial cultures (preserving bacteria, fungi, viruses, and other microorganisms), tissue cultures of
parts of plants and animals (including sperm storage), and gene libraries
(involving storage and replication of partial segments of plant or animal
DNA, or deoxyribonucleic acid).

Conservation areas are the preferred in situ (at the natural or original
place) means of protecting genetic resources. Ideally these include preserving the number and relative proportions of species and the genetic diver-

sity they represent, the physical features of the habitat, and all ecosystem processes. It is not always enough, however, to maintain the ecosystem which the threatened species inhabits. It is sometimes necessary to take an active interventionist position in order to save a species. Controversial strategies can include reintroduction of captive species into the wild, sometimes after they have been genetically manipulated. Direct management of the ecosystem may also be attempted by either lessening human exploitation and interference or by reducing the number of natural predators or competitors. However, management of a specific conservation area varies in terms of what is valued and how preservation is accomplished.

Crop Diversity

One area of keen interest that illustrates the issues involved with the preservation of any kind of genetic diversity is how to preserve crop germ plasm. Largely conserved in gene banks, crop germ plasm was historically protected by farmers who selected for success in differing environments and other useful traits. Traditionally cultivated varieties (landraces) diversified as people spread into new areas. Colonial expansion produced new varieties as farmers adapted to new conditions and previously separated plant species interbred; other species were lost when some societies declined and disappeared.

By the early 1900's field botanists and agronomists were expressing concern about the rapidly escalating loss of traditionally cultivated varieties. This loss accelerated after the 1940's as high yielding hybrids of cereal and vegetable crops replaced local landraces. Wild relatives of these landraces are also disappearing as their habitats are destroyed through human activity. Gene banks preserve both kinds of plants because, as argued by Nikolai I. Vavilov in 1926, crop plant improvement can best be accomplished by taking advantage of these preserved genetic stocks. Vavilov also noted that genetic variation for most cultivated species was concentrated in specific regions, his "centers of diversity," most of which are regions where crop species originated.

The vulnerability to parasites and climate of an agriculture that relies on one or a few varieties of crops necessitates the maintenance of adequate reserves of genetic material for breeding. In addition to the preservation of species known to be useful, many people advocate preservation of wild species for aesthetic reasons as well as for their unknown future potential.

Maintenance of Productivity

Farmers in developed nations change crop varieties every four to ten years in order to maintain consistent levels of food production. This necessitates an ongoing search for new breeds with higher yields and an ability to withstand several environmental challenges, including resistance to multiple pests and drought. Over time, older varieties mutate, become less popular at the marketplace, or are unable to adapt to new conditions. However, farmers from poorer nations are not always able to take advantage of the new breeds or afford the expensive support systems, including chemical fertilizers. Moreover, not all types of crops have benefited equally from conservation efforts.

Another tension between the world's poor and rich nations concerns ownership of genetic diversity. The Convention on Biodiversity, signed by 167 nations in 1992, states that genetic materials are under the sovereign control of the countries in which they are found. This policy is particularly controversial regarding medicinal plants, because "biodiversity prospecting" for new drugs has economically benefited either individuals or corporations based in the developed countries.

Joan C. Stevenson

See also: Adaptive radiation; Clines, hybrid zones, and introgression; Convergence and divergence; Evolution: definition and theories; Evolution of plants and climates; Extinctions and evolutionary explosions; Gene flow; Isolating mechanisms; Natural selection; Punctuated equilibrium vs. gradualism; Speciation.

Sources for Further Study

Hawkes, John G. *The Diversity of Crop Plants*. Cambridge, Mass.: Harvard University Press, 1983.

Fundamentals of Conservation Biology. Cambridge, Mass.: Blackwell Science, 1996.

Krattiger, Anatole F., et al., eds. *Widening Perspectives on Biodiversity*. Geneva, Switzerland: International Academy of the Environment, 1994.

Orians, Gordon H., et al., eds. *The Preservation and Valuation of Biological Resources*. Seattle: University of Washington Press, 1990.

GENETIC DRIFT

Types of ecology: Evolutionary ecology; Population ecology

Genetic drift refers to random changes in the genetic composition of a population. It is one of the evolutionary forces that cause biological evolution, the others being natural selection, mutation, and migration, or gene flow.

Drift occurs because the genetic variants, or alleles, present in a population are a random sample of the alleles that adults in the previous generation would have been predicted to pass on, where predictions are based on expected migration rates, expected mutation rates, and the direct effects of alleles on fitness. If this sample is small, then the genetic composition of the offspring population may deviate substantially from expectation, just by chance. This deviation is called genetic drift. Drift becomes increasingly important as population size decreases. The key feature of drift that distinguishes it from the other evolutionary forces is the unpredictable direction of evolutionary change.

Anything that generates fitness variation among individuals (that is, variation in the ability of individuals to survive and reproduce) will increase the magnitude of drift for all genes that do not themselves cause the fitness variation. Because of their indeterminate growth, plants often vary greatly in reproductive potential because of local environmental variation, and this magnifies genetic drift. For example, the magnitude of drift in most annual plants is more than doubled by size variation among adults. This makes sense if one considers that larger individuals contribute a larger number of offspring to the next generation, so any alleles they carry will tend to be overrepresented.

Fitness variation caused by selection will also increase the magnitude of drift at any gene not directly acted upon by the selection. If an individual has high fitness because it possesses one or more favorable alleles, then all other alleles it possesses will benefit. This is called genetic hitchhiking. This is a potent source of evolution because the direction of change at a hitchhiking gene will remain the same for multiple generations. However, it is not possible to predict in advance what that direction will be because where and when a favorable mutation will occur cannot be predicted.

The opportunities for drift to occur are greatly influenced by gene flow. Most terrestrial plants are characterized by highly localized dispersal. Thus, even in large, continuous populations, the pool of potential mates for an individual, and the pool of seeds that compete for establishment at a site, are all drawn from a small number of nearby individuals known as the neighborhood. If the neighborhood is sufficiently small, genetic drift will have a significant impact on its genetic composition.

For these and other reasons, population size alone is not sufficient to predict the magnitude of drift. The effective size of a population, N_e, is a number that is directly related to the magnitude of drift through a simple equation. Thus, N_e incorporates all characteristics of a population that influence drift.

Loss of Variability

The long-term consequence of drift is a loss of genetic variation. As alleles increase and decrease in frequency at random, some will be lost. In the absence of mutation and migration, such losses are permanent. Eventually, only one allele remains at each gene, which is said to be fixed. Thus, all else being equal, smaller populations are expected to harbor less genetic variation than larger populations.

An important way in which different plant populations are not equal is in their reproductive systems. With self-fertilization (selfing), or asexual reproduction, genetic hitchhiking becomes very important. In the extreme cases of 100 percent selfing or 100 percent asexual reproduction, hitchhiking will determine the fates of most alleles. Thus, as a new mutation spreads or is eliminated by selection, so too will most or all of the other alleles carried by the individual in which the mutation first arose. This is called a selective sweep, and the result is a significant reduction in genetic variation. Which alleles will be swept to fixation or elimination cannot be predicted in advance, so the loss of variation reflects a small N_e. Consistent with this expectation, most populations of flowering plants that reproduce partly or entirely by selfing contain significantly less genetic variation than populations of related species that do not self-fertilize.

Extinction

Mutations that decrease fitness greatly outnumber mutations that increase fitness. In a large population in which drift is weak, selection prevents most such mutations from becoming common. In very small populations, however, alleles that decrease fitness can drift to fixation, causing a decrease in average fitness. This is one manifestation of a phenomenon called inbreeding depression. In populations with very small N_e, this inbreeding depression can be significant enough to threaten the population with extinction. If a population remains small for many generations, mean fitness will continue to decline as new mutations become fixed by drift. When fitness declines to the point where offspring are no longer overproduced, population size will decrease further. Drift then becomes stronger, mutations are fixed faster, and the population heads down an accelerating trajectory toward extinction. This is called mutational meltdown.

Creative Potential

By itself, drift cannot lead to adaptation. However, drift can enhance the ability of selection to do so. Because of diploidy and sexual recombination, some types of mutations, either singly or in combinations, will increase fitness when common but not when rare. Genetic drift can cause such genetic variants to become sufficiently common for selection to promote their fixation. A likely example is the fixation of new structural arrangements of chromosomes that occurred frequently during the diversification of flowering plants. New chromosome arrangements are usually selected against when they are rare because they disrupt meiosis and reduce fertility. The initial spread of such a mutation can therefore only be caused by strong genetic drift, either in an isolated population of small effective size or in a larger population divided into small neighborhoods.

John S. Heywood

See also: Biodiversity; Clines, hybrid zones, and introgression; Extinctions and evolutionary explosions; Gene flow; Genetic diversity; Isolating mechanisms; Natural selection; Nonrandom mating, genetic drift, and mutation; Population genetics; Punctuated equilibrium vs. gradualism; Speciation.

Sources for Further Study

Denny, Mark, and Steven Gaines. *Chance in Biology: Using Probability to Explore Nature.* Princeton, N.J.: Princeton University Press, 2000.
Futuyma, Douglas J. *Evolutionary Biology.* 3d ed. Sunderland, Mass.: Sinauer Associates, 1998.

GENETICALLY MODIFIED FOODS

Types of ecology: Agricultural ecology; Chemical ecology;
Ecotoxicology; Evolutionary ecology

*Applications of genetic engineering in agriculture and the food industry could in-
crease world food supplies, reduce environmental problems associated with food
production, and enhance the nutritional values of certain foods. However, these
benefits are countered by food-safety concerns, the potential for ecosystem disruption, and fears of unforeseen consequences resulting from altering natural selection.*

Humans rely on plants and animals as food sources and have long
used microbes to produce foods such as cheese, bread, and fermented beverages. Conventional techniques such as cross-hybridization,
production of mutants, and selective breeding have resulted in new varieties of crop plants or improved livestock with altered genetics. However,
these methods are relatively slow and labor-intensive, are generally limited to intraspecies crosses, and involve a great deal of trial and error.

Transgenic Technology
Recombinant DNA techniques, which manipulate cells' deoxyribonucleic
acid (DNA), were developed in the 1970's and enabled researchers to make
specific, predetermined genetic changes in a variety of organisms. Because
the technology also allows for the transfer of genes across species and
kingdom barriers, an infinite number of novel genetic combinations are
possible. The first animals and plants containing genetic material from
other organisms (transgenics) were developed in the early 1980's. By 1985
the first field trials of plants engineered to be pest-resistant were conducted. In 1990 the U.S. Food and Drug Administration (FDA) approved
chymosin as the first substance produced by modified organisms to be
used in the food industry for dairy products such as cheese. That same year
the first transgenic cow was developed to produce human milk proteins
for infant formula. The well-publicized Flavr Savr tomato, modified to delay ripening and rotting, obtained FDA approval in 1994.

Goals and Uses
By the mid-1990's, more than one thousand genetically modified crop
plants were approved for field trials. The goals for altering food crop
plants by genetic engineering fall into three main categories: to create

plants that can adapt to specific environmental conditions to make better use of agricultural land, increase yields, or reduce losses; to increase nutritional value or flavor; and to alter harvesting, transport, storage, or processing properties for the food industry. Many genetically modified crops are sources of ingredients for processed foods and animal feed.

Herbicide-resistant plants, such as the Roundup Ready soybean, can be grown in the presence of glyphosphate, an herbicide that normally destroys all plants with which it comes in contact. Beans from these plants were approved for food-industry use in several countries, but there has been widespread protest by activists such as Jeremy Rifkin and environmental organizations such as Greenpeace. Frost-resistant fruit containing a fish antifreeze gene, insect-resistant plants with a bacterial gene that encodes for a pesticidal protein (*Bacillus thuringiensis*), and a viral disease-

Genetically Modified Crop Plants Unregulated by the U.S. Department of Agriculture

Crop	Patent Holder	Genetically Engineered Trait
Canola	AgrEvo	herbicide tolerance
Corn	AgrEvo	herbicide tolerance
	Ciba-Geigy	insect resistance
	DeKalb	herbicide tolerance; insect resistance
	Monsanto	herbicide tolerance; insect resistance
	Northrup King	insect resistance
Cotton	Calgene	herbicide tolerance; insect resistance
	DuPont	herbicide tolerance
	Monsanto	herbicide tolerance; insect resistance
Papaya	Cornell	virus resistance
Potato	Monsanto	insect resistance
Squash	Asgrow	virus resistance
	Upjohn	virus resistance
Soybean	AgrEvo	herbicide tolerance
	DuPont	altered oil profile
	Monsanto	herbicide tolerance
Tomato	Agritope	altered fruit ripening
	Calgene	altered fruit ripening
	Monsanto	altered fruit ripening
	Zeneca	altered chemical content in fruit

Source: U.S. Department of Agriculture and Plant Health Inspection Service (APHIS).

resistant squash are examples of other genetically modified food crops that have undergone field trials.

Scientists have also created plants that produce healthier unsaturated fats and oils rather than saturated ones. Genetic engineering has yielded coffee plants whose beans are caffeine-free without processing and tomatoes with altered pulp content for improved canned products. Genetically modified microbes are used for the production of food additives such as amino acid supplements, sweeteners, flavors, vitamins, and thickening agents. In some cases, these substances had to be obtained from slaughtered animals. Altered organisms are also used for improving fermentation processes in the food industry.

Ecological Implications

Food safety and quality are at the center of the genetically modified food controversy. Concerns include the possible introduction of new toxins or allergens into the diet and changes in the nutrient composition of foods. Proponents argue that food sources could be designed to have enhanced nutritional value.

A large percentage of crops worldwide are lost each year to drought, temperature extremes, and pests. Plants have already been engineered to exhibit frost, insect, disease, and drought resistance. Such alterations would increase yields and allow food to be grown in areas that are currently too dry or infertile, positively impacting the world food supply.

Environmental problems such as deforestation, erosion, pollution, and loss of biodiversity have all resulted, in part, from conventional agricultural practices. Use of genetically modified crops could allow better use of existing farmland and lead to a decreased reliance on pesticides and fertilizers. However, critics fear the creation of unpredictable ecological consequences as well. For example, "superweeds"—either the engineered plants or new plant varieties formed by the transfer of recombinant genes conferring various types of resistance to wild species—might compete with valuable plants and have the potential to destroy ecosystems and farmland unless stronger poisons were used for eradication. The transfer of genetic material to wild relatives (outcrossing, or "genetic pollution") might also lead to the development of new plant diseases.

Diane White Husic

See also: Biopesticides; Erosion and erosion control; Grazing and overgrazing; Integrated pest management; Multiple-use approach; Pesticides; Rangeland; Slash-and-burn agriculture; Soil; Soil contamination; Species loss.

Sources for Further Study

American Chemical Society. *Genetically Modified Foods: Safety Issues.* Washington, D.C.: Author, 1995.

Anderson, Luke. *Genetic Engineering, Food, and Our Environment.* White River Junction, Vt.: Chelsea Green, 1999.

Paredes-Lopez, Octavio, ed. *Molecular Biotechnology for Plant Food Production.* Lancaster, Pa.: Technomic, 1999.

Rissler, Jane. *The Ecological Risks of Engineered Crops.* Cambridge, Mass.: MIT Press, 1996.

Shannon, Thomas A., ed. *Genetic Engineering: A Documentary History.* Westport, Conn.: Greenwood Press, 1999.

Yount, Lisa. *Biotechnology and Genetic Engineering.* New York: Facts on File, 2000.

GEOCHEMICAL CYCLES

Types of ecology: Ecoenergetics; Ecosystem ecology; Global ecology

Geochemical cycles refer to the movement, or cycling, of elements through ecosystems and the biosphere. Both biotic (living) and abiotic (nonliving) components participate in these cycles, and their interdependency means that a change in one component of such a cycle will have an impact on all other components.

Geochemical cycles are generally considered to be those involving nutrient elements utilized by organisms in various ecosystems. Cycling involves both biological and chemical processes. While nearly all natural elements could be considered as being cycled through both abiotic and living systems, certain elements are most commonly described in such systems. These include carbon, nitrogen, phosphorus, and a variety of lesser elements (including iron, sulfur, and trace elements such as copper and mercury).

Although the cycling of elements is often thought of as occurring in a relatively rapid fashion, many of these elements spend long periods locked in abiotic systems. For example, carbon may be found in materials that require millions of years to cycle through ocean sediment back into the atmosphere. The fate of such elements depends on many factors, including their chemical properties and their ability to erode or return to the atmosphere. Some chemical elements, such as carbon, oxygen, and nitrogen, are incorporated into organisms from the atmosphere. Other elements, such as phosphorus, potassium, sulfur, and iron, are found mainly in rocks and sediments.

Carbon and Oxygen Cycles

The carbon and oxygen cycles are greatly dependent on each other. Molecular oxygen, which represents approximately 20 percent of the atmosphere, is used by organisms through a metabolic process called respiration. In these reactions, the oxygen reacts with reduced carbon compounds such as carbohydrates (sugars) and generates carbon dioxide (CO_2). Though carbon dioxide constitutes only a small proportion of the volume of the atmosphere (0.04 percent), it is in this form that it is used by primary producers such as plants. In the process of photosynthesis, utilizing sunlight as an energy source, plants and some microorganisms bind, or fix, the CO_2, converting the carbon again into carbohydrates, resulting in growth of the plant, or replication of the microorganism. The complex carbohydrates which are generated in photosynthesis serve as the food source for

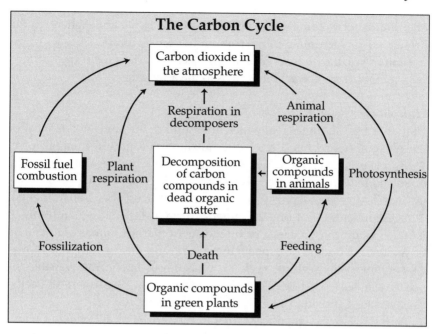

The Carbon Cycle

consumers—organisms such as animals (including humans) that eat the plants. The carbohydrates are then broken down, regenerating carbon dioxide. In a sense, the combinations of respiration and photosynthesis represent the cycle of life. Approximately 70 billion metric tons of carbon dioxide (10 percent of the total atmospheric CO_2) are fixed each year. The concentration of carbon dioxide in the atmosphere is a factor in regulating the temperature of Earth. Consequently, the release of large quantities of the gas into the atmosphere through the burning of fossil fuels could potentially alter the earth's climate.

Nitrogen Cycle

Nitrogen gas (N_2) represents 78 percent of the total volume of the atmosphere. However, because of the extreme stability of the bond between the two nitrogens in the gas, plants and animals are unable to use atmospheric nitrogen directly as a nutrient. Nitrogen-fixing bacteria in the soil and in the roots of leguminous plants (peas, clover) are able to convert the gaseous nitrogen into nitrites and nitrates, chemical forms that can be used by plants. Animals then obtain nitrogen by consuming the plants. The decomposition of nitrogen compounds results in the accumulation of ammonium (NH_4+) compounds in a process called ammonification. It is in this form that nitrogen is commonly found under conditions in which oxygen is lim-

289

ited. In this form, some of the nitrogen returns to the atmosphere. In the presence of oxygen, ammonium compounds are oxidized to nitrates (nitrification). Once the plant or animal has died, bacteria convert the nitrogen back into nitrogen gas, and it returns to the atmosphere.

Phosphorus Cycle

Unlike carbon and nitrogen, which are found in the atmosphere, most of the phosphorus required for biotic nutrition is found in mineral form. Phosphorus is relatively water insoluble in this form; it is only gradually dissolved in water. Available phosphorus is therefore often growth-limiting in soils (it is second only to nitrogen as the scarcest of the soil nutrients). Ocean sediments may bring the mineral to the surface through uplifting of land, as along coastal areas, or by means of marine animals. Enzymatic breakdown of organic phosphate by bacteria and the consumption of marine organisms by seabirds cycle the phosphorus into forms available for use by plants. Deposition of guano (bird feces) along the American Pacific coast has long provided a fertilizer rich in phosphorus.

Bacteria also play significant roles in the geochemical cycling of many other elements. Iron, despite its abundance in the earth's crust, is largely

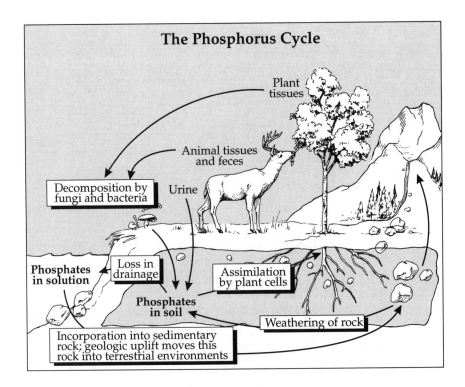

The Phosphorus Cycle

Plant tissues

Animal tissues and feces

Decomposition by fungi and bacteria

Urine

Phosphates in solution

Loss in drainage

Assimilation by plant cells

Phosphates in soil

Weathering of rock

Incorporation into sedimentary rock; geologic uplift moves this rock into terrestrial environments insoluble in water. Consequently, it is generally found in the form of precipitates of ferric (Fe^{+3}) compounds, seen as brown deposits in water. Acids are often formed as by-products in the formation of ferric compounds. The bacterial oxidation of pyrite (FeS_2) is a major factor in the leaching process of iron ores and in the formation of acid mine drainage. Likewise, much of the sulfur found in the earth's crust is in the form of pyrite and gypsum ($CaSO_4$). Weathering processes return much of the sulfur to water-soluble forms; in the absence of air, the bacterial reduction of sulfate (SO_4^{-2}) to forms such as hydrogen sulfide (H_2S) allows its return to the atmosphere. Since sulfide compounds are highly toxic to many organisms, bacterial reduction of sulfates is of major biogeochemical significance.

Richard Adler

See also: Balance of nature; Biomass related to energy; Communities: ecosystem interactions; Communities: structure; Food chains and webs; Herbivores; Hydrologic cycle; Nutrient cycles; Omnivores; Phytoplankton; Rain forests and the atmosphere; Trophic levels and ecological niches.

Sources for Further Study

Berner, Elizabeth K., and Robert A. Berner. *Global Environment: Water, Air, and Geochemical Cycles.* Upper Saddle River, N.J.: Prentice Hall, 1996.

Bolin, Bert. "The Carbon Cycle." *Scientific American* 223 (September, 1970).

Clark, F. E., and T. Rosswall, eds. *Terrestrial Nitrogen Cycles: Processes, Ecosystem Strategies, and Management Impacts.* Stockholm: Swedish Natural Science Research Council, 1981.

Heimann, Martin. *The Global Carbon Cycle.* New York: Springer-Verlag, 1993.

Kasischke, Eric S., and Brian J. Stocks, eds. *Fire, Climate Change, and Carbon Cycling in the Boreal Forest.* New York: Springer, 2000. Discusses the direct and indirect mechanisms by which fire and climate interact to influence carbon cycling in North American boreal forests.

Krebs, Charles J. *Ecology: The Experimental Analysis of Distribution and Abundance.* 5th ed. San Francisco: Benjamin Cummings, 2001.

Lasserre, P., and J. M. Martin, eds. *Biogeochemical Processes at the Land-Sea Boundary.* Amsterdam, N.Y.: Elsevier, 1986.

Pomeroy, Lawrence, ed. *Cycles of Essential Elements.* Stroudsburg, Pa.: Dowden, Hutchinson, & Ross, 1974.

Tiessen, Holm, ed. *Phosphorus in the Global Environment: Transfers, Cycles, and Management.* New York: Wiley, 1995.

GLOBAL WARMING

Type of ecology: Global ecology

Global warming is the term applied to rising global air temperatures. This rise in temperature has the potential to cause drastic changes in climate and weather patterns worldwide.

Greenhouse Effect

"Global warming" is the term for the rise in the earth's average temperature. Scientists know that the earth's average global temperature has been rising since the beginning of the Industrial Revolution in the second half of the eighteenth century. Increases in air temperature could alter precipitation patterns, change growing seasons, result in coastal flooding, and turn some areas into deserts. Scientists do not know the cause or causes of this phenomenon but believe that it could be part of a normal climate cycle or be caused by natural events or the activities of humankind.

When the ground is heated by sunlight, it gives off the heat as infrared radiation. The atmosphere absorbs the infrared radiation and keeps it from escaping into space. This is called the "greenhouse effect" because it was once believed that the glass panes of greenhouses acted similarly to the atmosphere, capturing the infrared radiation given off by Earth inside the greenhouse and not allowing it to pass through. Although it has subsequently been shown that greenhouses work by simply trapping the heated air, the name has stuck.

The atmosphere eventually releases its heat into space. The amount of heat stored in the atmosphere remains constant as long as the composition of gases in the atmosphere does not change. Some gases, including carbon dioxide, water vapor, and methane, store heat more efficiently than others and are called "greenhouse gases." If the composition of the atmosphere changes to include more of these greenhouse gases, the air retains more heat, and the atmosphere becomes warmer.

Global levels of greenhouse gases have been steadily increasing and in 1990 were more than 14 percent higher than they were in 1960. At the same time, the average global temperature has also been rising. Meteorological records show that from 1890 to the mid-1990's, the average global temperature rose by between 0.4 and 0.7 degree Celsius. About 0.2 degree Celsius of this temperature increase has occurred since 1950. In comparison, the difference between the average global temperature in the 1990's and in the last ice age is approximately 10 degrees Celsius, and it is estimated that a drop of as little as 4 to 5 degrees Celsius could trigger the formation of continental glaciers. Therefore, the rise in average temperature is significant and already beginning to cause some changes in the global climate. Documented changes include the melting of glaciers and rising sea levels as the ocean gets warmer and its waters expand. Measurements of plant activity indicate that the annual growing season has become approximately two weeks longer in the middle latitude regions.

Effects of Global Warming
A common misunderstanding is that global warming simply means that winters will be less cold and summers will be hotter, while everything else

U.S. Greenhouse Gas Emissions, 1990-1999

Type	1990	1994	1995	1996	1997	1998	1999
Carbon Dioxide (carbon)[1]	1,350.5	1,422.5	1,434.7	1,484.1	1,505.2	1,507.4	1,526.8
Methane Gas[1]	31.74	31.17	31.18	30.16	30.11	29.29	28.77
Nitrous Oxide[2]	1,168	1,310	1,257	1,246	1,226	1,223	1,224
Chlorofluorocarbons[2]	202	109	102	67	51	49	41
Halons[2]	2.8	2.7	2.9	3.0	3.0	3.0	3.0
Hydrofluorocarbons[2]							
HFC-23	3.0	3.0	2.0	3.0	3.0	3.4	2.6
HFC-125	(Z)	0.3	0.5	0.7	0.9	1.1	1.3
HFC-134a	1.0	6.3	14.3	19.0	23.5	26.9	30.3
HFC-143a	(Z)	0.1	0.1	0.2	0.3	0.5	0.7
Perfluorocarbons[2]							
CF-4	3	2	2	2	2	2	2
C-2F-6	1	—	1	1	1	1	1
C-4F-10	(Z)	(Z)	(Z)	(Z)	(Z)	(Z)	(Z)
Sulfur hexafluoride[2]	1	1	2	2	2	2	1

Source: Abridged from U.S. Energy Information Administration, *Emissions of Greenhouse Gases in the United States,* annual. From *Statistical Abstract of the United States: 2001* (Washington, D.C.: U.S. Census Bureau, 2001).

Note: Emission estimates were mandated by Congress through Section 1605(a) of the Energy Policy Act of 1992 (Title XVI). Gases that contain carbon can be measured either in terms of the full molecular weight of the gas or just in terms of their carbon content.

[1] In millions of metric tons.

[2] In thousands of metric tons.

Z. Less than 500 metric tons.

— Represents or rounds to zero

will be basically the same. Actually, the earth's weather system is very complicated, and higher global temperatures will result in significant changes in weather patterns. Most changes will be observed in the middle and upper latitudes, with equatorial regions witnessing fewer changes. The areas that will experience most of the changes include North America, Europe, and most of Asia. The Southern Hemisphere will experience less severe effects because it contains more water than the Northern Hemisphere does, and it takes more energy to heat water than land.

It is difficult to predict precisely what these changes will be, but observation of the changing climate and scientific studies allow researchers to make some rough estimates of the kinds of changes Earth will experience. Summers will be hotter, with more severe heat waves. Because hot air holds more moisture than cool air, rain will fall less frequently in the summer. Droughts can be expected to be more common and more severe. Through the late 1980's and into the early 1990's, annual temperatures climbed higher and higher, and summer heat waves became more frequent. This is a particularly troubling problem in areas where homes are typically built without air conditioning. The air in a closed-up house during a heat wave can reach temperatures well over 40 degrees Celsius. Because these temperatures exceed people's normal body temperature, which is about 36.5 degrees Celsius, it becomes very difficult for the body to cool down. For this reason, heat waves are particularly dangerous for the very young, the elderly, and people who are ill. More frequent heat waves will cause increases in the use of air conditioning, which requires more energy and will release additional greenhouse gases.

Global warming will also produce severe autumn rains. The overheated summer air will cool in the autumn and will no longer be able to hold all of the moisture it was storing. It will release the moisture as heavy rains, causing flooding. This phenomenon has already been observed, but not for a period of time that is scientifically significant. It is difficult to tell the difference between long-term changes and short-term fluctuations with only a few years of observations.

These changing rain patterns—droughts and severe autumn storms—will certainly have an effect on the earth's landscape. Some areas may be turned to deserts, and others may be transformed from plains to forests.

One of the strange aspects of global warming is that it is predicted to result in not only hotter periods during the summers but also colder periods during the winter. It takes energy to move large cold-air masses from the polar regions in winter, so it is possible that large winter storms will be colder, more violent, and more frequent. This pattern has been evident since about the mid-1970's. Winter storms have brought record low temperatures and enough snow to close cities for days. However, this time span is too short to determine whether this is a temporary phenomenon or a trend. It is possible that some smaller, less permanent event than global warming is responsible for the more severe winter storms. A shorter (although more severe) winter may create a surge in pest populations and diseases that are normally controlled by long winters.

Global warming will cause ocean levels to rise because water expands when heated and also because of the melting of glaciers on Greenland and Antarctica. The melting of the ice in the northern polar areas will not contribute to rising ocean levels because that ice, unlike the ice on Greenland and Antarctica, is already in the ocean. Just as the melting of the ice in a drink does not cause the level of the drink to rise, the melting of the northern oceanic ice sheets will not affect sea levels.

Causes of Global Warming

Although many scientists are convinced, based on the abundant evidence, that global warming is occurring, they are less sure of why. One significant factor, the rise in greenhouse gases, can be attributed to the activities of humankind. Burning forests to clear land and operating factories and automobiles produce carbon dioxide and water vapor. Livestock herds and rotting vegetation release methane, and fertilizers used on farms also release greenhouse gases. Power plants that consume fossil fuels such as coal, oil, or natural gas release massive amounts of carbon dioxide into the air. However, no one knows whether humankind's activities are the real or only reason for the increase in global temperature. The last ice age ended very recently in geologic terms, and a number of changes are still taking place as the globe recovers from the presence of huge ice sheets. It is possible that the world's climate is still warming up from the last ice age. Volcanoes are another major source of greenhouse gases, and the level of volcanic activity has been increasing since approximately the beginning of the Industrial Revolution.

Global warming could also be part of a natural cyclical change in the climate. Evidence indicates that the earth's climate varies between warmer and colder periods that last a few centuries. History provides many stories of dramatically changing climate, including one period from 1617 to 1650 that was so unusually cold that it is called the "Little Ice Age." Therefore, Earth may be merely experiencing another cyclical change in its climate.

Study of Global Warming

The problem with studying global warming is that no one can be sure of its extent, and some scientists debate whether it actually exists. They argue that the observed changes may only be a warm phase of a climatic cycle that will make the earth warmer for a few years, then cooler for a few years. If this is true, then the coming decades may see average temperatures leveling off or even dropping.

A major source of the confusion involves the way global warming is studied. It would seem easy to record temperatures for a number of years and then compare them. However, detailed records on the weather have not been kept for more than a few decades in many areas. Scientists are forced to rely on interpretations of historical accounts and the clues left in fossil records. The analysis of tree rings, sedimentary deposits, and even very old ice from deep within glaciers can provide data about the climate in the past.

In addition, the existing records must be reviewed carefully to identify local changes that may not reflect global ones. For example, as towns grow into cities, the temperature climbs simply because larger cities are warmer than smaller ones, a phenomenon known as the urban heat island effect. Measurements taken years ago in a more rural environment should be lower than those taken after the population around the measuring station increased. This problem can be overcome with balloons. By sending instruments high in the atmosphere on weather balloons, air temperatures can be measured without being affected by urbanization. Although data recorded this way show a consistent rise in global temperature, again such measurements go back only a few decades.

Measurements of the level of greenhouse gases in the atmosphere are also affected by urbanization. As a small town becomes a city, levels can be expected to rise. However, recording stations located in regions far removed from cities and factories also show an increase in the level of greenhouse gases. One station on the island of Hawaii has shown a rise in the amount of carbon dioxide present in the air since early 1958, with similar reports coming from stations in Point Barrow, Alaska, and Antarctica.

Another important variable in looking at global warming is sea surface temperature. Measurements can be skewed by local effects that have no impact on the global climate. One method used to make detailed measurements of seawater temperature is to broadcast a particular frequency of noise through the water and measure it at distant locations. The speed and frequency of the sound are affected by the temperature of the water. However, more accurate and more global data are becoming available through observations made from satellites placed in geosynchronous orbits, which can take a variety of detailed measurements of many conditions that are factors in global warming. These data will help immensely in enabling scientists to understand the global climate and the changes it is undergoing.

Christopher Keating

See also: Biodiversity; Biomes: determinants; Biomes: types; Biosphere concept; Ecosystems: definition and history; Ecosystems: studies; Geochemical cycles; Greenhouse effect; Hydrologic cycle; Nutrient cycles; Ozone depletion and ozone holes; Rain forests and the atmosphere; Trophic levels and ecological niches.

Sources for Further Study

Abrahamson, Dean Edwin, ed. *The Challenge of Global Warming*. Washington, D.C.: Island Press, 1989.

Berger, John J. *Beating the Heat: Why and How We Must Combat Global Warming*. Berkeley, Calif.: Berkeley Hills Books, 2000.

Gates, David M. *Climate Change and Its Biological Consequences*. Sunderland, Mass.: Sinauer Associates, 1993.

Houghton, John T. *Global Warming: The Complete Briefing*. 2d ed. New York: Cambridge University Press, 1997.

Nance, John J. *What Goes Up: The Global Assault on Our Atmosphere*. New York: William Morrow, 1991.

Rosenzweig, Cynthia. *Climate Change and the Global Harvest: Potential Impacts of the Greenhouse Effect on Agriculture*. New York: Oxford University Press, 1998.

Sommerville, Richard C. J. *The Forgiving Air: Understanding Environmental Change*. Berkeley: University of California Press, 1996.

Wyman, Richard L., ed. *Global Climate Change and Life on Earth*. New York: Chapman and Hall, 1991.

GRASSLANDS AND PRAIRIES

Types of ecology: Biomes; Ecosystem ecology

Grasslands, including the prairies of North America, are a biome characterized by the presence of low plants, mostly grasses, and are distinguished from woodlands, deserts, and tundra. They support a great variety of plants and animals. At present, the remaining grasslands provide grazing for livestock and wildlife.

Grasslands occupied vast areas of the world more than ten thousand years ago, before the development of agriculture, industrialization, and the subsequent explosive growth of the human population. They are characterized by the presence of low plants (mostly grasses), experience sparse to moderate rainfall, and are found in both temperate and tropical climatic zones. The main grasslands of the planet include the prairies of North America, the pampas of South America, the steppes of Eurasia, and the savannas of Africa.

Grasslands are intermediate between deserts and woodlands in terms of precipitation and biomass. The warmer tropical savannas average 60 to 150 centimeters (25 to 60 inches) of rain. The temperate grasslands range between twenty-five to seventy-five centimeters (ten to thirty inches) of precipitation, some of which may be in the form of snow. The biomass of grasslands, predominantly grasses, is quantitatively intermediate between that of deserts and woodlands, which produce 10 to 15 percent and 200 to 300 percent, respectively, of the amount of plant material. It should be recognized that the grassland biomes can be subdivided in terms of climate, plant species, and animal species. It should also be noted that grasslands do not always shift abruptly to deserts or woodlands, leading to gradations between them. In addition, grasslands do have scattered trees, often along streams or lakes, and low-lying brush.

Grasses have extensive root systems and the ability to become dormant. These permit them to survive low rainfall, including periodic droughts, or the winter cold typical of temperate regions. Furthermore, grasslands have always been subjected to periodic fires, but the deep root systems of grassland plants also permit them to regrow after fire. Grasses coevolved over millions of years with the grazing animals that depend on them for food. Ten thousand years ago, wild ancestors of cattle and horses, as well as antelope and deer, were on the Eurasian steppes; bison and pronghorn prospered on the North American prairies; wildebeest, gazelle, zebra, and buffalo dominated African savannas; and the kangaroo was the predominant grazer in Australia. Grazing is a symbiotic relationship, whereby animals gain their nourishment from plants, which in turn benefit from the activity. It removes vegetative matter, which is necessary in order for grasses to grow, facilitates seed dispersal, and disrupts mature plants, permitting young plants to take hold. Urine and feces from grazing animals recycle nutrients to the plants. The grassland ecosystem also includes other animals, including worms, insects, birds, reptiles, rodents, and predators. The grasses, grazing animals, and grassland carnivores, such as wolves or large cat species, constitute a food chain.

Humans have been an increasing presence in grassland areas, where more than 90 percent of contemporary crop production now occurs and much urbanization and industrialization have taken place. Remaining grassland areas are not used for crops, habitation, or industry because of inadequate water supplies or unsuitable terrain but instead are used for grazing domesticated or wild herbivores. In addition, many woodland areas around the world have been cleared and converted to grasslands for crops, livestock, living, or working.

The Prairies of North America

Originally stretching east from the Rocky Mountains to Indiana and Ohio, and from Alberta, Canada to Texas, the prairies were the major grassland of North America. The short-grass prairie extended about 200 miles (300 kilometers) east of the mountains, and the long-grass prairie bordered the deciduous forest along the eastern edge, while the mixed-grass prairie was between the two. Going from west to east, the amount of precipitation increases, causing changes in plant populations. The short-grass prairie receives only about 25 centimeters (10 inches) of precipitation each year, mostly as summer rain, and, as its name suggests, has short grass, less than 60 centimeters (2 feet) tall. Today, it is used primarily for grazing because the soil is shallow and unsuited for farming without irrigation. The mixed-grass prairie receives moderate precipitation, ranging from 35 to 60 centimeters (14 to 24 inches) and has medium-height grasses, ranging from 60 to 120 centimeters (2 to 4 feet) tall. Much of it is now used for growing wheat. The tall-grass prairie receives more than 60 centimeters (24 inches) of precipitation, mostly in the summer, and had grasses that grow to over 150 centimeters (5 feet) tall. It has rich soil and has been mostly converted to very productive cropland, primarily for corn and soybeans. The prairies experience very cold winters (down to –45 degrees Celsius, –50 degrees Fahrenheit) and very hot summers (up to 45 degrees Celsius, 110 degrees Fahrenheit). They are often windy and experience severe storms, blizzards in winter, thunderstorms and tornadoes in summer.

Grasses have extensive root systems and the ability to become dormant, permitting them to survive the low to moderate rainfall that characterizes regions on the dry sides of mountain ranges. The grassland ecosystem includes native grazers such as bison or, more often today, domesticated grazers such as cattle, as well as worms, insects, birds, reptiles, rodents, and predators such as wolves or large cats. (PhotoDisc)

Like other biomes, the prairies have a characteristic assortment of animals, herbivores that eat the plants and carnivores that prey on the herbivores. Before 1500 C.E., two ruminants, the bison (commonly but inaccurately called buffalo) and the pronghorn (not a true antelope), were the major grazers on the prairies. The prairie dog, a herbivorous rodent that burrows, lived in large communities on the prairies. The major predators were the wolf and coyote for bison and pronghorn, and the black-footed ferret and fox for prairie dogs. A variety of birds (herbivorous and carnivorous), reptiles, and insects also made their home on the prairie.

Overgrazing Grasslands

While grazing is of mutual benefit to plant and animal, overgrazing is ultimately detrimental to both the plant and animal populations, as well as the environment. Continued heavy grazing leads to deleterious consequences. Removal of leaf tips, even repeated, will not affect regeneration of grasses provided that the basal zone of the plant remains intact. While the upper half of the grass shoot can generally be eaten without deleterious consequences, ingesting the lower half, which sustains the roots and fuels regrowth, will eventually kill the plants. Overgrazing leads to denuding the land, to invasion by less nutritious plant species, to erosion due to de-creased absorption of rainwater, and to starvation of the animal species. Because the loss of plant cover changes the reflectance of the land, climate changes can follow and make it virtually impossible for plants to return, with desertification an ultimate consequence. It is not just the number of animals, but the timing of the grazing that can be detrimental. Grasses re-quire time to regenerate, and continuous grazing will inevitably kill them. Consumption too early in the spring can stunt their development.

Semiarid regions are particularly prone to overgrazing because of low and often unpredictable rainfall; regrettably, these are the areas of the world where much grazing has been relegated, because the moister grass-land areas have been converted to cropland. Overgrazing has contributed to environmental devastation worldwide. Excessive grazing by cattle, sheep, goats, and camels is partly responsible for the desert of the Middle East, ironically the site of domestication for many animals and plants. Un-controlled livestock grazing in the late 1800's and early 1900's negatively affected many areas of the American West, where sagebrush and juniper trees have invaded the grasslands. Livestock overgrazing has similarly devastated areas of Africa and Asia. In the early twenty-first century, feral horses in the American West and the Australian outback are damaging those environments. Overgrazing by wildlife can also be deleterious. The 1924 Kaibab Plateau deer disaster in the Grand Canyon National Park and Game Preserve is one such example, where removal of natural predators led to overpopulation, overgrazing, starvation, and large die-offs.

Riparian zones, the strips of land on either side of a river or stream, are particularly susceptible to overgrazing. Because animals naturally congre-gate in these areas with water, lush vegetation, and shade, they can seri-ously damage them by preventing grasses from regrowing and young trees from taking root, as well as trampling and compacting the soil and fouling the water course. The ecosystem can be devastated, threatening survival of plant and animal species and leading to serious erosion. While herding and fencing can be used to control animals in these areas, a less ex-pensive method is to disperse the location of water supplies and salt blocks to encourage movement away from rivers or streams. Grassland animals crave salt, and if deprived of it, will seek it out.

Grassland Management

Grassland areas need not deteriorate if properly managed, whether for livestock, wildlife, or both. Managing grasslands involves controlling the number of animals and enhancing their habitat. Carrying capacity, which is the number of healthy animals that can be grazed indefinitely on a given unit of land, must not be exceeded. Because of year-to-year changes in weather conditions and hence food availability, determining carrying capacity is not simple; worst-case estimates are preferred in order to minimize the chances of exceeding it. The goal should be a healthy grassland achieved by optimizing, not maximizing, the number of animals. For private land, optimizing livestock numbers is in the long-term interest of the landowner, although not always seen as such. For land that is publicly held, managed in common, or with unclear or disputed ownership, restricting animals to the optimum level is particularly difficult to achieve. Personal short-term benefit often leads to long-term disaster, described as the "tragedy of the commons" by biologist Garrett Hardin.

Appropriate management of grasslands involves controlling animal numbers and enhancing grassland plants. Restricting cattle and sheep is physically easy through herding and fencing, although it can be politically difficult and expensive. Much more problematic is controlling charismatic feral animals, such as horses, or wildlife, when natural predators have been eliminated and hunting is severely restricted. As for habitat improvement, the use of chemical, fire, mechanical, and biological approaches can increase carrying capacity for either domesticated or wild herbivores. Removing woody vegetation by burning or mechanical means will increase grass cover, fertilizing can stimulate grass growth, and reseeding with desirable species can enhance the habitat. Plants native to a particular region can be best for preserving that environment. Effective grassland management requires matching animals with the grasses on which they graze.

James L. Robinson

See also: Biomes: determinants; Biomes: types; Chaparral; Deserts; Forests; Habitats and biomes; Lakes and limnology; Marine biomes; Mediterranean scrub; Mountain ecosystems; Old-growth forests; Rain forests; Rain forests and the atmosphere; Rangeland; Reefs; Savannas and deciduous tropical forests; Taiga; Tundra and high-altitude biomes; Wetlands.

Sources for Further Study
Brown, Lauren. *The Audubon Society Nature Guides: Grasslands*. New York: Knopf, 1985.

Collinson, Alan. *Grasslands*. New York: Dillon Press, 1992.

Demarais, Stephen, and Paul R. Krausman, eds. *Ecology and Management of Large Mammals in North America*. Upper Saddle River, N.J.: Prentice Hall, 2000.

Humphreys, L. R. *The Evolving Science of Grassland Improvement*. New York: Cambridge University Press, 1997.

Joern, Anthony, and Kathleen H. Keeler, eds. *The Changing Prairie*. New York: Oxford University Press, 1995.

Pearson, C. J., and R. L. Ison. *Agronomy of Grassland Systems*. 2d ed. New York: Cambridge University Press, 1997.

Sampson, Fred B., and Fritz L. Knopf, eds. *Prairie Conservation*. Washington, D.C.: Island Press, 1996.

Steele, Philip. *Grasslands*. Minneapolis: Carolrhoda Books, 1997.

GRAZING AND OVERGRAZING

Types of ecology: Agricultural ecology; Restoration and conservation ecology

Animals that eat grass, or graze, can actually help the earth produce richer land cover and soil. When the land suffers ill effects because of too much grazing, over-grazing has occurred.

The effects of overgrazing occur where there are more grazing animals than the land and vegetation can support. Overgrazing has negatively affected regions of the United States, primarily in the Southwest. Areas that have been severely damaged by overgrazing typically show declining or endangered plant and animal species.

Herbivores are animals that feed on plant material, and grazers are herbivores that feed specifically on grass. Examples are horses, cows, antelope, rabbits, and grasshoppers. Overgrazing occurs when grazer populations exceed the carrying capacity of a specified area (the number of individual organisms the resources of a given area can support). In overgrazing conditions, there is insufficient food to support the animal population in question. Depending on the grazer's strategy, emigration or starvation will follow. Grasslands can handle, and even benefit from, normal grazing; only overgrazing adversely affects them.

Grasses' Defenses Against Grazing

Grasslands and grazers coevolved, so grasses can withstand grazing within the ecosystem's carrying capacity. All plants have a site of new cell growth called the meristem, where growth in height and girth occurs. Most plants have the meristem at the very top of the plant (the apical meristem). If a plant's apical meristem is removed, the plant dies.

If grasses had an apical meristem, grazers—and lawn mowers—would kill grasses. Grasses survive mowing and grazing because the meristem is located at the junction of the shoot and root, close to the ground. With the exception of sheep, grazers in North America do not disturb the meristem, and sheep do so only during overgrazing conditions. At proper levels of grazing, grazing actually stimulates grass to grow in height in an attempt to produce a flowering head for reproduction. Grazing also stimulates grass growth by removing older plant tissue at the top that is functioning at a lower photosynthetic rate.

Grazers

Mammalian grazers have high, crowned teeth with a great area for grinding to facilitate opening of plants' cell walls as a means to release nutrients. The cell wall is composed of cellulose, which is very difficult for grazers to digest. Two major digestive systems of grazing strategies have evolved to accommodate grazing. Ruminants, such as cows and sheep, evolved stomachs with four chambers to allow regurgitation in order to chew food twice to maximize cellulose breakdown. Intestinal bacteria digest the cellulose, releasing fatty acids that nourish the ruminants. Other grazers, such as rabbits and horses, house bacteria in the cecum, a pouch at the junction of the small and large intestines. These bacteria ferment the plant material ingested. The fermented products of the bacteria nourish these grazers.

Impacts in the Southwest

As previously mentioned, in the United States the negative effects of overgrazing are most intense in the Southwest. Some ecologists believe that one significant factor was the pattern of early European colonization of the area. Missions were abundant in the Southwest, and the missions owned

Grazers such as sheep form a symbiotic relationship with the grasses, getting their nourishment from these plants and in turn facilitating growth of new grass by removing excess vegetative matter, dispersing seed through their droppings, and recycling nutrients via urine and feces. Overgrazing occurs when the number of grazers exceeds the carrying capacity of the grassland. (PhotoDisc)

cattle that were rarely slaughtered, except on big feast days. Because Catholic missionaries received some financial support from their religious orders in Europe, mission cattle were not restrained as strictly as were those owned by cattle ranchers, whose sole livelihood came from raising and selling cattle. Mission cattle roamed greater distances and began the pattern of overgrazing in the Southwest. The impact of overgrazing was particularly intense because much of the Southwest has desertlike conditions. Extreme environmental conditions result in particularly fragile ecosystems. Hence the Southwest was, and is, vulnerable to the effects of overgrazing.

Another possible—though disputed—contribution to overgrazing may stem from the fact that much of the land in the Southwest is public land under jurisdiction of the Bureau of Land Management. This federal agency leases land to private concerns for the purpose of grazing cattle or sheep. Some observers feel that the bureau has a conflict of interest in that its primary source of income is money obtained from leasing public land under its jurisdiction. They suspect that the bureau has granted, and fear that it may continue to grant, grazing leases in regions threatened with or suffering from overgrazing.

Effects of Overgrazing

Overgrazing can lead to a number of ecological problems. Depletion of land cover leads to soil erosion and can ultimately cause desertification. Other possible results are the endangering of some species of grass and the creation of monocultures in regions where certain species have been removed. Desertification is the intensification and expansion of deserts at the expense of neighboring grasslands. When overgrazing occurs along desert perimeters, plant removal leads to decreased shading. Decreased shading increases the local air temperature. When the temperature increases, the air may no longer cool enough to release moisture in the form of dew. Dew is the primary source of precipitation in deserts, so without it, desert conditions intensify. Even a slight decrease in desert precipitation is serious. The result is hotter and drier conditions, which lead to further plant loss and potentially to monocultures.

Overgrazing of grasslands, combined with the existence of nonnative species in an ecosystem, can result in the endangerment of species of native grasses. At one time, cattle in the Southwest fed exclusively on native grasses. Then nonnative plant species arrived in the New World in the guts of cows shipped from Europe. They began to compete with the native grasses. European grass species have seeds with prickles and burs; southwestern native grasses do not, making them softer and more desirable to the cattle. Hence European grasses experienced little, if any, grazing, while the much more palatable southwestern native grasses were grazed to the point of overgrazing. The result was drastic decline or loss of native grassland species. In such cases animals dependent on native grassland species must emigrate or risk extinction. For example, many ecologists conjecture that the Coachella Valley kit fox in California is threatened because of the loss of grassland habitat upon which it is dependent.

Solutions

Desertification is usually considered irreversible, but the elimination of grazing along desert perimeters can help to prevent further desertification. One kind of attempt to reestablish native grass species involves controlled-burn programs. Nonnative grassland species do not appear to be as fire-resistant as native grass species. Controlled burn programs are therefore being used in some overgrazed grassland areas to try to eliminate nonnatives and reestablish native grass species. If successful, such programs will improve the health of the ecosystem.

Jessica O. Ellison

See also: Biopesticides; Desertification; Erosion and erosion control; Forest fires; Forest management; Forests; Genetically modified foods; Grazing and overgrazing; Integrated pest management; Multiple-use approach; Pesticides; Rangeland; Slash-and-burn agriculture; Soil; Soil contamination; Wildlife management.

Sources for Further Study

Hodgson, J., and A. W. Illius, eds. *The Ecology and Management of Grazing Systems.* Wallingford, Oxon, England: CAB International, 1996.

McBrien, Heather, et al. "A Case of Insect Grazing Affecting Plant Succession." *Ecology* 64, no. 5 (1983).

Sousa, Wayne P. "The Role of Disturbance in Natural Communities." *Annual Review Ecological Systems* 15, 1984.

_____. "Some Basic Principles of Grassland Fire Management." *Environmental Management* 3, no. 1 (1979).

WallisDeVries, Michiel F., Jan P. Bakker, Sipke E. Van Wieren, eds. *Grazing and Conservation Management.* Boston: Kluwer Academic, 1998.

GREENHOUSE EFFECT

Type of ecology: Global ecology

The greenhouse effect is a natural process of atmospheric warming in which solar energy that has been absorbed by the earth's surface is reradiated and then absorbed by particular atmospheric gases, primarily carbon dioxide and water vapor. Without this warming process, the atmosphere would be too cold to support life. Since 1880, however, the surface atmospheric temperature has been rising, paralleling a rise in the concentration of carbon dioxide and other gases produced by human activities.

Since 1880, carbon dioxide, along with several other gases—chloro-fluorocarbons (CFCs), methane, hydrofluorocarbons (HFCs), per-fluourocarbons (PFCs), sulfur hexafluoride, and nitrous oxide—have been increasing in concentration and have been identified as likely contributors to a rise in global surface temperature. These gases are called greenhouse gases. The temperature increase may lead to drastic changes in climate and food production as well as widespread coastal flooding. As a result, many scientists, organizations, and governments have called for curbs on the production of greenhouse gases. Since the predictions are not definite, however, debate continues about the costs of reducing the production of these gases without being sure of the benefits.

Global Warming

The greenhouse effect occurs because the gases in the atmosphere are able to absorb only particular wavelengths of energy. The atmosphere is largely transparent to short-wave solar radiation, so sunlight basically passes through the atmosphere to the earth's surface. Some is reflected or absorbed by clouds, some is reflected from the earth's surface, and some is absorbed by dust or the earth's surface. Only small amounts are actually absorbed by the atmosphere. Therefore, sunlight contributes very little to the direct heating of the atmosphere. On the other hand, the greenhouse gases are able to absorb long-wave, or infrared, radiation from the earth, thereby heating the earth's atmosphere.

Discussion of the greenhouse effect has been confused by terms that are imprecise and even inaccurate. For example, the atmosphere was believed to operate in a manner similar to a greenhouse, whose glass would let visible solar energy in but would also be a barrier preventing the heat energy from escaping. In actuality, the reason that the air remains warmer inside a greenhouse is probably because the glass prevents the warm air from mixing with the cooler outside air. Therefore the greenhouse effect could be more accurately called the "atmospheric effect," but the term greenhouse effect continues to be used.

Even though the greenhouse effect is necessary for life on earth, the term gained harmful connotations with the discovery of apparently increasing atmospheric temperatures and growing concentrations of greenhouse gases. The concern, however, is not with the greenhouse effect itself but rather with the intensification or enhancement of the greenhouse effect, presumably caused by increases in the level of gases in the atmosphere resulting from human activity, especially industrialization. Thus the term global warming is a more precise description of this presumed phenomenon.

The Greenhouse Effect

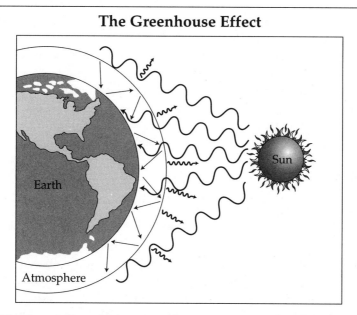

Clouds and atmospheric gases such as water vapor, carbon dioxide, methane, and nitrous oxide absorb part of the infrared radiation emitted by the earth's surface and reradiate part of it back to the earth. This process effectively reduces the amount of energy escaping into space and is popularly called the "greenhouse effect" because of its role in warming the lower atmosphere. The greenhouse effect has drawn worldwide attention because increasing concentrations of carbon dioxide from the burning of fossil fuels may result in a global warming of the atmosphere. Scientists know that the greenhouse analogy is incorrect. A greenhouse traps warm air within a glass building where it cannot mix with cooler air outside. In a real greenhouse, the trapping of air is more important in maintaining the temperature than is the trapping of infrared energy. In the atmosphere, air is free to mix and move about.

Greenhouse effect

Human Contributions

A variety of human activities appear to have contributed to global warming. Large areas of natural vegetation and forests have been cleared for agriculture. The crops may not be as efficient in absorbing carbon dioxide as the natural vegetation they replaced. Increased numbers of livestock have led to growing levels of methane. Several gases that appear to be intensifying global warming, including CFCs and nitrous oxides, also appear to be involved with ozone depletion. Stratospheric ozone shields the earth from solar ultraviolet (short-wave) radiation; therefore, if the concentration of these ozone-depleting gases continues to increase and the ozone shield is depleted, the amount of solar radiation reaching the earth's surface should increase. Thus, more solar energy would be intercepted by the earth's surface to be reradiated as long-wave radiation, which would presumably increase the temperature of the atmosphere.

However, whether there is a direct cause-and-effect relationship between increases in carbon dioxide and the other gases and surface temperature may be impossible to determine because the atmosphere's temperature has fluctuated widely over millions of years. Over the past 800,000 years, the earth has had several long periods of cold temperatures—during which thick ice sheets covered large portions of the earth—interspersed with shorter warm periods. Since the most recent retreat of the glaciers around 10,000 years ago, the earth has been relatively warm.

Problems of Prediction

How much the temperature of the earth might rise is not clear. So far, the temperature increase of around 1 degree Fahrenheit is within the range of normal (historic) trends. The possibility of global warming became a serious issue during the late twentieth century because the decades of the 1980's and the 1990's included some of the hottest years recorded for more than a century. On the other hand, warming has not been consistent since 1880, and for many years cooling occurred. The cooling might have resulted from the increase of another product of fossil fuel combustion, sulfur dioxide aerosols, which reflect sunlight, thus lessening the amount of solar energy entering the atmosphere. Similarly, in the early 1990's temperatures declined, perhaps because of ash and sulfur dioxide produced by large volcanic explosions during that period. In the late 1990's and early 2000's temperatures appeared to be rising again, thus indicating that products of volcanic explosions may have masked the process of global warming. The United States Environmental Protection Agency (EPA) states that the earth's average temperature will probably continue to increase because the greenhouse gases stay in the atmosphere longer than the aerosols.

310

Proper analysis of global warming is dependent on the collection of accurate temperature records from many locations around the world and over many years. Because human error is always possible, "official" temperature data may not be accurate. This possibility of inaccuracy compromises examination of past trends and predictions for the future. However, the use of satellites to monitor temperatures has probably increased the reliability of the data.

Predictions for the future are hampered in various ways, including lack of knowledge about all the components affecting atmospheric temperature. Therefore, computer programs cannot be sufficiently precise to make accurate predictions. A prime example is the relationship between ocean temperature and the atmosphere. As the temperature of the atmosphere increases, the oceans would absorb much of that heat. Therefore, the atmosphere might not warm as quickly as predicted. However, the carbon dioxide absorption capacity of oceans declines with temperature. Therefore, the oceans would be unable to absorb as much carbon dioxide as before, but exactly how much is unknown. Increased ocean temperatures might also lead to more plant growth, including phytoplankton. These plants would probably absorb carbon dioxide through photosynthesis. A warmer atmosphere could hold more water vapor, resulting in the potential for more clouds and more precipitation. Whether that precipitation would fall as snow or rain and where it would fall could also affect air temperatures. Air temperature could lower as more clouds might reflect more sunlight, or more clouds might absorb more infrared radiation.

To complicate matters, any change in temperature would probably not be uniform over the globe. Because land heats up more quickly than water, the Northern Hemisphere, with its much larger landmasses, would probably have greater temperature increases than the Southern Hemisphere. Similarly, ocean currents might change in both direction and temperature. These changes would affect air temperatures as well. In reflection of these complications, computer models of temperature change range widely in their estimates. Predicted increases range from 1.5 to 11 degrees Celsius (3 to 20 degrees Fahrenheit) over the early decades of the twenty-first century.

Mitigation Attempts

International conferences have been held, and international organizations have been established to research and minimize potential detriments of global warming. In 1988 the United Nations Environment Programme and the World Meteorological Organization established the International Panel on Climate Change (IPCC). The IPCC has conducted much research on climate change and is now considered an official advisory body on the climate change issue. In June, 1992, the United Nations Conference on Environment and Development, or Earth Summit, was held in Brazil. Participants devised the Framework Convention on Climate Change and considered the landmark international treaty. It required signatories to reduce and monitor their greenhouse gas emissions.

A more advanced agreement, the Kyoto Accords, was developed in December, 1997, by the United Nations Framework Convention on Climate Change. It set binding emission levels for all six greenhouse gases over a five-year period for the developed world. Developing countries do not have any emission targets. It also allows afforestation to be used to offset emissions targets. The Kyoto agreement includes the economic incentive of trading emissions targets. Some countries, because they have met their targets, would have excess permits, which they might be willing to sell to other countries that have not met their targets.

Margaret F. Boorstein

See also: Biodiversity; Biomes: determinants; Biomes: types; Biosphere concept; Geochemical cycles; Global warming; Hydrologic cycle; Ozone depletion and ozone holes; Rain forests and the atmosphere.

Sources for Further Study
Berger, John J. *Beating the Heat: Why and How We Must Combat Global Warming.* Berkeley, Calif.: Berkeley Hills Books, 2000.
Flavin, Christopher, Odil Tunali, and Jane A. Peterson, et al., eds. *Climate of Hope: New Strategies for Stabilizing the World's Atmosphere.* Washington, D.C.: Worldwatch Institute, 1996.
Graedel, Thomas E. *Atmosphere, Climate, and Change.* New York: Scientific American Library/W. H. Freeman, 1997.
Houghton, John T. *Global Warming: The Complete Briefing.* 2d ed. New York: Cambridge University Press, 1997.
Kondratsev, Kirill, and Arthur P. Cracknell. *Observing Global Climate Change.* London: Taylor and Francis, 1998.
McKibben, Bill. *The End of Nature.* 10th anniversary ed. New York: Anchor Books, 1999.
Rosenzweig, Cynthia. *Climate Change and the Global Harvest: Potential Impacts of the Greenhouse Effect on Agriculture.* New York: Oxford University Press, 1998.
Sommerville, Richard C. J. *The Forgiving Air: Understanding Environmental Change.* Berkeley: University of California Press, 1996.

HABITATS AND BIOMES

Types of ecology: Biomes; Ecosystem ecology

The biosphere is the sum total of all habitats on earth that can be occupied by living organisms. Descriptive and experimental studies of habitat components allow scientists to predict how various organisms will respond to changes in their environment, whether caused by humans or nature.

Life in the form of individual organisms composed of one or more living cells is found in a vast array of different places on earth, each with its own distinctive types of organisms. Life on earth has been classified by scientists into units called species, whose individuals appear similar, have the same role in the environment, and breed only among themselves. The space in which each species lives is called its habitat.

Habitats, Communities, Ecosystems, and Biomes

The term "habitat" can refer to specific places with varying degrees of accuracy. For example, rainbow trout can be found in North America from Canada to Mexico, but more specifically they are found in freshwater streams and lakes, with an average temperature below 70 degrees Fahrenheit and a large oxygen supply. The former example describes the macrohabitat of the rainbow trout, which is a broad, easily recognized area. The latter example describes its microhabitat—the specific part of its macrohabitat in which it is found. Similarly, the macrohabitat of one species can refer to small or large areas of its habitat. The macrohabitat of rainbow trout may refer to the habitat of a local population, the entire range of the species, or (most often) an area intermediate to those extremes. While "habitat," therefore, refers to the place an organism lives, it is not a precise term unless a well-defined microhabitat is intended.

The total population of each species has one or more local populations, which are all the individuals in a specific geographic area that share a common gene pool; that is, they commonly interbreed. For example, rainbow trout of two adjacent states will not normally interbreed unless they are part of local populations that are very close to each other. The entire geographic distribution of a species, its range, may be composed of many local populations.

On a larger organizational scale, there is more than only one local population of one species in any habitat. Indeed, it is natural and necessary for many species to live together in an area, each with its own micro- and macrohabitat. The habitat of each local population of each species overlaps the habitat of many others. This collective association of populations in one general area is termed a community, which may consist of thousands of species of animals, plants, fungi, bacteria, and other one-celled organisms.

Groups of communities that are relatively self-sufficient in terms of both recycling nutrients and the flow of energy among them are called ecosystems. An example of an ecosystem could be a broad region of forest community interspersed with meadows and stream communities that share a common geographic area. Some ecosystems are widely distributed across the surface of the earth and are easily recognizable as similar ecosystems known as biomes—deserts, for example. Biomes are usually named for the dominant plant types, which have very similar shapes and macrohabitats. Thus, similar types of organisms inhabit them, though not necessarily ones of the same species. These biomes are easily mapped on the continental scale and represent a broad approach to the distribution of organisms on the face of the earth. One of the more consistent biomes is the northern coniferous forest, which stretches across Canada and northern Eurasia in a latitudinal belt. Here are found needle-leafed evergreen conifer trees adapted to dry, cold, windy conditions in which the soil is frozen during the long winter.

North American Biomes

The biomes of North America, from north to south, are the polar ice cap, the Arctic tundra, the northern coniferous forest; then, at similar middle latitudes, eastern deciduous forest, prairie grassland, or desert; and last, subtropical rain forest near the equator. Complicating factors that determine the actual distribution of the biomes are altitude, annual rainfall, topography, and major weather patterns. These latter factors, which influence the survival of the living, or biotic, parts of the biome, are called abiotic factors. These are the physical components of the environment for a community of organisms.

The polar ice cap is a hostile place with little evidence of life on the surface except for polar bears and sea mammals that depend on marine animals for food. A distinctive characteristic of the Arctic tundra, just south of the polar ice cap, is its flat topography and permafrost, or permanently frozen soil. Only the top meter or so thaws during the brief Arctic summer to support low-growing mosses, grasses, and the dominant lichens known as reindeer moss. Well-known animals found there are the caribou, muskox, lemming, snowy owl, and Arctic fox.

The northern coniferous forest is dominated by tall conifer trees. Familiar animals include the snowshoe hare, lynx, and porcupine. This biome stretches east to west across Canada and south into the Great Lakes region of the United States. It is also found at the higher elevations of the Rocky Mountains and the western coastal mountain ranges. Its upper elevation limit is the "treeline," above which only low-growing grasses and herbaceous plants grow in an alpine tundra community similar to the Arctic tundra. In mountain ranges, the change in biomes with altitude mimics the biome changes with increasing latitude, with tundra being the highest or northernmost.

Approximately the eastern half of the United States was once covered with the eastern deciduous forest biome, named for the dominant broadleaved trees that shed their leaves in the fall. This biome receives more than 75 centimeters (about 30 inches) of rainfall each year and has a rich diversity of bird species, such as the familiar warblers, chickadees, nuthatches, and woodpeckers. Familiar mammals include the white-tailed deer, cottontail rabbit, and wild turkey. The Great Plains, between the Mississippi River and the Rocky Mountains, receives 25 to 75 centimeters (about 10 to 30 inches) of rain annually to support an open grassland biome often called the prairie. The many grass species that dominate this biome once supported vast herds of bison and, in the western parts, pronghorn antelope. Seasonal drought and periodic fires are common features of grasslands.

The land between the Rockies and the western coastal mountain ranges is a cold type of desert biome; three types of hotter deserts are found from western Texas west to California and south into Mexico. Deserts receive fewer than 25 centimeters (10 inches) of rainfall annually. The hot deserts are dominated by many cactus species and short, thorny shrubs and trees, whereas sagebrush, grass, and small conifer trees dominate the cold desert. These deserts have many lizard and snake species, including poisonous rattlesnakes and the Gila monster. The animals often have nocturnal habits to avoid the hot, dry daytime.

Southern Mexico and the Yucatan Peninsula are covered by evergreen, broad-leaved trees in the tropical rain-forest biome, which receives more than 200 centimeters (almost 80 inches) of rain per year. Many tree-dwelling animals, such as howler monkeys and tree frogs, spend most of their lives in the tree canopy, seldom reaching the ground.

Aquatic biomes can be broadly categorized into freshwater, marine, and estuarine biomes. Freshwater lakes, reservoirs, and other still-water environments are called lentic, in contrast to lotic, or running-water, environments. Lentic communities are often dominated by planktonic organisms, small, drifting (often transparent) microscopic algae, and the small animals that feed on them. These, in turn, support larger invertebrates and fish. Lotic environments depend more on algae that are attached to the bottoms of streams, but they support equally diverse animal communities. The marine biome is separated into coastal and pelagic, or open-water, environments, which have plant and animal communities somewhat similar to lotic and lentic freshwater environments, respectively. The estuarine biome is a mixing zone where rivers empty into the ocean. These areas have a diverse assemblage of freshwater and marine organisms.

The Biosphere
All the biomes together, both terrestrial and aquatic, constitute the biosphere, which by definition is all the places on earth where life is found. Organisms that live in a biome must interact with one another and must successfully overcome and exploit their abiotic environment. The severity and moderation of the abiotic environments determine whether life can exist in that microhabitat. Such things as minimum and maximum daily and annual temperature, humidity, solar radiation, rainfall, and wind speed directly affect which types of organisms are able to survive. Amazingly, few places on earth are so hostile that no life exists there. An example would be the boiling geyser pools at Yellowstone National Park, but even there, as the water temperature cools at the edges to about 75 degrees Celsius, bacterial colonies begin to appear. There is abundant life in the top meter or two of soil, with plant roots penetrating to twenty-two meters or more in extreme cases. Similarly, the mud and sand bottoms of lakes and oceans contain a rich diversity of life. Birds, bats, and insects exploit the airspace above land and sea up to a height of about 1,200 meters (nearly 4,000 feet), with bacterial and fungal spores being found much higher. Thus, the biosphere generally extends about 10 to 15 meters (33 to 50 feet) below the surface of the earth and about 1,200 meters above it. Beyond that, conditions are too hostile. A common analogy is that if the earth were a basketball, the biosphere would constitute only the thin outer layer.

Studying Habitats and Biomes
Abiotic habitat requirements for a local population or even for an entire species can be determined in the laboratory by testing its range of tolerance for each factor. For example, temperature can be regulated in a laboratory experiment to determine the minimum and maximum survival temperatures as well as an optimal range. The same can be done with humidity, light, shelter, and substrate type: "Substrate preference" refers to the solid or liquid matter in which an organism grows and/or moves—for example, soil or rock. The combination of all ranges of tolerance for abiotic factors should describe a population's actual or potential microhabitat within a community. Furthermore, laboratory experiments can theoretically indi-

cate how much environmental change each population can tolerate before it begins to migrate or die.

Methods to study the interaction of populations with one another or even the interaction of individuals within one local population are much more complicated and are difficult or impossible to bring into a laboratory setting. These studies most often require collecting field data on distribution, abundance, food habits or nutrient requirements, reproduction and death rates, and behavior in order to describe the relationships between individuals and populations within a community. Later stages of these field investigations could involve experimental manipulations in which scientists purposely change one factor, then observe the population or community response. Often, natural events such as a fire, drought, or flood can provide a disturbance in lieu of human manipulation.

There are obvious limits to how much scientists should tinker with the biosphere merely to see how it works. Populations and even communities in a local area can be manipulated and observed, but it is not practical or advisable to manipulate whole ecosystems or biomes. To a limited extent, scientists can document apparent changes caused by civilization, pollution, and long-term climatic changes. This information, along with population- and community-level data, can be used to construct a mathematical model of a population or community. The model can then be used to predict the changes that would happen if a certain event were to occur. These predictions merely represent the "best guesses" of scientists, based on the knowledge available. Population ecologists often construct reasonably accurate population models that can predict population fluctuations based on changes in food supply, abiotic factors, or habitat. As models begin to encompass communities, ecosystems, and biomes, however, their knowledge bases and predictive powers decline rapidly. Perhaps the most complicating factor in building and testing these large-scale models is that natural changes seldom occur one at a time. Thus, scientists must attempt to build cumulative-effect models that are capable of incorporating multiple changes into a predicted outcome.

James F. Fowler

See also: Biomes: determinants; Biomes: types; Biosphere concept; Chaparral; Deserts; Ecosystems: definition and history; Ecosystems: studies; Forests; Grasslands and prairies; Lakes and limnology; Marine biomes; Mediterranean scrub; Mountain ecosystems; Old-growth forests; Rain forests; Rain forests and the atmosphere; Rangeland; Reefs; Savannas and deciduous tropical forests; Taiga; Tundra and high-altitude biomes; Wetlands.

Sources for Further Study

Allaby, Michael. *Biomes of the World*. 9 vols. Danbury, Conn.: Grolier International, 1999.

Bradbury, Ian K. *The Biosphere*. New York: Belhaven Press, 1991.

Cox, George W. *Conservation Biology: Concepts and Applications*. 2d ed. Dubuque, Iowa: Wm. C. Brown, 1997.

Hanks, Sharon La Bonde. *Ecology and the Biosphere: Principles and Problems*. Delray Beach, Fla.: St. Lucie Press, 1996.

Luoma, Jon R. *The Hidden Forest: The Biography of an Ecosystem*. New York: Henry Holt, 1999.

Miller, G. Tyler, Jr. *Living in the Environment*. 12th ed. Belmont, Calif.: Thomson Learning, 2001.

Sutton, Ann, and Myron Sutton. *Eastern Forests*. New York: Alfred A. Knopf, 1986.

Wetzel, Robert G. *Limnology*. 3d ed. San Diego: Academic Press, 2001.

HABITUATION AND SENSITIZATION

Type of ecology: Behavioral ecology

Habituation is learning to ignore irrelevant stimuli that previously produced a reaction. Results of habituation studies have been used to explain, predict, and control behavior of humans and other organisms.

Habituation is a simple form of nonassociative learning that has been demonstrated in organisms as diverse as protozoans, insects, *Nereis* (clam worms), birds, and humans. The habituated organism learns to ignore irrelevant, repetitive stimuli which, prior to habituation, would have produced a response. With each presentation of the habituating stimulus, the responsiveness of the organism decreases toward the zero, nonresponse level. If habituation training continues after the zero-response level, the habituation period is prolonged. Habituation to a particular stimulus naturally and gradually disappears unless the training continues. If training is resumed after habituation has disappeared, habituation occurs more rapidly in the second training series than in the first. Habituation is important for survival of the individual. Many stimuli are continuously impinging upon it: Some are important, others are not. Important stimuli require an immediate response, but those which result in neither punishment nor reward may be safely ignored.

Stimulus and Response

When a new stimulus is presented (when a sudden change in the environment occurs), the organism—be it bird, beast, or human—exhibits the "startle" or "orientation" response. In essence, it stops, looks, and listens. If the stimulus is repeated and is followed by neither reward nor punishment, the organism will pay less and less attention to it. When this happens, habituation has occurred, and the organism can now respond to and deal with other stimuli. On the other hand, if, during habituation learning, a painful consequence follows a previously nonconsequential stimulus, the organism has been sensitized to that stimulus and will respond to it even more strongly than it did before the learning sessions, whether they are occurring in the laboratory or in the field.

Young birds must learn to tell the difference between and respond differently to a falling leaf and a descending predator. A young predatory bird must learn to ignore reactions of its prey which pose no danger, reactions that the predator initially feared.

A theory known as the dual-process habituation-sensitization theory was formulated in 1966 and revised in 1973. It establishes criteria for both habituation and sensitization. Criteria for habituation (similar to those proposed by E. N. Sokolov in 1960) are that habituation will develop rapidly; the frequency of stimulation determines the degree of habituation; if stimulation stops for a period of time, habituation will disappear; the stronger the stimulus, the slower the rate of habituation; the frequency of stimulation is more important than the strength of the stimulus; rest periods between habituation series increase the degree of habituation; and the organism will generalize and therefore exhibit habituation to an entire class of similar stimuli. Stimulus generalization can be measured: If a different stimulus is used in the second habituation series, habituation occurs more rapidly; it indicates generalization.

Sensitization
Sensitization, a very strong response to a very painful, injurious, or harmful stimulus, is not limited to stimulus-response circuits but involves the entire organism. After sensitization, the individual may respond more strongly to the habituating stimulus than it did prior to the start of habituation training.

There are eight assumptions about sensitization in the dual-process theory. Sensitization does not occur in stimulus-response circuits but involves the entire organism. Sensitization increases during the early stages of habituation training but later decreases. The stronger the sensitizing stimulus and the longer the exposure to it the greater the sensitization; weaker stimuli may fail to produce any sensitization. Even without any external intervention, sensitization will decrease and disappear. Increasing the frequency of sensitization stimulation causes a decrease in sensitization. Sensitization will extend to similar stimuli. Dishabituation, the loss of habituation, is an example of sensitization. Sensitization may be time-related, occurring only at certain times of the day or year.

According to the dual-process theory, the response of an organism to a stimulus will be determined by the relative strengths of habituation and sensitization. Charles Darwin, the father of evolution, observed and described habituation, although he did not use the term. He noted that the birds of the Galápagos Islands were not disturbed by the presence of the giant tortoises, *Amblyrhynchus*; they disregarded them just as the magpies in England, which Darwin called "shy" birds, disregarded cows and horses grazing nearby. Both the giant tortoises of the Galápagos Islands and the grazing horses and cows of England were stimuli which, though present, would not produce profit or loss for the birds; therefore, they could be ignored.

The Neurology of Stimulus Response

Within the bodies of vertebrates is a part of the nervous system called the reticular network or reticular activating system; it has been suggested that the reticular network is largely responsible for habituation. It extends from the medulla through the midbrain to the thalamus of the forebrain. (The thalamus functions as the relay and integration center for impulses to and from the cerebrum of the forebrain.) Because it is composed of a huge number of interconnecting neurons and links all parts of the body, the reticular network functions as an evaluating, coordinating, and alarm center. It monitors incoming message impulses. Important ones are permitted to continue to the cerebral cortex, the higher brain. Messages from the cerebral cortex are coordinated and dispatched to the appropriate areas.

During sleep, many neurons of the reticular network stop functioning. Those that remain operational may inhibit response to unimportant stimuli (habituation) or cause hyperresponsiveness (sensitization). The cat who is accustomed to the sound of kitchen cabinets opening will sleep through a human's dinner being prepared (habituation) but will charge into the kitchen when the she hears the sound of the cat food container opening (sensitization).

Researcher E. N. Sokolov concluded that the "orientation response" (which can be equated with sensitization) and habituation are the result of the functioning of the reticular network. According to Sokolov, habituation results in the formation of models within the reticular activating system. Incoming messages that match the model are disregarded by the organism, but those that differ trigger alerting reactions throughout the body, thus justifying the term "alerting system" as a synonym for the reticular network. Habituation to a very strong stimulus would take a long time. Repetition of this strong stimulus would cause an even stronger defensive reflex and would require an even longer habituation period.

The Role of Neurotransmitters

Neurotransmitters are chemical messengers that enable nerve impulses to be carried across the synapse, the narrow gap between neurons. They transmit impulses from the presynaptic axon to the postsynaptic dendrite(s). E. R. Kandell, in experiments with *Aplysia* (the sea hare, a large mollusk), demonstrated that as a habituation training series continues, smaller amounts of the neurotransmitter acetylcholine are released from the axon of the presynaptic sensory neuron. On the other hand, after sensitization, this neuron released larger amounts of acetylcholine because of the presence of serotonin, a neurotransmitter secreted by a facilitory interneuron. When a sensitizing stimulus is very strong, it usually generates an impulse within the control center—a ganglion, a neuron, or the brain. The control center then transmits an impulse to a facilitory interneuron, causing the facilitory interneuron to secrete serotonin.

Increased levels of acetylcholine secretion by the sensory neuron result from two different stimuli: direct stimulation of the sensory neurons of the siphon or serotonin from the facilitory interneuron. Facilitory interneurons synapse with sensory neurons in the siphon. Serotonin discharged from facilitory interneurons causes the sensory neurons to produce and secrete more acetylcholine.

On the molecular level, the difference between habituation and adaptation—the failure of the sensory neuron to respond—is very evident. The habituated sensory neuron has a neurotransmitter in its axon but is unable to secrete it and thereby enable the impulse to be transmitted across the synapse. The adapted sensory neuron, by contrast, has exhausted its current supply of neurotransmitter. Until new molecules of neurotransmitter are synthesized within the sensory neuron, none is available for release.

In 1988, Emilie A. Marcus, Thomas G. Nolen, Catherine H. Rankin, and Thomas J. Carew published the multiprocess theory to explain dishabituation and sensitization in the sea hare, *Aplysia*. On the basis of their experiments using habituated sea hares that were subjected to different stimuli, they concluded that dishabituation and sensitization do not always occur together; further, they decided, there are three factors to be considered: dishabituation, sensitization, and inhibition.

Habituation Studies

Habituation studies have utilized a wide variety of approaches, ranging from the observation of intact organisms carrying out their normal activities in their natural surroundings to the laboratory observation of individual nerve cells. With different types of studies, very different aspects of habituation and sensitization can be investigated. Surveying the animal kingdom in 1930, G. Humphrey concluded that habituation-like behavior exists at all levels of life, from the simple one-celled protozoans to the multicelled, complex mammals.

E. N. Sokolov, a compatriot of Ivan P. Pavlov, used human subjects in the laboratory. In 1960, he reported on the results of his studies, which involved sensory integration, the makeup of the orientation reflex (which he credited Pavlov with introducing in 1910), a neuronal model and its role in the orientation reflex, and the way that this neuronal model could be used to explain the conditioned reflex. Sokolov measured changes in the diameter of blood vessels in the head and finger, changes in electrical waves within the brain, and changes in electrical conductivity of the skin. By lowering the intensity of a tone to which human subjects had been habituated, Sokolov demonstrated that habituation was not the result of fatigue, because subjects responded to the lower-intensity tone with the startle or orientation reflex just as they would when a new stimulus was introduced. Sokolov concluded that the orientation response (which is related to sensitization) and habituation are the result of the functioning of the reticular network of the brain and central nervous system. Sokolov emphasized that the orientation response was produced after only the first few exposures to a particular stimulus, and it increased the discrimination ability of internal organizers. The orientation response was an alerting command. Heat, cold, electric shock, and sound were the major stimuli that he used in these studies.

E. R. Kandell used the sea hare, *Aplysia*, in his habituation-sensitization studies. *Aplysia* is a large sluglike mollusk, with a sheetlike, shell-producing body covering, the mantle. *Aplysia* has a relatively simple nervous system and an easily visible gill-withdrawal reflex. (The gill is withdrawn into the mantle shelf.) Early habituation-sensitization experiments dealt with withdrawal or absence of gill withdrawal. Later experiments measured electrical changes that occurred within the nerve cells that controlled gill movement. These were followed by studies which demonstrated that the gap (synapse) between the receptor nerve cell (sensory neuron) and the muscle-moving nerve cell (motor neuron) was the site where habituation and dishabituation occurred and that neurohormones such as acetylcholine and serotonin played essential roles in these processes. Kandell called the synapse the "seat of learning."

Charles Sherrington used spinal animals in which the connection between the brain and the spinal nerve cord had been severed. Sherrington demonstrated that habituation-sensitization could occur within the spinal nerve cord even without the participation of the brain. Pharmaceuticals have also been used in habituation-sensitization studies. Michael Davis and Sandra File used neurotransmitters such as serotonin and norepinephrine to study modification of the startle (orientation) response.

Habituation studies conducted in the laboratory enable researchers to control variables such as genetic makeup, previous experiences, diet, and the positioning of subject and stimulus; however, they lack many of the background stimuli present in the field. In her field studies of the chimpanzees of the Gombe, Jane Goodall used the principles of habituation to decrease the distance between herself and the wild champanzees until she was able to come close enough to touch and be accepted by them. The field-experimental approach capitalizes on the best of both laboratory and field techniques. In this approach, a representative group of organisms that are in their natural state and habitat are subjected to specific, known stimuli.

Learning to Survive

Habituation is necessary for survival. Many stimuli are constantly impinging upon all living things; since it is biologically impossible to respond simultaneously to all of them, those which are important must be dealt with immediately. It may be a matter of life or death. Those which are unimportant or irrelevant must be ignored.

Cell physiologists and neurobiologists have studied the chemical and electrical changes that occur between one nerve cell and another and between nerve and muscle cells. The results of those studies have been useful in understanding and controlling these interactions as well as in providing insights for therapies. Psychologists utilize the fruits of habituation studies to understand and predict, modify, and control the behavior of intact organisms. For example, knowing that bulls serving as sperm donors habituate to one cow or model and stop discharging sperm into it, the animal psychologist can advise the semen collector to use a different cow or model or simply to move it to another place—even as close as a few yards away.

Conservationists and wildlife protectionists can apply the principles of habituation to wild animals, which must live in increasingly closer contact with one another and with humans, so that both animal and human populations can survive and thrive. For example, black-backed gulls, when establishing their nesting sites, are very territorial. Males which enter the territory of another male gull are rapidly and viciously attacked. After territorial boundaries are established, however, the males in contiguous territories soon exhibit "friendly enemy" behavior: They are tolerant of the proximity of other males that remain within their territorial boundaries. This has been observed in other birds as well as in fighting fish.

Walter Lener

See also: Altruism; Communication; Defense mechanisms; Displays; Ethology; Herbivores; Hierarchies; Insect societies; Isolating mechanisms; Mammalian social systems; Migration; Mimicry; Omnivores; Pheromones; Poisonous animals; Predation; Reproductive strategies; Territoriality and aggression.

Sources for Further Study

Alcock, John. *Animal Behavior: An Evolutionary Approach*. 6th ed. Sunderland, Mass.: Sinauer Associates, 1998.

Alkon, Daniel L. "Learning in a Marine Snail." *Scientific American* 249 (July, 1983): 70-84.

Barash, David P. *Sociobiology and Behavior*. New York: Elsevier, 1982.

Drickamer, Lee C., Stephen H. Vessey, and Doug Meikle. *Animal Behavior: Mechanisms, Ecology, Evolution*. 4th ed. Dubuque, Iowa: Wm. C. Brown, 1996.

Eckert, Roger, and David Randall. *Animal Physiology*. 4th ed. San Francisco: W. H. Freeman, 1997.

Gould, James L. *Ethology: The Mechanisms and Evolution of Behavior*. New York: W. W. Norton, 1980.

Gould, James L., and Peter Marler. "Learning by Instinct." *Scientific American* 256 (January, 1987): 74-85.

Halliday, Tim, ed. *Animal Behavior*. Norman: University of Oklahoma Press, 1994.

Klopfer, P. H., and J. P. Hailman. *An Introduction to Animal Behavior: Ethology's First Century*. Englewood Cliffs, N.J.: Prentice-Hall, 1967.

Slater, P. J. B., and T. R. Halliday, eds. *Behavior and Evolution*. New York: Cambridge University Press, 1994.

HERBIVORES

Types of ecology: Behavioral ecology; Ecoenergetics

Herbivores, animals which eat only plants, include insects and other arthropods, fish, birds, and mammals. They keep plants from overgrowing and are food for carnivores or omnivores.

Herbivores are animals whose diets consist entirely of plants. They occupy one of the major trophic levels and have two ecological functions. First, they eat plants and keep them from overgrowing. Second, they are food for carnivores, which subsist almost entirely upon their flesh, and omnivores, which eat both plants and animals. Herbivores live on land or in oceans, lakes, and rivers. They can be insects, other arthropods, fish, birds, or mammals.

Insect Herbivores

Insects are the largest animal class, with approximately one million species. Fossils show their emergence 400 million years ago. Insects occur worldwide, from pole to pole, on land and in fresh or salt water. They are the best developed invertebrates, except for some mollusks. They mature by metamorphosis, passing through at least two dissimilar stages before adulthood. Metamorphosis can take up to twenty years or may be complete a week after an egg is laid.

Many insects are herbivores. Some feed on many different plants, others depend on one plant variety or a specific plant portion, such as leaves or stems. Relationships between insects and the plants they eat are frequently necessary for plant growth and reproduction. Among the insect herbivores are grasshoppers and social insects such as bees.

Artiodactyls

Artiodactyls are another type of herbivore—hoofed mammals, including cattle, pigs, goats, giraffes, deer, antelope, and hippopotamuses. Artiodactyls walk on two toes. Their ancestors had five, but evolution removed the first toe and the second and fifth toes are vestigial. Each support toe—the third and fourth—ends in a hoof. The hippopotamus, unique among artiodactyls, stands on four toes of equal size and width.

What makes artiodactyls herbivorous is the fact that they lack upper incisor and canine teeth, whereas pads in their upper jaws help the lower teeth grind food. Domesticated artiodactyls include bovids such as cattle, sheep, and goats. Many artiodactyls, such as antelope, cattle, deer, goats, and giraffes, are ruminants. They chew and swallow vegetation, which enters the stomach for partial digestion, is regurgitated, chewed again, and reenters the stomach for more digestion. This maximizes nutrient intake from food. Members of the deer family, which include approximately 40 species from the seven-foot-tall moose to the one-foot-tall pudu, are hoofed ruminants that inhabit many continents and biomes: Asia, Europe, the Americas, and North Africa; woods, prairies, swamps, mountains, and tundra. These animals eat the twigs, leaves, bark, and buds of bushes and saplings, and grasses.

Most antelope, a group of approximately 150 ruminant species, are African, although some are European or Asian. They live on plains, marshes, deserts, and forests, eating grass, twigs, buds, leaves, and bark. In Asia, Siberian saigas and goat antelope (takin) inhabit mountain ranges. Chamois goat antelope live in Europe's Alps.There are no true antelope in North America, where their closest relatives are pronghorns and Rocky Mountain goats (goat-antelope with both goat and antelope anatomic features). The smallest antelope, the dik-dik, is rabbit-sized. Elands, the largest antelope, are ox-sized.

Giraffes and hippos are artiodactyls that inhabit the dry, tree-scattered land south of the Sahara in Africa. Giraffes rarely graze and can go for

Deer are herbivores, eating grasses and the tender buds, shoots, bark, and twigs of trees. The boundary between forest and field offers them the widest range of food choices. (PhotoDisc)

months without drinking, getting most of their water from the leaves they eat, because it is difficult for them to reach the ground or the surface of a river with their mouths. By contrast, the short-legged, stocky hippos are semiaquatic, spending most daylight hours nearly submerged eating aquatic plants. At night they eat land plants.

Aquatic Herbivores
Fish are aquatic vertebrates, having gills, scales, and fins. They include rays, lampreys, sharks, lungfish, and bony fish. The earliest vertebrates, 500 million years ago, were fish. They comprise more than 50 percent of all vertebrates and have several propulsive fins: dorsal fins along the central back; caudal fins at tail ends; and paired pectoral and pelvic fins on sides and belly. Fish inhabit lakes, oceans, and rivers, even in Arctic and Antarctic areas. Most marine fish are tropical. The greatest diversity of freshwater species is found in African and rain forest streams.

Ecological Importance
It is clear that wild herbivores are ecologically important to food chains. This is because they eat plants, preventing their overgrowth, and they are eaten by carnivores and omnivores. Domesticated herbivores—cattle, sheep and goats, used for human sustenance—account for three to four billion living creatures. Future production of better strains of domesticated herbivores via recombinant deoxyribonucleic acid (DNA) research may cut the numbers of such animals killed to meet human needs. Appropriate species conservation should maintain the present balance of nature and sustain the number of wild herbivore species living on earth.

Sanford S. Singer

See also: Balance of nature; Food chains and webs; Omnivores; Predation; Trophic levels and ecological niches.

Sources for Further Study
Gerlach, Duane, Sally Atwater, and Judith Schnell. *Deer*. Mechanicsburg, Pa.: Stackpole Books, 1994.
Gullan, P. J., and P. S. Cranston. *Insects: An Outline of Entomology*. 2d ed. Malden, Mass.: Blackwell Science, 2000.
Olsen, Sandra L., ed. *Horses Through Time*. Boulder, Colo.: Roberts Rinehart, 1996.
Rath, Sara. *The Complete Cow*. Stillwater, Minn.: Voyageur Press, 1990.
Shoshani, Jeheskel, and Frank Knight. *Elephants: Majestic Creatures of the Wild*. Rev. ed. New York: Checkmark Books, 2000.

HIERARCHIES

Type of ecology: Behavioral ecology

Hierarchies, systems of establishing dominance and subordination, are important in maintaining social order in many species of animals.

All animal species strive for their share of fitness. In this struggle for reproductive success, individuals that make up a population often compete for essential resources such as food, mates, or nesting sites. In many species, competition over resources may lead to fighting among individuals. Fighting, however, can be costly to the individuals involved. The loser may suffer real injury or even death, and the winner has to expend energy and still may suffer an injury. In order to prevent constant fighting over resources, many animal species have adopted a system of what sociobiologists call a dominance hierarchy or social hierarchy. The dominance hierarchy is a set of aggression-submission relationships among the animals of a population. With an established system of dominance, the subordinate individuals will acquiesce rather than compete with the dominant individuals for resources.

Dominance and Subordinance

To be dominant is to have the priority of access to the essential resources of life and reproduction. In almost all cases, the superior dominant animals will displace the subordinates from food, mates, and nest sites. In the matter of obtaining food, for example, wood pigeons are flock feeders. The dominant pigeons are always found near the center of the flock when feeding and feed more quickly than the subordinate birds at the edge of the flock. The birds at the edge of the flock accumulate less food and often obtain just enough to sustain them through the night. Among sheep and reindeer, the lowest-ranking females are also the worst-fed animals and among the poorest of mothers. Baby pigs compete for teat position on the mother and once established will maintain that position until weaning. Those piglets that gain access to the most anterior teats will weigh more at weaning than those who have to settle for posterior teat positions. In gaining access to mates, one study with laboratory mice has shown that while the dominant males constituted only one third of the male population, they sired 92 percent of the offspring.

Life is still not all that hopeless for the subordinates. Oftentimes the loser in the battle for dominance is given a second chance, and in some of the more social species, the subordinate only has to await its turn to rise in the hierarchy. In some species, cooperation among subordinate groups, especially kin groups, can lead to the formation of a new colony and a new opportunity to establish dominance. In other species, it may well be advantageous for the subordinate to stay with the group. For example, individual baboons and macaques will not survive very long if they are away from the group's sleeping area, and they will have no opportunity to reproduce. It has been shown that even a low-ranking male eats well if he is part of a troop, and he may occasionally have the opportunity to mate. In addition, the dominant male will eventually lose prowess, and the subordinate will have a chance to move up in the dominance hierarchy.

Types of Hierarchies

The simplest possible type of hierarchy is a despotism, in which one individual rules over all other members of the group and no rank distinctions are made among the subordinates. Hierarchies more frequently contain multiple ranks in a more or less linear fashion. An alpha individual dominates all others, a beta individual is subordinate to the alpha but dominates all others, and so on down to the omega individual at the bottom, who is dominated by all of the others. Sometimes, the network is complicated by triangular or other circular relationships in which two or three individuals might be at the same dominance level. Such relationships appear to be less stable than despotisms or linear orders.

Nested hierarchies are often observed in some animal species. Societies that are divided into groups can display dominance both within and between the various components. For example, white-fronted geese establish a rank order of several subgroups including parents, mated pairs without young, and free juveniles. These hierarchies are superimposed over the hierarchy within each of the subgroups. In wild turkeys, brothers establish a rank order among their brotherhood, but each brotherhood competes for dominance with other brotherhoods on the display grounds prior to mating.

Formation and Maintenance of Hierarchies

Hierarchies are formed during the initial encounters between animals through repeated threats and fighting, but once the issue of dominance has been determined, each individual gives way to its superiors with little or no hostile exchange. Life in the group may eventually become so peaceful that the existence of ranking is hidden from the observer until some crisis occurs to force a confrontation. For example, a troop of baboons can go for hours without engaging in sufficient hostile exchanges to reveal their ranking, but in a moment of crisis such as a quarrel over food the hierarchy will suddenly be evident. Some species are organized in absolute dominance hierarchies, in which the rank orders remain constant regardless of the circumstances. Status within an absolute dominance hierarchy changes only when individuals move up or down in rank through additional interaction with their rivals. Other animal societies are arranged in relative dominance hierarchies. In these arrangements, such as with crowded domestic house cats, even the highest-ranking individuals acquiesce to subordinates when the latter approach a point that would normally be too close to their personal sleeping space.

The stable, peaceful hierarchy is often supported by status signs. In other words, the mere actions of the dominant individual advertise his dominance to the other individuals. The leading male in a wolf pack can control his subordinates without a display of excessive hostility in the great majority of cases. He advertises his dominance by the way he holds his head, ears, and tail, and the confident face-forward manner in which he approaches other members of his pack. In a similar manner, the dominant rhesus monkey advertises his status by an elaborate posture which includes elevated head and tail, lowered testicles, and slow, deliberate body movements accompanied by an unhesitating but measured scrutiny of other monkeys he encounters. Animals use not only visual signals to advertise dominance but also acoustic and chemical signals. For example, dominant European rabbits use a mandibular secretion to mark their territory.

Uses of Hierarchies

A stable dominance hierarchy presents a potentially effective united front against strangers. Since a stranger represents a threat to the status of each individual in the group, he is treated as an outsider. When expelling an intruder, cooperation among individuals within the group reaches a maximum. Chicken producers have long been aware of this phenomenon. If a new bird is introduced to the flock, it will be subjected to attacks for many days and be forced down to the lowest status unless it is exceptionally vigorous. Most often, it will simply die with very little show of fighting back. An intruder among a flock of Canada geese will be met with the full range of threat displays and repeated mass approaches and retreats.

In some primate societies, the dominant animals use their status to stop fighting among subordinates. This behavior has been observed in rhesus and pig-tailed macaques and in spider monkeys. This behavior has been observed even in animal societies, such as squirrel monkeys, that do not exhibit dominance behavior. Because of the power of the dominant individual, relative peace is observed in animal societies organized by despotisms, such as hornets, paper wasps, bumblebees, and crowded territorial fish and lizards. Fighting increases significantly among the equally ranked subordinates as they vie for the dominant position when the dominant animal is removed.

Young males are routinely excluded from the group in a wide range of aggressively organized mammalian societies such as baboons, langur monkeys, macaques, elephant seals, and harem-keeping ungulates. At best, these young males are tolerated around the fringes of the group, but many are forced out of the group and either join bachelor herds or wander as solitary nomads. As would be expected, these young males are the most aggressive and troublesome members of the society. They compete with one another for dominance within their group and often unite into separate bands that work together to reduce the power of the dominant males. Males in the two groups show different behaviors. Among the Japanese macaques, the dominant males stay calm and aloof when introduced to a new object so as to not risk loss of their status, but the females and young males will explore new areas and examine new objects.

D. R. Gossett

See also: Altruism; Communication; Defense mechanisms; Displays; Ethology; Habituation and sensitization; Herbivores; Insect societies; Isolating mechanisms; Mammalian social systems; Migration; Mimicry; Omnivores; Pheromones; Poisonous animals; Predation; Reproductive strategies; Territoriality and aggression.

Sources for Further Study

Barash, David P. *Sociobiology and Behavior*. 2d ed. New York: Elsevier, 1982.

Campbell, Neil A., Lawrence G. Mitchell, and Jane B. Reece. *Biology: Concepts and Connections*. 3d ed. San Francisco: Benjamin/Cummings, 2000.

Feldhamer, G. A., L. C. Drickamer, S. H. Vessey, and J. F. Merritt. *Mammalogy*. Boston: WCB/McGraw-Hill, 1999.

Ridley, Mark. *Animal Behavior: An Introduction to Behavioral Mechanisms, Development, and Ecology*. 2d ed. Boston: Blackwell Scientific Publications, 1995.

Wilson, Edward O. *Sociobiology: The Abridged Edition*. Cambridge, Mass.: Belknap Press of Harvard University Press, 1980.

Wittenberger, James F. *Animal Social Behavior*. Boston: Duxbury Press, 1981.

HUMAN POPULATION GROWTH

Type of ecology: Population ecology

Since the Industrial Revolution of the nineteenth century, human populations have experienced a period of explosive growth. Overpopulation now poses a real threat to plant and animal life, ecosystems, and the long-term sustainability of the earth's current ecological balance.

Just eleven thousand years ago, only about five million humans lived on the earth. The initial population growth was slow, largely because of the way humans lived—by hunting and gathering. Such a mobile lifestyle limited the size of families for practical reasons. When simple means of birth control, often abstention from sex, failed, a woman would elect abortion or, more commonly, infanticide to limit her family size. Furthermore, a high mortality rate among the very young, the old, the ill, and the disabled acted as a natural barrier to rapid population growth.

Agricultural Revolution

It took more than two million years—from the earliest animal considered to be human, *Homo habilis*—or about 100,000 years from the time modern human beings, *Homo sapiens sapiens*, migrated out of Africa into the rest of the world, for the world's population to reach one billion. The second billion was added in about one hundred years, the third billion in fifty years, the fourth in fifteen years, the fifth billion in twelve years. By the close of the twentieth century, the world's population was exceeding six billion.

This explosion had become possible with the development of agriculture. A hunting-gathering lifestyle requires a nomadic existence over a large range of territory, which makes the establishment of infrastructures, such as permanent housing and long-range food stores, impractical. Agricultural societies, by contrast, can support more people in a limited area and, because settlements are permanent, can build infrastructures over time and therefore minimize efforts directed to basic subsistence, such as the erecting of shelters. Moreover, when humans became sedentary, some limits on family size were lifted. With the development of agriculture, children became an asset to their families by helping with farming and other chores.

Starting about eleven thousand years ago (5 million people), humans began to cultivate such plants as barley, lentils, wheat, and peas in the

Middle East—an area that today extends from Lebanon and Syria in the northwest eastward through Iraq to Iran. In doing so, human beings began to have a profound ecological impact as well. In cultivating and caring for these crops, early farmers changed the characteristics of these plants, making them higher yielding, more nutritious, and easier to harvest. Agriculture spread and first reached Europe approximately six thousand years ago. By the beginning of the common era (1 C.E.), human population had grown to about 130 million, distributed all over the earth.

Agriculture might also have originated independently in Africa in one or more centers. Many crops were domesticated there, including yams, okra, coffee, and cotton. In Asia, agriculture based on staples such as rice and soybeans and many other crops such as citrus, mangos, taro, and bananas was developed. Agriculture was developed independently in the New World. It began as early as nine thousand years ago in Mexico and Peru. Christopher Columbus and his followers found many new crops to bring back to the Old World, including corn, kidney beans, lima beans, tomatoes, tobacco, chili peppers, potatoes, sweet potatoes, pumpkins and squashes, avocados, cacao, and the major cultivated species of cotton.

Ecological Impact

For the last five to six centuries, important staple crops have been cultivated throughout the world. Wheat, rice, and corn, which provide 60 percent of the calories people consume, are cultivated wherever they will grow. Other crops, including spices and herbs, have also been brought under cultivation. One of the results of the agricultural developments—particularly pronounced since the Green Revolution of high-yield crops in the mid-twentieth century—has been a tendency toward "monoculture," or the reduction of diversity, in crop plants worldwide.

The growing population has also changed the landscape, distribution, and diversity of plants dramatically. Clear-cutting and deforestation have driven many species (both plant and animal) to extinction. Relatively little has been done to develop agricultural practices suitable for tropical regions. As a result, the tropics are being devastated ecologically, with an estimated 20 percent of the world's species likely to be lost by the mid-twenty-first century.

Industrial Revolution

By 1650, the world population had reached 500 million. The process of industrialization had begun, bringing about profound changes in the lives of humans and their interactions with the natural world. With improved living standards, lower death rates, and prolonged life expectancies, human

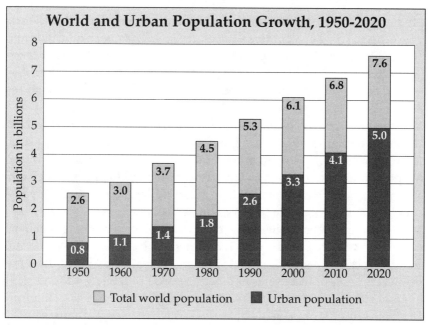

World and Urban Population Growth, 1950-2020

Sources: Data are from U.S. Bureau of the Census International Data Base and John Clarke, "Population and the Environment: Complex Interrelationships," in *Population and the Environment* (Oxford, England: Oxford University Press, 1995), edited by Bryan Cartledge.

Note: The world's population passed 6 billion in the year 2000.

population grew exponentially. By 1999, there were about 6 billion people, compared with 2.5 billion in 1950. By 2002, the world population was well on its way to 7 billion, with an annual growth rate of nearly 100 million. The processes of industrialization and exponential population growth combined to multiply the ecological effects of human activity on the biosphere. Human use of fossil fuels, fertilizers, pesticides, global trading pracitices, agribusiness practices—all these large-scale activities began to exhibit large-scale effects on both the abiotic and biotic components of ecosystems everywhere.

Threats to Sustainability

Without effective measures of control, the human population could exceed the earth's carrying capacity. Humans are, at present, estimated to consume about 40 percent of the total net products generated via photosynthesis by plants. Human activities have reduced the productivity of earth's forests and grasslands by 12 percent. Each year, millions of acres of once-productive land are turned into desert through overgrazing and deforestation, especially in developing countries. As a result of overfertilization and aggressive practices in agriculture, topsoil is lost at an annual rate of 24 billion metric tons. Collectively, these practices caused the destruction of 40 million acres of rain forest each year during the 1960's and 1970's and the extinction of enormous numbers of species.

Through technological innovation and aggressive practices in agriculture, a 2.6-fold increase in world grain production has been achieved since 1950. However, this increase in food output is not nearly enough to feed the population. Based upon an estimate by the World Bank and the Food and Agriculture Organization of the United Nations, one out of every five people is living in absolute poverty, unable to obtain food, shelter, or clothing dependably. About one out of every ten people receives less than 80 percent of the daily caloric intake recommended by the United Nations. In countries such as Bangladesh and Haiti and in regions as East Africa, humans are dying in increasing numbers because of the lack of food. This food shortage may stem from drought, soil depletion, or soil loss; more often, famine results from inequitable distribution of resources among populations. Situations exacerbated by a growing population also pose threats

By the end of the twentieth century, the world's human population had passed the 6 billion mark, and most of that population lived in or near large urban centers typified by this street in Barcelona, Spain: with high-rises and other structures designed for high-density living and working accommodations. (PhotoDisc)

to the environment, aggravating the problems of acid rains, toxic and hazardous wastes, water shortages, topsoil erosion, ozone layer punctuation, greenhouse effects, and groundwater contamination.

Ming Y. Zheng

See also: Acid deposition; Biological invasions; Biomagnification; Biopesticides; Deforestation; Erosion and erosion control; Eutrophication; Genetically modified foods; Grazing and overgrazing; Integrated pest management; Invasive plants; Landscape ecology; Multiple-use approach; Ocean pollution and oil spills; Ozone depletion and ozone holes; Pesticides; Pollution effects; Population growth; Rangeland; Slash-and-burn agriculture; Soil; Soil contamination; Urban and suburban wildlife; Waste management; Wildlife management.

Sources for Further Study

Brown, L. *State of the World 2000*. New York: W. W. Norton, 2000.

Heiser, C. B., Jr. *Seed to Civilization: The Story of Food*. 2d ed. San Francisco: W. H. Freeman, 1981.

National Research Council. *Lost Crops of the Incas: Little-Known Plants of the Andes with Promise for Worldwide Cultivation*. Washington, D.C.: National Academy Press, 1990.

Weiner, J. *The Next One Hundred Years: Shaping the Future of Our Living Earth*. New York: Bantam Books, 1990.

HYDROLOGIC CYCLE

Types of ecology: Ecoenergetics; Ecosystem ecology; Global ecology

The hydrologic cycle, one of the most important geochemical cycles with both short- and long-range impacts on the biosphere, is a continuous system through which water circulates through vegetation, in the atmosphere, in the ground, on land, and in surface water such as rivers and oceans. The sun and earth's gravity provide the energy that drives the hydrologic cycle.

The total amount of water on earth is an estimated 1.36 billion cubic kilometers. Of this water, 97.2 percent is found in the earth's oceans. The ice caps and glaciers contain 2.15 percent of the earth's water. The remainder, 0.65 percent, is divided among rivers (0.0001 percent), freshwater and saline lakes (0.017 percent), groundwater (0.61 percent), soil moisture (0.005 percent), the atmosphere (0.001 percent), and the biosphere and groundwater below 4,000 meters (0.0169 percent). While the percentages of water appear to be small for these water reservoirs, the total volume of water contained in each is immense.

Evaporation

Evaporation is the process whereby a liquid or solid is changed to a gas. Heat causes water molecules to become increasingly energized and to move more rapidly, weakening the chemical force that binds them together. Eventually, as the temperature increases, water molecules move from the ocean's surface into the overlying air. The rate of evaporation is influenced by radiation, temperature, humidity, and wind velocity.

Each year, about 320,000 cubic kilometers of water evaporate from oceans. It is estimated that an additional 60,000 cubic kilometers of water evaporate from rivers, streams, and lakes or are transpired by plants each year. A total of about 380,000 cubic kilometers of water is evapotranspired from the earth's surface every year.

Condensation and Precipitation

Wind may transport the moisture-laden air long distances. The amount of water vapor the air can hold depends upon the temperature: The higher the temperature, the more vapor the air can hold. As air is lifted and cooled at higher altitudes, the vapor in it condenses to form droplets of water. Condensation is aided by small dust and other particles in the atmosphere. As droplets collide and coalesce, raindrops begin to form, and precipitation begins. Most precipitation events are the result of three causal factors: frontal precipitation, or the lifting of an air mass over a moving weather front; convectional precipitation related to the uneven heating of the earth's surface, causing warm air masses to rise and cool; and orographic precipitation, resulting from a moving air mass being forced to move upward over a mountain range, cooling the air as it rises.

Each year, about 284,000 cubic kilometers of precipitation fall on the world's oceans. This water has completed its cycle and is ready to begin a new cycle. Approximately 96,000 cubic kilometers of precipitation fall upon the land surface each year. This precipitation follows a number of different pathways in the hydrologic cycle. It is estimated that 60,000 cubic kilometers evaporate from the surface of lakes or streams or transpire directly back into the atmosphere. The remainder, about 36,000 cubic kilometers, is intercepted by human structures or vegetation, infiltrates the soil or bedrock, or becomes surface runoff.

Interception
In cities, the amount of water intercepted by human structures may approach 100 percent. However, much urban water is collected in storm sewers or drains that lead to a surface drainage system or is spread over the land surface to infiltrate the subsoil. Interception loss from vegetation depends upon interception capacity (the ability of the vegetation to collect and retain falling precipitation), wind speed (the higher the wind speed, the greater the rate of evaporation), and rainfall duration (the interception loss will decrease with the duration of rainfall, as the vegetative canopy will become saturated with water after a period of time). Broad-leaf forests may intercept 15 to 25 percent of annual precipitation, and a bluegrass lawn may intercept 15 to 20 percent of precipitation during a growing season.

Transpiration
Plants are continuously extracting soil moisture and passing it into the atmosphere through a process called transpiration. Moisture is drawn into the plant rootlet through osmotic pressure. The water moves through the plant to the leaves, where it is passed into the atmosphere through the leaf openings, or stomata. The plant uses less than 1 percent of the soil moisture in its metabolism; thus, transpiration is responsible for most water vapor loss from the land in the hydrologic cycle. For example, an oak tree may transpire 151,200 liters per year.

Overland Flow and Infiltration
When the amount of rainfall is greater than the earth's ability to absorb it,

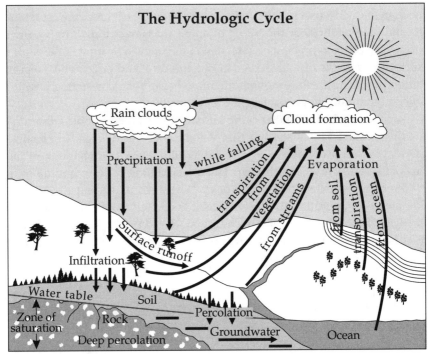

The Hydrologic Cycle

Source: U.S. Department of Agriculture, *Yearbook of Agriculture* (Washington, D.C.: Government Printing Office, 1955).

excess water begins to run off, a process termed overland flow. Overland flow begins only if the precipitation rate exceeds the infiltration capacity of the soil. Infiltration occurs when water sinks into the soil surface or into fractures of rocks; the amount varies according to the characteristics of the soil or rock and the nature of the vegetative cover. Sandy soils have higher infiltration rates than clay rock soils. Nonporous rock has an infiltration rate of zero, and all precipitation that reaches it becomes runoff. The presence of vegetation impedes surface runoff and increases the potential for infiltration to occur.

Water infiltrating the soil or bedrock encounters two forces: capillary force and gravitational force. A capillary force is the tendency of the water in the subsurface to adhere to the surface of soil or sediment particles. Capillary forces are responsible for the soil moisture a few inches below the land surface.

The water that continues to move downward under the force of gravity through the pores, cracks, and fissures of rocks or sediments will eventually enter a zone of water saturation. This source of underground water is called an aquifer—a rock or soil layer that is porous and permeable enough to hold and transport water. The top of this aquifer, or saturated zone, is the water table. This water is moving slowly toward a point where it is discharged to a lake, spring, or stream. Groundwater that augments the flow of a stream is called base flow. Base flow enables streams to continue to flow during droughts and winter months. Groundwater may flow directly into the oceans along coastlines.

When the infiltration capacity of the earth's surface is exceeded, overland flow begins. Broad, thin sheets of water a few millimeters thick are called sheet flow. After flowing a few meters, the sheets break up into threads of current that flow in tiny channels called rills. The rills coalesce into gullies and, finally, into streams and rivers. Some evaporation losses occur from the stream surface, but much of the water is returned to the oceans, thus completing the hydrologic cycle.

Residence Time
Residence time refers to how long a molecule of water will remain in various components of the hydrologic cycle. The average length of time that a water molecule stays in the atmosphere is about one week. Two weeks is the average residence time for a water molecule in a river, and ten years in a lake. It would take four thousand years for all the water molecules in the oceans to be recycled. Groundwater may require anywhere from a few weeks to thousands of years to move through the cycle. This time period suggests that every water molecule has been recycled millions of times.

Methods of Study
Several techniques are used to gather data on water in the hydrologic cycle. These data help scientists determine the water budget for different geographic areas. Together, these data enable scientists to estimate the total water budget of the earth's hydrologic cycle.

Scientists have developed a vast array of mathematical equations and instruments to collect data on the hydrologic cycle. Variations in temperature, precipitation, evapotranspiration, solar radiation, vegetative cover, soil and bedrock type, and other factors must be evaluated to understand the local or global hydrologic cycle.

Precipitation is an extremely variable phenomenon. The United States has some thirteen thousand precipitation stations equipped with rain gauges, placed strategically to compensate for wind and splash losses. Precipitation falling on a given area is determined using a rain-gauge network of uniform density to determine the arithmetic mean for rainfall in the area. The amount of water in a snowpack is estimated by snow surveys.

The depth and water content of the snowpack are measured and the extent of the snow cover mapped using satellite photography.

The amount of precipitation lost by interception can be measured and evaluated. Most often, interception is determined by measuring the amount above the vegetative canopy and at the earth's surface. The difference is what is lost to interception.

The volume of water flowing by a given point at a given time in an open stream channel is called discharge. Discharge is determined by measuring the velocity of water in the stream channel, using a current meter. The cross-sectional area of the stream channel is determined at a specific point and multiplied by the stream velocity. Automated stream-gauging stations are located on most streams to supply data for various hydrologic investigations.

The U.S. National Weather Service maintains about five hundred stations using metal pans, mimicking reservoirs, to measure free-water evaporation. Water depths of 17 to 20 centimeters are maintained in the pans. Errors may result from splashing by raindrops or birds. Because the pans will heat and cool more rapidly than will a natural reservoir, a pan coefficient is employed to compensate for this phenomenon. The wind velocity is also determined. A lake evaporation nomograph determines daily lake evaporation. The mean daily temperature, wind velocity, solar radiation, and mean daily dew point are all used in the calculation.

The amount of evapotranspiration can be measured using a lysimeter, a large container holding soil and living plants. The lysimeter is set outside, and the initial soil moisture is determined. All precipitation or irrigation is measured accurately. Changes in the soil moisture storage determine the amount of evapotranspiration.

Samuel F. Huffman

See also: Balance of nature; Biomass related to energy; Food chains and webs; Geochemical cycles; Herbivores; Nutrient cycles; Omnivores; Phytoplankton; Rain forests and the atmosphere; Trophic levels and ecological niches.

Sources for Further Study
Berner, Elizabeth K., and Robert A. Berner. *Global Environment: Water, Air, and Geochemical Cycles.* Upper Saddle River, N.J.: Prentice Hall, 1996.

Moore, J. W. *Balancing the Needs of Water Use.* New York: Springer-Verlag, 1989.

Viessman, Warren, Jr., and Gary L. Lewis. *Introduction to Hydrology.* 4th ed. Glenview, Ill.: HarperCollins, 1995.

INSECT SOCIETIES

Types of ecology: Behavioral ecology; Population ecology

Ants, termites, and many kinds of bees and wasps live in complex groups known as insect societies. Studies of such societies have enriched scientific knowledge about some of the most successful species on earth and have provided insights into the biological basis of social behavior in other animals.

Many of the most robust, thriving species today owe their success in great part to benefits that they reap from living in organized groups or societies. Nowhere are the benefits of group living more clearly illustrated than among the social insects. Edward O. Wilson, one of the foremost authorities on insect societies, estimates that more than twelve thousand species of social insects exist in the world today. This number is equivalent to all the species of known birds and mammals combined. Although insect societies have reached their pinnacle in bees, wasps, ants, and termites, many insects show intermediate degrees of social organization—providing insights regarding the probable paths of the evolution of sociality.

Scientists estimate that eusociality has evolved at least twelve times: once in the *Isoptera*, or termites, and eleven separate times in the *Hymenoptera*, comprising ants, wasps, and bees. In addition, one group of aphids has been found to have a sterile soldier caste. Although the eusocial species represent diverse groups, they all show a high degree of social organization and possess numerous similarities, particularly with regard to division of labor, cooperative brood care, and communication among individuals.

Ants

The organization of a typical ant colony is representative, with minor modifications, of all insect societies. A newly mated queen, or reproductive female, will start a new ant colony. Alone, she digs the first nest chambers and lays the first batch of eggs. These give rise to grublike larvae, which are unable to care for themselves and must be nourished from the queen's own body reserves. When the larvae have reached full size, they undergo metamorphosis and emerge as the first generation of worker ants. These workers—all sterile females—take over all the colony maintenance duties, including foraging outside the nest for food, defending the nest, and cleaning and feeding both the new brood and the queen, which subsequently becomes essentially an egg-laying machine. For a number of generations, all eggs develop into workers and the colony grows. Often, several types of workers can be recognized. Besides the initial small workers, or minor workers, many ant species produce larger forms known as major workers, or soldiers. These are often highly modified, with large heads and jaws, well suited for defending the nest and foraging for large prey. Food may include small insects, sugary secretions of plants or sap-feeding insects, or other scavenged foods. After several years, when the colony is large enough, some of the eggs develop into larger larvae that will mature into new reproductive forms: queens and males. Males arise from unfertilized eggs, while new queens are produced in response to changes in larval nutrition and environmental factors. These sexual forms swarm out of the nest in a synchronized fashion to mate and found new colonies of their own.

Bees and Wasps

With minor modifications, the same pattern occurs in bees and wasps. Workers of both bees and wasps are also always sterile females, but they differ from ants in that they normally possess functional wings and lack a fully differentiated soldier caste. Wasps, like ants, are primarily predators and scavengers; bees, however, have specialized on pollen and plant nectar as foods, transforming the latter into honey that is fed to both nestmates and brood. The bias toward females reflects a feature of the biology of the *Hymenoptera* that is believed to underlie their tendency to form complex societies. All ants, wasps, and bees have an unusual form of sex determination in which fertilized eggs give rise to females and unfertilized eggs develop into males. This type of sex determination, known as haplodiploidy, generates an asymmetry in the degree of relatedness among nestmates. As a consequence, sisters are more closely related to their sisters than they are to their own offspring or their brothers. Scientists believe that this provided an evolutionary predisposition for workers to give up their own personal reproduction in order to raise sisters—a form of natural selection known as kin selection.

Termites

The termites, or *Isoptera*, differ from the social *Hymenoptera* in a number of ways. They derive from a much more primitive group of insects and have been described as little more than "social cockroaches." Instead of the strong female bias characteristic of the ants, bees, and wasps, termites have regular sex determination; thus, workers have a fifty-fifty sex ratio. Additionally, termite development lacks complete metamorphosis. Rather, the young termites resemble adults in form from their earliest stages. As a consequence of these differences, immature forms can function as workers from an early age, and—at least among the lower termites—they regularly do so.

Termites also differ from *Hymenoptera* in their major mode of feeding. Instead of feeding on insects or flowers, all termites feed on plant material rich in cellulose. Cellulose is a structural carbohydrate held together by chemical bonds that most animals lack enzymes to digest. Termites have formed intimate evolutionary relationships with specialized microorganisms—predominantly flagellate protozoans and some spirochete bacteria—that have the enzymes necessary to degrade cellulose and release its food energy. The microorganisms live in the gut of the termite. Because these symbionts are lost with each molt, immature termites are dependent upon gaining new ones from their nestmates. They do this by feeding on fluids excreted or regurgitated by other individuals, a process known as trophallaxis. This essential exchange of materials also includes, along with food, certain nonfood substances known as pheromones.

Pheromones, by definition, are chemicals produced by one individual of a species that affect the behavior or development of other individuals of the same species that come in contact with them. Pheromones are well documented throughout the insect world, and they play a key role in communication between members of nonsocial or subsocial species. Moth mating attractants provide a well-studied example. Pheromones are nowhere better developed than among the social insects. They not only appear to influence caste development in the *Hymenoptera* and termites but also permit immediate communication among individuals. Among workers of the fire ant (*Solenopsis saevissima*), chemical signals have been implicated in controlling recognition of nestmates, grooming, clustering, digging, feeding, attraction or formation of aggregations, trail following, and alarm behavior. Nearly a dozen different glands have been identified which produce some chemical in the *Hymenoptera*, although the exact function of many of these chemicals remains unknown.

In addition to chemical communication, social insects may share information in at least three other ways: by tactile contact, such as stroking or grasping; by producing sounds, including buzzing of wings; and by employing visual cues. Through combinations of these senses, individuals can communicate complex information to nestmates. Indeed, social insects epitomize the development of nonhuman language. One such language, the "dance" language of bees, which was unraveled by Karl von Frisch and his students, provides one of the best-studied examples of animal behavior. In the waggle dance, a returning forager communicates the location of a food resource by dancing on the comb in the midst of its nestmates. It can accurately indicate the direction of the flower patch by incorporating the relative angles between the sun, the hive, and the food. Information about distance, or more precisely the energy expended to reach the food source, is communicated in the length of the run. Workers following the dance are able to leave the nest and fly directly to the food source, for distances in excess of one thousand meters.

Benefits of Cooperation

Living in cooperative groups has provided social insects with opportunities not available to their solitary counterparts. Not only can more individuals cooperate in performing a given task, but also several quite different tasks may be carried out simultaneously. The benefits from such cooperation are considerable. For example, group foraging allows social insects to increase the range of foods they can exploit. By acting as a unit, species such as army ants can capture large insects and even fledgling birds.

A second benefit of group living is in nest building. Shelter is a primary need for all animals. Most solitary species use naturally occurring shelters or, at best, build simple nests. By cooperating and sharing the effort, social insects are able to build nests that are quite elaborate, containing several kinds of chambers. Wasps and bees build combs, or rows of special cells, for rearing brood and storing food. Subterranean termites can construct mounds more than six meters high, while others build intricate covered nests in trees. Mound-building ants may cover their nests with a thatch that resembles, in both form and function, the thatched roofs of old European dwellings. Colonial nesting provides two additional benefits. First, it enhances defense. By literally putting all of their eggs in one basket, social insects can centralize and share the guard duties. The effectiveness of this approach is attested by one's hesitation to stir up a hornet's nest.

Nest construction also provides the potential to maintain homeostasis, the ability to regulate the environment within a desirable range. Virtually all living creatures maintain homeostasis within their bodies, but very few animals have evolved the ability to maintain a constant external living environment. In this respect, insect societies are similar to human societies. Workers adjust their activities to maintain the living environment within optimal limits. Bees, for example, can closely regulate the internal temperature of a hive. When temperatures fall below 18 degrees Celsius, they begin to cluster together, forming a warm cover of living bees to protect the vulnerable brood stages. To cool the hive in hot weather, workers initially circulate air by beating their wings. If further cooling is needed, they resort to evaporative cooling by regurgitating water throughout the nest. This

Social insects such as bees, wasps, termites, and ants are known for combining their efforts to build protective structures that allow them to live and reproduce in relative safety. This mound, for example, houses thousands of ants.
(Corbis)

water evaporates with wing fanning and serves to cool the entire hive. Other social insects rely on different but equally effective methods. Some ants, and especially termites, build their nests as mounds in the ground, with different temperatures existing at different depths. The mound nests of the African termite, *Macrotermes natalensis,* are an impressive engineering feat. They are designed to regulate both temperature and air flow through complex passages and chambers, with the mound itself serving as a sophisticated cooling tower.

Finally, group living allows the coordination of the efforts of individuals to accomplish complex tasks normally restricted to the higher vertebrates. The similarities between insect societies and human society are striking. An insect society is often referred to as a superorganism, reflecting the remarkable degree of coordination between individual insects. Individual workers have been likened to cells in a body, and castes to tissues or organs that perform specialized functions. Insect societies are not immortal; however, they often persist in a single location for periods similar to the life spans of much larger animals. The social insects have one of the most highly developed symbolic languages outside human cultures. Further, social insects have evolved complex and often mutually beneficial interactions with other species to a degree unknown except among human beings. Bees are inseparably linked with the flowers they feed upon and pollinate. Ants have actually developed agriculture of a sort with their fungus gardens and herds of tended aphids. On a more sobering note, ants are the only nonhuman animals that are known to wage war. These striking similarities with human societies have led researchers to study social insects to learn about the biological basis of social behavior and have led to the development of a new branch of science known as sociobiology.

Studying Insect Societies

Because of the diversity of questions that investigators have addressed regarding insect societies, many methods of scientific inquiry have been employed. In Karl von Frisch's experiments, for example, basic behavioral observations were coupled with simple but elegant experimental design to unravel the dance language of bees. The bees were raised in an observation colony. This was essentially a large hive housed between plates of glass so that an observer could watch the behavior of individual bees. Researchers followed specific workers by marking them with small numbers placed on the abdomen or thorax. Sometimes the entire observation hive was placed within a small, darkened shed to simulate more closely the conditions within a natural hive.

Bees learned to find an artificial "flower"—a glass dish filled with a sugar solution. Brightly colored backgrounds and odors such as peppermint oil were added to the sugars to provide specific cues for the bees to associate with the reward. Feeding stations were set up at fixed distances; observers could follow the exact movements of known individuals both at the feeder and at the hive. In this way, von Frisch was able to describe several types of dances (the round dance for near food sources, the waggle dance for feeding stations that were farther from the hive) and show that a returning bee could share information regarding the location and quality of a source with her nestmates. Scientists subsequently have developed robot bees that can be operated by remote control to perform different combinations of dance behaviors. This allows them to determine which parts of the dance actually convey the coded information.

The investigation of forms of chemical communication requires application of a variety of techniques. Chromatography is useful for identifying the minute amounts of chemical pheromones with which insects communicate. Chromatography (which literally means "writing with color") is particularly suitable for separating mixtures of similar materials. A solution of the mixture is allowed to flow over the surface of a porous solid material. Since each component of the mixture will flow at a slightly different rate, eventually they will become separated or spaced out on the solid material. Once the components of the pheromone have been separated and identified, their activity is assessed separately and in combination using living insects. Such bioassays allow researchers to determine exactly which fractions of the chemical generate the highest response.

Other biochemical techniques, such as electrophoresis, have been used to determine subtle behavioral differences, such as kin discrimination among hive mates. Each individual carries a complement of enzymes or proteins that catalyze biological reactions in the body. The structure of such enzymes is determined by the genetic makeup of the individual, and it varies among individuals. Because enzyme structure is inheritable, however, much as eye color is, the degree of similarities between the enzymes can be used as a measure of how closely related two individuals are. The amino acids composing the enzyme differ in their electrical charges, so different forms can be separated using the technique of electrophoresis. When a liquid containing their enzymes is subjected to an electrical field, the proteins with the highest negative charge will move farthest toward the positive pole. This provides a tool to distinguish close genetic relatives for use in conjunction with behavioral observations to test, for example, whether workers can discriminate full sisters from half sisters, or relatives from nonrelatives, as kin selection theory would predict.

The Success of Social Insects

Social insects are among the most successful groups of animals throughout the world, especially in the tropics. Although the number of species is low when compared to all insects (twelve thousand out of more than a million species), their relative contribution to the community may be unduly large. In Peru, for example, ants may make up more than 50 percent of the individual insects collected at any site.

The study of social insects has provided scientists with new ways of looking at social behavior in all animals. Charles Darwin described the evolution of sterile workers in the social insects as the greatest obstacle to his theory of evolution by natural selection. In attempts to explain this seeming paradox, William D. Hamilton closely examined the social *Hymenoptera*, where sociality had evolved eleven separate times. Realizing that the haplodiploid form of sex determination led to sisters being more closely related to one another than they would be to their own young, Hamilton developed a far-reaching new theory of social evolution: kin selection, or selection acting on groups of closely related individuals. This theory, which provides insights into the evolution of many kinds of seemingly altruistic behaviors, arose primarily from his perceptions regarding the asymmetrical relatedness of nestmates in the social *Hymenoptera*. These insects, then, should be credited with providing the model system that has led to a subdiscipline of behavioral ecology known as sociobiology, the study of the biological basis of social behavior. Moreover, given their central roles in critical ecological processes such as nutrient cycling and pollination, it would be hard to imagine life without them.

Catherine M. Bristow

See also: Altruism; Animal-plant interactions; Biopesticides; Coevolution; Communication; Communities: ecosystem interactions; Ethology; Mammalian social systems; Mimicry; Pollination; Symbiosis.

Sources for Further Study

Crozier, Ross H., and Pekka Pamilo. *Evolution of Social Insect Colonies: Sex Allocation and Kin Selection.* New York: Oxford University Press, 1996.

Frisch, Karl von. *The Dance Language and Orientation of Bees.* Translated by L. E. Chadwick. Cambridge, Mass.: Belknap Press, 1967.

Gordon, Deborah M. *Ants at Work: How an Insect Society Is Organized.* New York: W. W. Norton, 1999.

Hoelldobler, Bert, and Edward O. Wilson. *Journey to the Ants: A Story of Scientific Exploration.* Cambridge, Mass.: Belknap Press, 1994.

Ito, Yoshiaki. *Behavior and Social Evolution of Wasps: The Communal Aggregation Hypothesis.* New York: Oxford University Press, 1993.

Moffett, Mark W. "Samurai Aphids: Survival Under Siege." *National Geographic* 176 (September, 1989): 406-422.

Prestwich, Glenn D. "The Chemical Defenses of Termites." *Scientific American* 249 (August, 1983): 78-87.

Wilson, Edward O. *The Insect Societies.* Cambridge, Mass.: The Belknap Press of Harvard University Press, 1971.

_____. *Sociobiology: The New Synthesis.* Cambridge, Mass.: The Belknap Press of Harvard University Press, 1975.

_____. *Success and Dominance in Ecosystems: The Case of the Social Insects.* Oldendorf/Luhe, Federal Republic of Germany: Ecology Institute, 1990.

INTEGRATED PEST MANAGEMENT

Types of ecology: Agricultural ecology; Ecotoxicology; Restoration and conservation ecology

Integrated pest management (IPM) is the practice of integrating insect, animal, or plant management tactics, such as chemical control, cultural control, biological control, and plant resistance, to maintain pest populations below damaging levels in the most economical and environmentally responsible manner.

In the past, pest management strategies in agriculture focused primarily on eliminating all of a particular pest organism from a given field or area. These strategies depended on the use of chemical pesticides to kill all the pest organisms. Prior to the twentieth century, farmers used naturally occurring compounds such as kerosine or pyrethrum for this purpose. During the second half of the twentieth century, synthetic pesticides began playing a prominent role in controlling crop pests.

Chemical Effects

After 1939 the use of pesticides such as dichloro-diphenyl-trichloroethane (DDT) was so successful in controlling pest populations that farmers began to substitute a heavy dependence on pesticides for sound pest management strategies. Soon pests in high-value crops became resistant to one pesticide after another. In addition, outbreaks of secondary pests occurred because either they developed resistance to the pesticides or the pesticides killed their natural enemies. Among birds ingesting DDT in the food chain, eggshells were so thin as to render many eggs unviable, reducing the bird population. Such impacts of DDT supplied the impetus for chemical companies to develop new pesticides, to which the pests also eventually developed resistance.

Rationale for IPM

Certain pests have developed resistance to all federally registered materials designed to control them. In addition, many pesticides are toxic to humans, wildlife, and other nontarget organisms and therefore contribute to environmental pollution. For these reasons, and because it is very expensive for chemical companies to put a new pesticide on the market, many producers began looking at alternative strategies such as IPM for managing pests. The driving forces behind the development of IPM programs are concern about the contamination of groundwater and other nontarget sites, adverse effects on nontarget organisms, and development of pesticide resistance. Pesticides will probably continue to play a vital role in pest management, even in IPM, but it is believed that the role will be greatly diminished over time.

An agricultural ecosystem consists of the crop environment and its surrounding habitat. The interactions among soil, water, weather, plants, and animals in this ecosystem are rarely constant enough to provide the ecological stability of nonagricultural ecosystems. Nevertheless, it is possible to use IPM to manage most pests in an economically efficient and environmentally friendly manner. IPM programs have been successfully implemented in the cropping of cotton and potatoes, and they are being developed for other crops.

Developing IPM Programs
There are generally three stages of development associated with IPM programs, and the speed at which a program progresses through these stages is dependent on the existing knowledge of the agricultural ecosystem and the level of sophistication desired. The first phase is referred to as the pesticide management phase. The implementation of this phase requires that the farmer know the relationship between pest densities and the resulting damage to crops so that the pesticide is not applied excessively. In other words, farmers do not have to kill all the pests all the time. They must use pesticides only when the economic damage caused by a number of pest organisms present on a given crop exceeds the cost of using a pesticide. This practice alone can reduce the number of chemical applications by as much as half.

The second phase is called the cultural management phase. Implementation of this phase requires knowledge of the pest's biology and its relationship to the cropping system. Cultural management includes such practices as delaying planting times, rotating crops, altering harvest dates, and planting resistant cultivars. It is necessary to understand pest responses to other species as well as abiotic factors, such as temperature and humidity, in the environment. If farmers know the factors that control population growth of a particular pest, they may be able to reduce the impact of that pest on a crop. For example, if a particular pest requires short days to complete development, farmers might be able to harvest the crop before the pest has a chance to develop.

The third phase is the biological control phase, which involves the use of biological organisms rather than chemicals to control pests. This is the most difficult phase to implement because farmers must understand not only the pest's biology but also the biology of the pest's natural enemies and the degree of effectiveness with which these agents control the pest.

In general, it is not possible to rely completely on biological control methods. A major requirement in using biological agents is to have sufficient numbers of the control agent present at the same time that the pest population is at its peak. It is sometimes possible to change the planting dates so that the populations of the pests and the biological control agents are synchronized. Also, there is often more than one pest species present at the same time within the same crop, and it is extremely difficult to control simultaneously two pests with biological agents.

D. R. Gossett

See also: Biomagnification; Biopesticides; Genetically modified foods; Pesticides; Pollution effects; Soil contamination.

Sources for Further Study

Pedigo, Larry P. *Entomology and Pest Management.* 4th ed. Upper Saddle River, N.J.: Prentice Hall, 2002.
Romoser, William S., and John G. Stoffolano. *The Science of Entomology.* 4th ed. Boston: McGraw-Hill, 1998.

INVASIVE PLANTS

Types of ecology: Ecosystem ecology; Ecotoxicology

Nonnative (also termed "exotic") fungi and plants that can outcompete native species are called invasive plants. Invasive plants cause irreversible changes to ecosystems, threaten plant and animal species, and cost billions of dollars to control.

Between the damage they cause and the cost of control efforts, invasive plants cost the United States more than $140 billion every year. For example, nearly half of the threatened and endangered plant species listed for the United States in 1999, 400 of 958, were in peril because of competition from invasive species. Thus, invasive plants are capable of causing irreparable changes in ecosystems.

Most invasive species in the United States originated in Asia or Europe. The seeds or spores of these plants are accidentally transported into new habitats by humans, leaving the plants' natural enemies and competitors behind. Without natural biological controls, the alien species can thrive and outcompete the native flora, driving the native plants toward extinction and creating a near monoculture of the invader.

Invasive plants are weedy species that grow rapidly, produce large numbers of long-lived seeds, and frequently have perennial roots, or rhizomes, that enhance asexual propagation. Invasive plants have a variety of effects on invaded ecosystems. Many invasive species deplete soil moisture and nutrient levels, either by growing more vigorously than native plants early in the growing season or by being more tolerant of reduced levels of water and nutrients than are natives. Some invasive species produce toxic chemicals (allelopathy) that are released into the soil and inhibit the growth of competitors. By outcompeting native plants, the invader decreases species diversity as it replaces many native species. As a result, animal species dependent on native flora are also affected. Fungi and seed plants are among the most disruptive invasive plants in the United States today.

Control Methods

Invasive species are carried to new habitats, either in or on machinery or organisms, and are usually transported by humans, so prevention is the most cost-effective method of control. Once an invasive species has entered an area, plant quarantine is an effective first line of defense. For example, living plants and animals brought into the United States must pass inspection by the U.S. Department of Agriculture Animal and Plant Health Inspection Service (APHIS) to ensure that they are not carrying potentially invasive species. Particular care is taken to ensure that imports from known areas of infestation are clean of seeds, spores, or propagules.

The next most effective strategy is detection and control of small infestations. When there is a known threat of invasion, the affected area should be surveyed periodically and individual plants removed by hand or, in extreme cases, by "spot-spraying" herbicide. Eradication is possible when the infestation is small.

Once an invasive species becomes established, the only means of management are expensive chemical or biological controls which, at best, will only minimize damage. A variety of chemicals may be used to kill invasive plants. Most chemicals, however, affect a broad spectrum of plants, including native species. Biological controls, including natural enemies from the invasive plant's native ecosystem, can be more specific but may also be capable of displacing native species and becoming "invaders."

Fungi
Many of the most serious plant pathogens are invasive species introduced into the Americas since the beginning of European settlement. Two classic examples are Dutch elm disease, caused by the fungi *Ophiostoma ulmi* and *Ophiostoma novo-ulmi* and chestnut blight, caused by the fungus *Cryphonectria parasitica*. At the beginning of the twentieth century, the most common street tree growing in the cities of the eastern United States was the

This Salvinia molesta *plant floats on water surfaces, multiplying rapidly. Found over an increasingly wide range in the United States, it is known to clog waterways and crowd out native plants.* (AP/Wide World Photos)

American elm. About 1910, the European bark beetle was introduced into the United States. It was not until the 1930's that Dutch elm disease was observed in Ohio and a few eastern states. The fungal spores are carried by the beetles, which burrow under the elm bark. The native elm has little resistance to this fungus, whose spores rapidly germinate and form extensive mycelia within the phloem of the host tree, killing it within a few years. After its initial contact, the fungus spread throughout the cities and forests of the East and gradually westward, so that by 1990 nearly all the native American elm trees in the United States had been killed.

American chestnut was also one of the early dominant trees of the eastern U.S. forest. In addition to providing edible fruit, the chestnut became a commercially important timber tree. Chestnut blight fungus was first reported in 1904 on chestnut trees in the New York Zoological Garden and quickly began to spread. This infestation led directly to passage of the Plant Quarantine Act of 1912, the forerunner of APHIS. By 1950 most native chestnut trees were reduced to minor understory shrubs. Biological control using virus strains first isolated in Italy show promise for controlling the blight.

Terrestrial Green Plants

Virtually all the plants commonly called weeds are foreign invaders that are difficult, if not impossible, to control. Some of the most severe include Canada thistle (*Circium arvense*), leafy spurge (*Euphorbia esula*), and purple loosestrife (*Lythrum salicaria*).

Canada thistle is the most widespread and difficult species of thistle to control. It was introduced to Canada from Europe in the 1600's and in 1795 was listed as a noxious weed in Vermont. It is now found in most of the United States as well as in Canada. Single herbicide applications do not provide long-term control, and there are no effective biological controls that do not also attack native species.

Leafy spurge was first reported in Newbury, Massachusetts, in 1827, where it arrived in ship ballast. By 1900 it had reached the West Coast, and it now thrives in more than half the states and in Canada. Thirteen species of insects are approved for biological control, and several herbicides can be used to control infestations effectively. Sheep and goats will browse on spurge.

Purple loosestrife was introduced into the United States as an ornamental plant in the early 1800's and became established in New England by 1830. Its early spread into the Great Lakes region was by barge and other canal traffic. Rapid expansion of the pest, particularly in the West, occurred after 1940, primarily due to the plant's "escape" from ornamental cultivation into irrigation projects. It is now found in all the lower forty-eight states except Florida. At present, there are no effective controls.

Aquatic Green Plants
Invasive plants are not limited to the terrestrial habitat or to vascular plants. One dramatic example is the alga *Caulerpa taxifolia,* the so-called killer alga. This attractive tropical alga was found to be easy to grow in salt-water aquaria and useful as a secondary food source for herbivorous tropical fish. It began to be used this way at the Oceanographic Museum of Monaco in 1982. Two years later, a meter-square patch was found growing in the Mediterranean Sea, visible from a window of the museum. By 1990 the alga had reached France, and by 1995 it could be found from Spain to Croatia. *Caulerpa* produces a number of toxins that inhibit foraging by native fish, and it is a prolific vegetative reproducer. Fragments of the alga, stuck on an anchor for example, can start a new infestation wherever the anchor is next dropped. This species has been discovered in Southern California, and a related species has become dominant in Sydney Harbor, Australia. Other aquatic invasive plants in the United States include the mosquito fern (*Azolla*), the Eurasian water milfoil (*Myriophyllum*), and the water hyacinth (*Eichhornia crassipes*).

Marshall D. Sundberg

See also: Biological invasions; Biomagnification; Eutrophication; Genetically modified foods; Pesticides; Phytoplankton; Pollution effects; Waste management.

Sources for Further Study
Meinesz, Alexandre. *Killer Algae.* Chicago: University of Chicago Press, 1999.
Randall, John M., and Janet Marinelli, eds. *Invasive Plants: Weeds of the Global Garden.* Brooklyn, N.Y.: Brooklyn Botanic Garden, 1996.
Sheley, Roger L., and Janet K. Petroff. *Biology and Management of Noxious Rangeland Weeds.* Corvalis, Oreg.: Oregon State University Press, 1999.

ISOLATING MECHANISMS

Types of ecology: Behavioral ecology; Ecosystem ecology; Evolutionary ecology; Speciation

Isolating mechanisms act to prevent interbreeding and the exchange of genes between species. The establishment of isolating mechanisms between populations is a critical step in the formation of new species and ensuring biodiversity.

Isolating mechanisms (reproductive isolating mechanisms) prevent interbreeding between species. The term, which was first used by Theodosius Dobzhansky in 1937 in his landmark book *Genetics and the Origin of Species*, refers to mechanisms that are genetically influenced and intrinsic. Geographic isolation can prevent interbreeding between populations, but it is an extrinsic factor and therefore does not qualify as an isolating mechanism. Isolating mechanisms function only between sexually reproducing species. They have no applicability to forms that reproduce only by asexual means, as by mitotic fission, stoloniferous or vegetative reproduction, or egg development without fertilization (parthenogenesis in animals). Obligatory self-fertilization in hermaphrodites (mainly a phenomenon in plants, rare in animals) is a distortion of the sexual process that produces essentially the same results as asexual reproduction. Many lower animals and protists regularly employ both asexual and sexual means of reproduction, and the significance of isolating mechanisms in such forms is essentially the same as in normal sexual species.

Premating Mechanisms

Reproductive isolating mechanisms are usually classified into two main groups. Premating (prezygotic) mechanisms operate prior to mating, or the release of gametes, and, therefore, do not result in a wastage of the reproductive potential of the individual. Postmating (postzygotic) mechanisms come into play after mating, or the release of gametes, and could result in a loss of the genetic contribution of the individual to the next generation. This distinction is also important in the theoretical sense in that natural selection should favor genes that promote premating isolation; those that do not presumably would be lost more often through mismatings (assuming that hybrids are not produced, or are sterile or inferior), and this could lead to a reinforcement of premating isolation.

Ethological (behavioral) isolation is the most important category of premating isolation in animals. The selection of a mate and the mating process depends upon the response of both partners to various sensory cues, any one of which may be species-specific. Although one kind of sensory stimulus may be emphasized, different cues may come into play at different stages of the pairing process. Visual signals provided by color, pattern, or method of display are often of particular importance in diurnal animals such as birds, many lizards, certain spiders, and fish. Sounds, as in male mating calls, are often important in nocturnal breeders such as crickets or frogs but are also important in birds. Mate discrimination based on chemical signals or odors (pheromones) is of fundamental importance in many different kinds of animals, especially those where visual cues or sound are not emphasized; chemical cues also are often important in aquatic animals with external fertilization. Tactile stimuli (touch) often play an important role in courtship once contact is established between the sexes. Even electrical signals appear to be utilized in some electrogenic fish.

Ecological (habitat) isolation often plays an important role. Different forms may be adapted to different habitats in the same general area and may meet only infrequently at the time of reproduction. One species of deer mouse, for example, may frequent woods, while another is found in old fields; one fish species spawns in riffles, while another spawns in still pools. This type of isolation, although frequent and widespread, is often incomplete as the different forms may come together in transitional habitats. The importance of ecological isolation, however, is attested by the fact that instances in which hybrid swarms are produced between forms that normally remain distinct have often been found to be the result of disruption of the environment, usually by humans. Mechanical isolation is a less-important type of premating isolation, but it can function in some combinations. Two related animal species, for example, may be mismatched because of differences in size, proportions, or structure of genitalia.

Finally, temporal differences often contribute to premating isolation. The commonest type of temporal isolation is seasonal isolation: Species may reproduce at different times of the year. A species of toad in the eastern United States, for example, breeds in the early spring, while a related species breeds in the late spring, with only a short period of overlap. Differences can also involve the time of day, whereby one species may mate at night and another during the day. Such differences, as in the case of ecological isolation, are often incomplete but may be an important component of premating isolation.

Postmating Mechanisms

If premating mechanisms fail, postmating mechanisms can come into play. If gametes are released, there still may be a failure of fertilization (intersterility). Spermatozoa may fail to penetrate the egg, or even with penetration there may be no fusion of the egg and sperm nucleus. Fertilization failure is almost universal between remotely related species (as from different families or above) and occasionally occurs even between closely related forms.

If fertilization does take place, other postmating mechanisms may operate. The hybrid may be inviable (F1 or zygotic inviability). Embryonic development may be abnormal, and the embryo may die at some stage, or the offspring may be defective. In other cases, development may be essentially normal, but the hybrid may be ill-adapted to survive in any available habitat or cannot compete for a mate (hybrid adaptive inferiority). Even if hybrids are produced, they may be partially to totally sterile (hybrid sterility). Hybrids between closely related forms are more likely to be fertile than those between more distantly related species, but the correlation is an inexact one. The causes for hybrid sterility are complex and can involve genetic factors, differences in gene arrangements on the chromosomes that disrupt normal chromosomal pairing and segregation at meiosis, and incompatibilities between cytoplasmic factors and the chromosomes. If the hybrids are fertile and interbreed or backcross to one of the parental forms, a more subtle phenomenon known as hybrid breakdown sometimes occurs. It takes the form of reduced fertility or reduced viability in the offspring. The basis for hybrid breakdown is poorly understood but may result from an imbalance of gene complexes contributed by the two species.

It should be emphasized that in most cases of reproductive isolation that have been carefully studied, more than one kind of isolating mechanism has been found to be present. Even though one type is clearly of paramount importance, it is usually supplemented by others, and should it fail, others may come into play. In this sense, reproductive isolation can be viewed as a fail-safe system. A striking difference in the overall pattern of reproductive isolation between animals and plants, however, is the much greater importance of premating isolation in animals and the emphasis on postmating mechanisms in plants. Ethological isolation, taken together with other premating mechanisms, is highly effective in animals, and postmating factors usually function only as a last resort.

Field Studies and Experimental Studies

Field studies have often been employed in the investigation of some types of premating isolating mechanisms. Differences in such things as breeding times, factors associated with onset of breeding activity, and differences in habitat distribution or selection of a breeding site are all subject to direct field observation. Comparative studies of courtship behavior in the field or laboratory often provide clues as to the types of sensory signals that may be important in the separation of related species.

Mating discrimination experiments carried on in the laboratory have often been employed to provide more precise information on the role played by different odors, colors, or patterns, courtship rituals, or sounds in mate selection. Certain pheromones, for example, which act as sexual attractants, have been shown to be highly species-specific in some insects. The presence or absence of certain colors or their presentation has been shown experimentally to be important in mate discrimination in vertebrates as diverse as fish, lizards, and birds. Call discrimination experiments, in which a receptive female is given a choice between recorded calls of males of her own and another species, have demonstrated the critical importance of mating call differences in reproductive isolation in frogs and toads. Synthetically generated calls have sometimes been used to pinpoint the precise call component responsible for the difference in response.

Studies on postmating isolating mechanisms have most often involved laboratory crosses in which the degree of intersterility, hybrid sterility, or hybrid inviability can be analyzed under controlled conditions. In instances in which artificial crosses are not feasible, natural hybrids sometimes occur and can be tested. The identification of natural backcross products can attest incomplete postmating, as well as premating isolation. Instances of extensive natural hybridization are of special interest and have often been subjected to particularly close scrutiny. Such cases often throw light on factors that can lead to a breakdown of reproductive isolation. Also, as natural hybridization more often occurs between marginally differentiated forms in earlier stages of speciation, new insights into the process of species formation can sometimes be obtained. Finally, such studies may yield information on the evolutionary role of hybridization, including introgressive hybridization, the leakage of genes from one species into another. Morphological analysis has long been used in such cases, and chromosomal studies are sometimes appropriate. In recent years, allozyme analysis by gel electrophoresis has become a routine tool in estimates of gene exchange, and molecular analysis of nuclear deoxyribonucleic acid (DNA), mitochondrial DNA, have been useful. As mitochondria are normally passed on only maternally, their DNA can also be used to identify cases in which females of only one of the two species has been involved in the breakdown of reproductive isolation.

Investigations of the role of natural selection in the development and reinforcement of reproductive isolation have employed two different approaches. One has involved the measurement of geographic variation in the degree of difference in some signal character (call, color, or pattern, for example) thought to function in premating isolation between two species that have overlapping ranges. If the difference is consistently greater within the zone of overlap (reproductive character displacement), an argument can be made for the operation of reinforcement. Another approach has involved laboratory simulations, usually with the fruit fly Drosophila, in which some type of selective pressure is exerted against offspring produced by crosses between different stocks, and measurement is made of the frequency of mismatings through successive generations. The results of such studies to this time are contradictory, and the role of selection with regard to development of reproductive isolation requires further study.

Enhancing Reproductive Efficiency
The efficiency of reproduction in most animals is enhanced immeasurably by premating isolating mechanisms. Clearly, in animals a random testing of potential mates without regard to type is totally unacceptable for most species in terms of reproductive capacity and time and energy resources. Premating isolation in this sense is a major factor in promoting species diversity in animal communities.

Both premating and postmating isolating mechanisms are also critical to the maintenance of species diversity in that they act to protect the genetic integrity of each form: A species cannot maintain its identity without barriers that prevent the free exchange of genes with other species. Furthermore, a species functions as the primary unit of adaptation. Every species in a community has its own unique combination of adaptive features that enable it to exploit the resources of its environment and to coexist with other species with a minimum of competition. The diversity of different species that can coexist in the same area depends upon the unique "niche" that each occupies; adaptive features that determine that niche are based on the unique genetic constitution of each species, and this genetic constitution is protected through reproductive isolation.

The development of reproductive isolating mechanisms is also critical to the formation of new species (speciation), and ultimately to the development of new organic diversity. The most widely accepted, objective, and theoretically operational concept for a sexual species is the biological species concept. Such a species can be defined as population or group of populations, members of which are potentially capable of interbreeding but which are reproductively isolated from other species. The origin of new species, therefore, depends upon the development of reproductive isolating mechanisms between populations. A major focus of research in evolutionary biology and systematics has been, and continues to be, on the various factors that influence the development of reproductive isolating mechanisms.

John S. Mecham

See also: Adaptive radiation; Biodiversity; Clines, hybrid zones, and introgression; Communication; Convergence and divergence; Evolution: definition and theories; Gene flow; Mammalian social systems; Natural selection; Nonrandom mating, genetic drift, and mutation; Punctuated equilibrium vs. gradualism; Reproductive strategies; Speciation.

Sources for Further Study

Dobzhansky, Theodosius. *Genetics of the Evolutionary Process*. New York: Columbia University Press, 1970.

Dobzhansky, Theodosius, Francisco J. Ayala, G. Ledyard Stebbins, and James W. Valentine. *Evolution*. San Francisco: W. H. Freeman, 1977.

Mayr, Ernst. *Populations, Species, and Evolution*. Cambridge, Mass.: Harvard University Press, 1970.

LAKES AND LIMNOLOGY

Types of ecology: Aquatic and marine ecology; Biomes; Ecosystem ecology

Lakes are inland bodies of water that fill depressions in the earth's surface. They are generally too deep to allow vegetation to cover the entire surface and may be fresh or saline. The study of the physical, chemical, climatological, biological, and ecological aspects of lakes is known as limnology.

Geological Origin of Lakes

Several geologic mechanisms can create the closed basins that are needed to impound water and produce lakes. The most important of these mechanisms include glaciers, landslides, volcanoes, rivers, subsidence, and tectonic processes.

Continental glaciers formed thousands of lakes by the damming of stream valleys with moraine materials. Glaciers also scoured depressions in softer bedrock, and these later filled with water to form lakes. Depressions called kettles formed when buried ice blocks melted. Mountain glaciers also produce numerous small, high alpine lakes by plucking away bedrock. The bowl-shaped depressions that occur as a result of this plucking are called cirques; lakes that occupy cirques are called tarns. Sometimes a mountain glacier moves down a valley and carves a series of depressions along the valley that, from above, look like a row of beads along a string. When these depressions later fill with water, the lakes are called paternoster lakes, the name coming from their similarity to beads on a rosary.

Landslides sometimes form natural dams across stream valleys. Large lakes then pond up behind the dam. Volcanoes may produce lava flows that dam stream valleys and produce lakes. A volcanic explosion crater may fill with water and make a lake. After an eruption, the area around the eruption vent may collapse to form a depression called a caldera. Some calderas, such as Crater Lake in Oregon, fill with water. Rivers produce lakes along their valleys when a tight loop of a meandering channel finally is eroded through and leaves behind an oxbow lake, isolated from the main channel. Sediment may accumulate at the mouth of a stream, and the resulting delta may build, bridging across irregularities in the shoreline to create a brackish coastal lake.

Natural subsidence creates closed basins in areas underlain by soluble limestones or evaporite deposits. As the underlying limestone is dissolved away, the earth above collapses to form a cavity (sinkhole), which later fills with water. Finally, large-scale (tectonic) downwarping of the earth's crust produces some very large lakes. Large basins form when the crust warps or sinks downward in response to deep forces. The subsidence produces very large closed basins that can hold water. A few immense lakes owe their origins to tectonic downwarping.

Sedimentation

With few exceptions, most lakes exist in relatively small depressions and serve as the catch basins for sediment from the entire watershed around them. The natural process of sedimentation ensures that most lakes fill with sediment before very long periods of geologic time have passed. Lakes with areas of only a few square kilometers or less will fill within a few tens of thousands of years. Very large lakes, the inland seas, may endure for more than ten million years. Human-made lakes and reservoirs have unusually high sediment-fill rates in comparison with most natural lakes. Human-made lakes fill with sediment within a few decades to a few centuries.

Lake sediments come from four sources: allogenic clastic materials that are washed in from the surrounding watershed; endogenic chemical precipitates that are produced from dissolved substances in the lake waters;

Lakes offer rich habitats for complex ecosystems comprising organisms that live on land, in water, or in both environments. Depending on their elevation, lakes can support wildlife typical of taiga, such as freshwater fish and bears, or lower elevations, such as migratory seabirds. (Digital Stock)

endogenic biogenic organic materials produced by plants and animals living in the lake; and airborne substances, such as dust and pollen, transported to the lake in the atmosphere.

Allogenic clastic materials are mostly minerals; they are produced when rocks and soils in the drainage basin are weathered by mechanical and chemical processes to yield small particles. These particles are moved downslope by gravity and running water to enter streams, which then transport them to the lake. Clastic materials also enter the lake via waves, which erode the materials from the shoreline, and via landslides that directly enter the lake. In winter, ice formed on the lake can expand and push its way a few centimeters to 1 meter or so onto the shore. There, the ice may pick up large particles, such as gravel and cobbles. When spring thaw comes, waves can remove that ice, together with its enclosed particles, and float it out onto the lake. The process by which the large particles are transported out on the lake is called ice-rafting. As the ice melts, the large clastic particles drop to the bottom; they are termed dropstones when found in lake sediments. A landslide into a lake or a flood on a stream that feeds into the lake can produce water heavily laden with sediment. The sediment-laden water is more dense than clean water and therefore can rush down and across the lake bottom at speeds sufficient to carry even coarse sand far out into the lake. These types of deposits are called turbidite deposits.

Endogenic chemical precipitates in freshwater lakes commonly consist of carbonate minerals (calcite, aragonite, or dolomite) and mineraloids that consist of oxides and hydroxides of iron, manganese, and aluminum. In some saline and brine lakes, the main sediments may be carbonates, together with sulfates such as gypsum (hydrated calcium sulfate), thenardite (sodium sulfate), or epsomite (hydrated magnesium sulfate), or with chlorides such as halite (sodium chloride) or more complex salts. Of the endogenic precipitates, calcite is the most abundant. Its precipitation represents a balance between the composition of the atmosphere and that of the lake water.

Diatoms are distinctive microscopic algae that produce a frustule (a kind of shell) made of silica glass that is highly resistant to weathering. When seen under a high-powered microscope, diatom frustules appear to be artwork—beautiful and highly ornate saucer- and pen-shaped works of glass. A tiny spot of lake sediment may contain millions.

A lake's sediment may contain from less than 1 percent to more than 90 percent organic materials, depending upon the type of lake. Most organic matter in lake sediments is produced within the lake by plankton and consists of compounds such as carbohydrates, proteins, oils, and waxes that are made up of organic carbon, hydrogen, nitrogen, and oxygen, with a little phosphorus. Plankton, with an approximate bulk composition of 36 percent carbon, 7 percent hydrogen, 50 percent oxygen, 6 percent nitrogen, and 1 percent phosphorus (by weight), includes microscopic plants (phytoplankton) and microscopic animals (zooplankton) that live in the water column. Lakes that are very high in nutrients (eutrophic lakes) commonly have heavy blooms of algae, which contribute much organic matter to the bottom sediment. Terrestrial (land-derived) organic material such as leaves, bark, and twigs form a minor part of the organic matter found in most lakes. Terrestrial organic material is higher in carbon and lower in hydrogen, nitrogen, and phosphorus than is planktonic organic matter.

Airborne substances usually constitute only a tiny fraction of lake sediment. The most important material is pollen and spores. Pollen usually constitutes less than 1 percent of the total sediments, but that tiny amount is a very useful component for learning about the recent climates of the earth. Pollen is among the most durable of all natural materials. It survives attack by air, water, and even strong acids and bases. Therefore, it remains in the sediment through geologic time. As pollen accumulates in the bottom sediment, the lake serves as a kind of recorder for the vegetation that exists around it at a given time. By taking a long core of the bottom sediment from certain types of lakes, a geologist may look at the pollen changes that have occurred through time and reconstruct the history of the climate and vegetation in an area.

Volcanic ash thrown into the atmosphere during eruptions enters lakes and forms a discrete layer of ash on the lake bottom. When Mount St. Helens erupted in 1980, it deposited several centimeters of ash in lakes more than 160 kilometers east of the volcano. Geologists have used layers of ash in lakes to reconstruct the history of volcanic eruptions in some areas. Although dust storms contribute sediment to lakes, such storms are usually too infrequent in most areas to contribute significant amounts.

Water Circulation

Lake waters are driven into circulation by temperature-induced density changes and wind. Most freshwater lakes in temperate climates circulate completely twice each year; they are termed dimictic lakes. Circulation exerts a profound influence on water chemistry of the lake and the amount and type of sediment present within the water column. During summer stratification, the lake is thermally stratified into three zones. The upper layer of warm water (epilimnion) floats above the denser cold water and prevents wind-driven circulation from penetrating much below the epilimnion. The epilimnion is usually in circulation, is rich in oxygen (from algal photosynthesis and diffusion from the atmosphere), and is well

Riparian ecosystems vary widely, depending on the course of the river. Upper courses, at higher and colder elevations, typically run faster and are highly oxygenated, supporting trout, bass, salmon, and other freshwater fish and their predators, but not as many grasses, for example, as middle or lower courses and estuaries. (PhotoDisc)

lighted. This layer is where summer blooms of green and blue-green algae occur and calcite precipitation begins. The middle layer (thermocline) is a transition zone in which the water cools downward at a rate of greater than 1 degree Celsius per meter. The bottom layer (hypolimnion) is cold, dark, stagnant, and usually poor in oxygen. There, bacteria decompose the bottom sediment and release phosphorus, manganese, iron, silica, and other constituents into the hypolimnion.

Sediment deposited in summer includes a large amount of organic matter, clastic materials washed in during summer rainstorms, and endogenic carbonate minerals produced within the lake. The most common carbonate mineral is calcite (calcium carbonate). The regular deposition of calcite in the summer is an example of cyclic sedimentation, a sedimentary event that occurs at regular time intervals. This event occurs yearly in the summer season and takes place in the upper 2 or 3 meters of water. On satellite photos, it is even possible to see the summer events as whitings on large lakes, such as Lake Michigan.

As the sediment falls through the water column in summer, it passes through the thermocline into the hypolimnion and onto the lake bottom.

As it sits on the bottom during the summer months, bacteria, particularly anaerobic bacteria (those that thrive in oxygen-poor environments), begin to decompose the organic matter. As this occurs, the dissolved carbon dioxide increases in the hypolimnion. If enough carbon dioxide is produced, the hypolimnion becomes slightly acidic, and calcite and other carbonates that fell to the bottom begin to dissolve. The acidic conditions also release dissolved phosphorus, calcium, iron, and manganese into the hypolimnion, as well as some trace metals. Clastic minerals such as quartz, feldspar, and clay minerals are not affected in such brief seasonal processes, but some silica from biogenic material such as diatom frustules can dissolve and enrich the hypolimnion in silica. As summer progresses, the hypolimnion becomes more and more enriched in dissolved metals and nutrients.

Autumn circulation begins when the water temperature cools and the density of the epilimnion increases until it reaches the same temperature and density as the deep water. Thereafter, there is no stratification to prevent the wind from circulating the entire lake. When this happens, the cold, stagnant hypolimnion, now rich in dissolved substances, is swept into circulation with the rest of the lake water. The dissolved materials from the hypolimnion are mixed into a well-oxygenated water column. Iron and manganese that formerly were present in dissolved form now oxidize to form tiny solid particles of manganese oxides, iron oxides, and hydroxides. The sediment therefore becomes enriched in iron, manganese, or both during the autumn overturn, the amount of enrichment depending upon the amount of dissolved iron and manganese that accumulated during summer in the hypolimnion. Dissolved silica is also swept from the hypolimnion into the entire water column. In the upper water column, where sunlight and dissolved silica become present in great abundance, diatom blooms occur. The diatoms convert the dissolved silica into solid opaline frustules.

As circulation proceeds, the currents may sweep over the lake bottom and actually resuspend 1 centimeter or more of sediment from the bottom and margins of the lake. The amount of resuspension that occurs each year in freshwater lakes is primarily the result of the shape of the lake basin. A lake that has a large surface area and is very shallow permits wind to keep the lake in constant circulation over long periods of the year.

As winter stratification comes, an ice cover forms over the lake and prevents any wind-induced circulation. Because the circulation is what keeps the lake sediment in suspension, most sediment quickly falls to the bottom; sedimentation then is minimal through the rest of winter. If light can penetrate the ice and snow, some algae and diatoms can utilize this weak light, present in the layer of water just below the ice, to reproduce. Their settling remains contribute small amounts of organic matter and diatom frustules. At the lake bottom, the most dense water (that at 4 degrees Celsius) accumulates. As in summer, some dissolved nutrients and metals can build up in this deep layer, but because the bacteria that are active in releasing these substances from the sediment are refrigerated, they work slowly, and not as much dissolved material builds up in the bottom waters.

When spring circulation begins, the ice at the surface melts, and the lake again goes into wind-driven circulation. Oxidation of iron and manganese occurs (as in autumn), although the amounts of dissolved materials available are likely to be less in spring. Once again, nutrients such as phosphorus and silica are circulated out of the dark bottom waters and become available to produce blooms of phytoplankton. Spring rains often hasten the melting, and runoff from rain and snowmelt in the drainage basin washes clastic materials into the lake. The period of spring thaw is likely to be the time of year when the maximum amount of new allogenic (externally derived) sediment enters the lake.

Spring diatom blooms continue until summer stratification prevents further replenishment of silica to the epilimnion. Thereafter, the diatoms are succeeded by summer blooms of green algae, closely followed by blooms of blue-green algae. Silica is usually the limiting nutrient for diatoms; phosphorus is the limiting nutrient for green and blue-green algae.

Diagenesis

After sediments are buried, changes occur; this process of change after burial is termed diagenesis. Physical changes include compaction and dewatering. Bacteria decompose much organic matter and produce gases such as methane, hydrogen sulfide, and carbon dioxide. The "rotten-egg" odor of black lake sediments, often noticed on boat anchors, is the odor of hydrogen sulfide. After long periods of time, minerals such as quartz or calcite slowly fill the pores remaining after compaction.

One of the first diagenetic minerals to form is pyrite (iron sulfide). Much pyrite occurs in microscopic spherical bodies that look like raspberries; these particles, called framboids, are probably formed by bacteria in areas with low oxygen within a few weeks. In fact, the black color of some lake muds and oozes results as much from iron sulfides as from organic matter. Other diagenetic changes include the conversion of mineraloid particles containing phosphorus into phosphate minerals such as vivianite and apatite. Manganese oxides may be converted into manganese carbonates (rhodochrosite). Freshwater manganese oxide nodules may form in high-energy environments such as Grand Traverse Bay in Lake Michigan.

Lake Ecosystems

Freshwater and saline lakes account for 0.009 and 0.008 percent of the total amount of water in the world, respectively. Although this is a minute fraction of the world's water—almost all of it is in the oceans and in glaciers—lakes are an extremely valuable resource. In terms of ecosystems, lakes are divided into a pelagial (open-water) zone and a littoral (shore) zone where macrovegetation grows. Sediments free of vegetation that occur below the pelagial zone are in the profundal zone.

The renewal times for freshwater and saline lakes range from 1 to 100 years and 10 to 1,000 years, respectively. The length of time varies directly with lake volume and average depth, and indirectly with a lake's rate of discharge. The rate of renewal, or turnover time, for lakes is much less than that of oceans and glacial ice, which is measured in thousands of years.

Eutrophication

The aging of a lake by biological enrichment is known as eutrophication. The water in young lakes is cold and clear, with minimal amounts of plant and animal life. The lake is then in the oligotrophic state. As time goes on, streams that flow into the lake bring in nutrients such as nitrates and phosphates, which encourage aquatic plant growth. As the fertility in the lake increases, the plant and animal life increases, and organic remains start accumulating on the bottom. The lake is now becoming eutrophic. Silt and organic debris continue to accumulate over time, slowly making the lake shallower. Marsh plants that thrive in shallow water start expanding and gradually fill in the original lake basin. Eventually the lake becomes a bog and then dry land.

This natural aging of a lake can take thousands of years, depending upon the size of the lake, the local climate, and other factors. However, human activities can substantially accelerate the eutrophication process. Among the problems caused by humans are the pollution of lakes by nutrients from agricultural runoff and poorly treated wastewater from municipalities and industries. The nutrients encourage algal growth, which clogs the lake and removes dissolved oxygen from the water. The oxygen is needed for other forms of aquatic life. The lake has now entered a hyper-eutrophic state as declining levels of dissolved oxygen result in incomplete oxidation of plant remains, a situation that eventually causes the death of the lake as a functioning aquatic ecosystem. In a real sense, the lake chokes itself to death.

Research Methods

Scientists who study lakes (limnologists) must study all the natural sciences—physics, chemistry, biology, meteorology, and geology—because lakes are complex systems that include biological communities, changing water chemistry, geological processes, and interaction among water, sunlight, and the atmosphere. Many who study ecology become limnologists, and vice versa.

Limnologists study modern lake sediments by collecting samples from the water column in sediment traps (cylinders and funnels into which the suspended sediment settles over periods of days or weeks) or by filtering large quantities of lake water. Living material is often sampled with a plankton net. Older sediments that have accumulated on the bottom are collected with dredges and by piston coring, which involves pushing a sharpened hollow tube (usually about 2.5 centimeters in diameter) downward into the sediment. Cores are valuable because they preserve the sediment in the order in which it was deposited, from oldest at the bottom to youngest at the top. Once the sample is collected, it is often frozen and taken to the laboratory. There, pollen and organisms may be examined by microscopy, minerals may be determined by X-ray diffraction, and chemical analyses may be made.

Varves are thin laminae that are deposited by cyclic processes. In freshwater lakes, each varve represents one year's deposit; it consists of a couplet with a dark layer of organic matter deposited in winter and a light-colored layer of calcite deposited in summer. Varves are deposited in lakes where annual circulations cannot resuspend bottom sediment and therefore cannot mix it to destroy the annual lamination. Some lakes that are small and very deep may produce varved sediments; Elk Lake in Minnesota is an example. In other lakes, the accumulation of dissolved salts on the bottom eventually produces a dense layer (monimolimnion), which prevents disturbance of the bottom by circulation in the overlying fresher waters. Soap Lake in Washington State is an example. Because each varve couplet represents one year, a geologist may core the sediments from a varved lake and count the couplets to determine the age of the sediment in any part of the core. The pollen, the chemistry, the diatoms, and other constituents may then be carefully examined to deduce what the lake was like during a given time period. The study is much like solving a mystery from a variety of clues. Eventually, the history of climate changes of the area may be learned from the study of lake varves.

Edward B. Nuhfer and Robert Hordon

See also: Acid deposition; Ecosystems: definition and history; Eutrophication; Habitats and biomes; Marine biomes; Ocean pollution and oil spills; Reefs; Wetlands.

Sources for Further Study

Bramwell, Martyn. *Rivers and Lakes*. London: Franklin Watts, 1986.

Burgis, Mary J., and Pat Morris. *The Natural History of Lakes*. New York: Cambridge University Press, 1987.

Cole, Gerald A. *Textbook of Limnology*. 4th ed. Prospect Heights, Ill.: Waveland Press, 1994.

Fraser, Andrew S., Michael Meybeck, and Edwin D. Ongley. *Water Quality of World River Basins*. 14th ed. New York: United Nations Publications, 1998.

Håkanson, Lars, and M. Jansson. *Principles of Lake Sedimentology*. New York: Springer-Verlag, 1983.

Horne, Alexander J., and Charles R. Goldman. *Limnology*. 2d ed. New York: McGraw-Hill, 1994.

Imberger, Jeorg, ed. *Physical Processes in Lakes* and Oceans. Washington, D.C.: American Geophysical Union, 1998.

Lerman, Abraham, Dieter M. Imboden, and Joel R. Gat, eds. *Physics and Chemistry of Lakes*. New York: Springer-Verlag, 1995.

Thornton, Kent W., Bruce L. Kimmel, and Forrest E. Payne, eds. *Reservoir Limnology: Ecological Perspectives*. New York: John Wiley, 1990.

U.S. Environmental Protection Agency. *The Great Lakes: An Environmental Atlas and Resource Book*. Chicago: Great Lakes Program Office, 1995.

Wetzel, Robert G. *Limnology*. 3d ed. San Diego: Academic Press, 2001.

LANDSCAPE ECOLOGY

Type of ecology: Landscape ecology

Humans live in natural landscapes that they have modified and managed to suit their own needs of shelter, security, aesthetics, and usefulness. The science of managing the habitat components of modified landscapes is called landscape ecology, a burgeoning field concerned with preserving the naturalness of modified landscapes while minimizing the negative impact of human intrusion in natural habitats within these landscapes.

K nowledge of landscape ecology is fundamentally important for a number of reasons: natural resource planning, residential and commercial development, wildlife conservation, forestry, agriculture, soil science, ecological function, sociobiology, and the structure of urban and suburban habitats are all elements of landscape ecology of continuing environmental and economic interest to humans.

Structurally, landscapes share four components: patches of habitat, corridors that connect patches of habitat, the background matrix that includes patches and corridors, and the wildlife that inhabits the landscape. Humans and natural processes constantly modify each of these landscape components.

From an ecological standpoint, humans most impact landscapes by fragmenting large blocks of natural habitat into landscapes consisting of patches of natural habitat, patches of modified habitat, and developed areas. For example, a forested landscape may be transformed into a mosaic of woodland patches, farms, residences, parks, preserves, and greenbelts—the whole typically crossed and gridworked by artificial corridors of roadways, power lines, and gas lines along with natural corridors such as rivers, streams, and fence rows.

Some important management issues in landscape ecology include preserving existing wildlife and wildlife habitats, maintaining and improving wildlife habitat and biodiversity, minimizing detrimental impacts of habitat fragmentation, creating natural corridors for wildlife movement between habitats, and creating and managing wildlife parks and preserves in human modified landscapes.

Landscape Fragmentation
The splitting of a contiguous area of natural landscape into two or more smaller blocks of habitat is called landscape fragmentation. The smaller

habitat parcels that result are called fragments or patches. As human populations grow they appropriate more and more of the natural landscape to accommodate their immediate needs for housing and agriculture, thus the rate of landscape fragmentation continues to increase. Because human populations are projected to grow for several more decades, at least, landscape fragmentation is considered to be a global wildlife and wildlife habitat issue of immediate and serious concern.

Natural processes such as floods, ice storms, winds, and landslides contribute to landscape fragmentation, but the vast majority of fragmentation occurs through human activities. Logging, agriculture, roadways, rail lines, power lines, gas lines, trails, the construction of houses, housing clusters, commercial and industrial developments are all some of the ways in which humans fragment natural habitats.

The Patchwork Mosaic
Habitat fragmentation ultimately results in a landscape mosaic comprising a patchwork of artificial and natural habitats strewn across the landscape often in haphazard and unplanned fashion. Natural habitats, slightly modified habitats, and totally altered habitats and the corridors that connect them juxtaposition one another. The interrelationships of these natural and modified patches with respect to size, shape, and connectivity can have profound impacts that collectively tend to limit natural habitats within the landscape and the wildlife that occur in those natural habitats. Some of the more important of these constraints include the decline or elimination of habitat specialists or area-sensitive species, the invasion of natural habitat patches by domestic, exotic, or alien species, the increased mortality of natural species caused by vehicles or predation by pets and feral animals, and the disturbance of reproductive efforts and other natural behaviors by the range of activities that invariably accompany human intrusion into and modification of, natural landscapes.

Initially, landscape fragmentation increases the number and variety of habitat patches within a landscape, thereby increasing habitat diversity. The many small patches provide "stepping stones" that promote species dispersal and interchange from one patch to another which in turn promotes gene flow between patches. Small patches may also provide habitat for certain familiar species, such as the American robin (*Turdus migratorius*) and gray catbird (*Dumetella carolinensis*), that use the edges of small patches for foraging, nesting, or refuges. Also, a large number of small patches may reduce erosion and ultimately loss of natural habitats that so often accompanies human activity and intrusion.

However, continued landscape fragmentation results in ever smaller

Landscape ecologists concern themselves with the interactions of the human built landscapes and wildlife habitat. This squirrel, found living on a golf course in Florida, has been tagged with a radio-transmitting collar, which will help researchers to determine its range and habits. The goal is to design urban settings that will allow wildlife and humans to coexist with minimal negative impacts. (AP/Wide World Photos)

habitat patches that necessarily support fewer species and smaller wildlife populations, thereby increasing the risk of extinction of remaining species. As habitat patches shrink in size or disappear, the remaining habitat patches become increasingly isolated, inhibiting wildlife interchange between patches and making recolonization of the more remote habitat patches increasingly difficult.

As habitat patches diminish in size, species populations begin to disappear. The first to go are the species that need large areas for their activities such as predators that require large territories for their food base. Thus, as patches shrink to a critical point the area cannot support sufficient prey populations to sustain viable populations of larger predators, such as northern goshawks (*Accipiter gentilis*) or red-shouldered hawk (*Buto lineatus*).

Area-sensitive species and interior species are also increasingly placed at risk. Area-sensitive species such as the Carolina chickadee (*Poecile atricapilla*) and eastern wood-peewee (*Contopus virens*) require large areas of undisturbed habitat for nesting, foraging, and other activities. As habitat patches decline in size, the numbers of these species dramatically decrease. Interior species such as the ovenbird (*Seiurus aurocapillus*) select habitats deep within the interior of natural habitats. As the size of habitat patches decreases, there is insufficient amount of core area needed to sustain their activities. The decline in populations of many neotropical migrants that nest in woodlands of Eastern North America is attributed to habitat fragmentation.

The transformation of natural landscapes into human and natural patches almost always creates a modified habitat suitable for intrusive species that are tolerant of humans and human-modified landscapes, such as the house sparrow (*Passer domesticus*), Eurasian starling (*Sturnis vulgaris*), raccoon (*Procyon lotor*), and opossum (*Didelphis virginiana*), all of which are behaviorally adapted to exploit the modified landscapes. Landscape modification also provides an environment for pets that inevitably accompany human intrusions, such as house cats (*Felis catus*) and dogs (*Canis familiaris*), both of which may predate natural wildlife species. Both intrusive and domesticated species reduce biodiversity either through outright predation or by competition.

A critical concern to landscape ecologists involves the ratio of edge habitat to interior habitat in habitat patches resulting from fragmentation. As habitat patches get smaller, the ratio of edge habitat to interior habitat of each patch increases. Greater edge habitat provides favorable habitat for game and other edge species, but interior species and area-sensitive species are placed at increased risk because predators, scavengers, and parasitic species can now penetrate farther into the interior of the habitat patch. Some of the more familiar and important avian edge species in terms of their impacts on interior species include the brown-head cowbird (*Molothrus ater*) and the blue jay (*Cyanocitta cristata*). Two of the many species of mammals that frequent edge habitat and may become nuisance species in human-modified landscapes include the raccoon and the opossum.

The potential impact of edge species on interior species is exemplified by a small blackbird with a distinctive brown head called the brown-headed cowbird. The cowbird is a brood parasite that frequents open woodland, grasslands, and forest edge. Unlike other birds, the cowbird does not build a nest but rather deposits its eggs in the nests of other birds, especially flycatchers, vireos, warblers, orioles, and finches. Female brown-headed cowbirds visit nests of potential host species during construction. Once the host bird lays its clutch, the female cowbird visits the nest and removes one or more of the eggs, either by eating them or by dropping them over the edge of the nest cup. She always leaves one egg—otherwise the host might not return. She then deposits a single egg in the host nest. The host female returns and incubates the clutch, which now includes the cowbird egg. The cowbird egg hatches a day or two before the other eggs in the nest and the young grows quickly, demanding and receiving a large portion of the food that the host adult birds bring to the nestlings. The impact of cowbird brood parasitism increases with increasing landscape fragmentation. As patches get smaller the ratio of edge habitat to core habitat in a given habitat patch increases, exposing more of the interior to cowbird activity. Cowbirds penetrate deeper into habitats to parasitize more nests of interior species. In fact, increased cowbird brood parasitism accompanying habitat fragmentation is one reason that interior woodland species such as wood warblers are thought to be declining.

Corridors and Connectivity

Corridors are links of natural habitats that connect landscape patches to one another. Examples of corridors include areas of vegetation left undeveloped by human disturbance such as belts of vegetation along rivers and streams, vegetated strips along roadways, railways, and drainage ditches, vegetation strips beneath power lines and above gas lines, and linear extensions of vegetated areas such as windbreaks, fence rows, and hedgerows.

Corridors are critically important elements of landscape ecology that help maintain stability and diversity of wildlife within each of the natural patches by providing dispersal routes for plants and animals between habitat patches. The ability of wildlife to use these corridors to disperse from one habitat patch to another is called connectivity. Larger patches that support larger species populations serve as the source of individuals that disperse along corridors to repopulate or augment species populations in smaller patches, where species populations are low, are declining in numbers, or have been extirpated.

Corridors of natural vegetation promote movement of organisms from one habitat patch by reducing their vulnerability during dispersal. Corridors also facilitate gene flow between wildlife populations in each patch, reducing the change of inbreeding depression. Conversely, lack of corridors or low corridor connectivity puts dispersing wildlife at greater risk. For example, if vegetated corridors are lacking woodland, animals must cross open areas to disperse from one woodland habitat patch to another, increasing their vulnerability to predators. Some animals, such as certain amphibians and reptiles, may be unable to disperse from one habitat patch to another if connecting corridors are not available.

Corridor width is also an important consideration in landscape ecology. Narrow corridors increase the potential for intrusion of domestic animals such as cats and dogs along and into the corridor, making wildlife dispersal more difficult and more dangerous. Narrow corridors may also contribute to poaching of larger animals, which are more visible and therefore more vulnerable during their dispersal within the confines of narrow corridors. However, low connectivity also decreases the rate of spread of invasive species and pests, reduces the dispersal of pollutants, thereby enhancing survival of interior species that occupy patches.

Some types of natural corridors also have distinct ecological character-istics that contribute to the overall species and landscape biodiversity in a given landscape. For example, belts of vegetation along rivers and streams form a distinctive ecological habitat called a riparian community that is in-habited by riparian species as well as edge and dispersal species.

Parks and Preserves as Landscape Components

Parks and preserves are natural units of landscape that have been deliber-ately set aside as protected parcels in an attempt to maintain ecologically functioning communities of plants and animals. Their size is usually a con-sequence of how valuable the land is that is being set aside, the value that humans place on wildlife and wildlife communities, the types of land that can be set aside, and the economic sacrifices that humans are willing to make to ensure that natural wildlife communities are protected.

Much of the debate over size of parks and preserves has focused on whether to establish a single large preserve or several small preserves. This debate is sometimes called the SLOSS controversy, the acronym standing for "single larger or several smaller" preserves.

Large preserves are favored by many ecologists because they provide greater amounts of natural habitat that can provide the resource base needed to support greater wildlife diversity and also larger populations of each plant and animal species. A larger species population reduces the risk of extinction or extirpation due to chance events such as skewed sex ratios or poor reproductive rates in a given year or set of years. A major draw-back of a single large preserve, however, is that all of the wildlife of an area is concentrated within a limited landscape that can be destroyed by a sin-gle catastrophe such as a severe rainstorm, floods, fires, ice storms, heavy snowfall, landslides, and other natural catastrophes. Furthermore, concen-tration of wildlife populations in a single large preserve exposes all of the members of a species population to risk from pollutants, chance appear-ance of competitors, or invasion of exotic species.

The alternate choice of creating several small preserves distributes members of a species population among several preserves rather than concentrating them all in a single large preserve. This greatly reduces the chance that a catastrophe will eliminate the wildlife population, as it is ex-tremely unlikely that a chance catastrophe will impact all of the smaller preserves simultaneously. However, the smaller resource base of a smaller preserve automatically limits numbers of individuals in each species pop-ulation, thereby greatly increasing risk of extirpation of a given species population in a smaller preserve due to chance events. For example, small populations face increased risks of genetic inbreeding, the chance accumulation of deleterious mutations, and genetic drift. Environmentally, smaller species populations are also more susceptible to disease, the introduction of predators, and the possibility of a skewed sex ratio that results in low reproduction in a given generation. To compensate for the increased risk of wildlife extinction in smaller preserves, landscape ecologists must provide corridors to connect preserves with one another, thereby promoting wildlife movement between preserves.

Managing and maintaining parks and preserves and their connecting corridors depends on several factors. Virtually all parks and preserves in urban and suburban landscapes require developing and implementing management strategies to ensure that disturbances from adjacent human-modified habitats is minimized. Rare habitats or habitats that contain rare or endangered wildlife must be managed differently from open-space parcels that contain common and widespread species. In some locales, parks and preserves may best be managed under the multiple-use concept, functioning simultaneously as wildlife preserves and as areas that provide recreational opportunities for hiking, biking, bird-watching, and wildlife appreciation.

Dwight G. Smith

Sources for Further Study

Forman, Richard T. T., and M. Godron. *Landscape Ecology.* New York: John Wiley & Sons, 1986.

Forman, Richard T. T. *Land Mosaics: The Ecology of Landscapes and Regions.* New York: Cambridge University Press, 1995.

Gardner, R. H., R. V. O'Neill, and M. G. Turner. *Landscape Ecology in Theory and Practice.* New York: Springer, 2001.

Garner, H. F. *The Origin of Landscapes: A Synthesis of Geomorphology.* New York: Oxford University Press, 1974.

Gergel, S. E., and Monica G. Turner, eds. *Learning Landscape Ecology: A Practical Guide to Concepts and Techniques.* New York: Springer, 2001.

Gutzwiller, K. J. *Applying Landscape Ecology in Biological Conservation.* New York: Springer, 2002.

Jongman, R. H. G., C. J. F. Ter Braak, and P. F. R. Van Tongeren. *Data Analysis in Communities and Landscape Ecology.* New York: Cambridge University Press, 1995.

LICHENS

Type of ecology: Community ecology

Lichens are an example of a very specialized "ecosystem" composed of two distinct species, a fungus and a photosynthetic alga (or bacterium) that have coevolved to live in a symbiotic relationship with each other in a community that grows on rocks, trees, and other substrates.

Lichens are classified as members of the kingdom *Fungi*, with most being placed under the phyla *Ascomycota* and *Basidiomycota*. It is estimated there are seventeen thousand species of lichen, representatives of which have been found nearly everywhere in the world.

Symbiosis

Symbiosis is an extreme form of an ecological relationship or mutualism between members of different species, in which each partner in the union derives benefits from the other. In symbiotic unions, the partners are so dependent on each other they can no longer independently survive.

In lichens, the fungal (mycobiont) symbiont provides protection, while the green-algal or cyanobacterial (photobiont) symbiont provides sugars, created by photosynthesis. It is often suggested that the fungus in lichen species might also pass water and nutrients to the photobiont, but this function is less well documented. This special relationship allows lichens to survive in many environments, such as hot deserts and frozen Arctic tundra, that are inhospitable to most other life-forms. As a result, the lichen whole is greater than the sum of its parts. While in nature lichen partners always exist together, under laboratory conditions it is possible to take the lichen apart and grow the two partners separately.

Anatomy

Whereas in most plant species the anatomy of the organism is identified with structures associated with a single vegetative body, the "lichen body" is more aptly described as a colony of cells that share a variety of associations with one another that vary from one species of lichen to the next. In some species of lichen, fungal and algal cells merely coexist. *Coenogonium leprieurii*, for example, is a lichen that lives in low-light tropical and subtropical forests in which the filamentous green-algal partner (*Trebouxia*) is dominant.

In most lichen species, however, the relationship between the symbiotic partners is more intimate, with the lichen body appearing to be a single entity. In these species the algal symbiont has no cell walls and is penetrated by filaments, or haustoria, from the fungal symbiont. The haustoria pass sugars from the algal cell to the fungal cell and may have a role in the transportation of water and nutrients from the fungal cell back to the algal cell. This integration is so complete that many naturalists prior to the nineteenth century mistakenly classified lichens as mosses.

In most lichen species it is nevertheless possible, with a good magnifying device, to identify several distinct regions of the thallus or lichen body. The outermost region is the cortex, a compacted layer composed of short, thick hyphae (widely dilated filaments) of the fungal symbionts that protect the lichen from abiotic factors in its environment. These hyphae extend downward into a second region, the photobiont layer, where they surround the algal symbionts. Below this is a third region, the medulla, composed of a loosely woven network of hyphae.

Underneath this is a fourth region, the undercortex, that is similar in appearance and structure to the cortex. The bottom of the lichen body is com-

Lichens are one of nature's best examples of symbiosis, a mutualism in which two different organisms have coevolved to become mutually dependent. Lichens are composed of two species which are partners: a fungal mycobiont and a bacterial or algal photobiont. In most lichen species, the lichen body appears to be a single entity. Here, a yellow lichen grows on a dead tree trunk. (AP/Wide World Photos)

posed of rhizines, rootlike structures composed of bundles of hyphae that attach the lichen to its substrate (the rock, bark, or other support on which it resides). This arrangement of regions into layers serves to prevent water loss. Many species can survive complete desiccation, coming back to life when water becomes available again. The cortex also contains pseudo-cyphellae, which are pores that allow for the exchange of gases necessary for photosynthesis.

Life Cycle
Lichens typically live for ten years or more, and in some species the lichen body can survive for more than a hundred years. Reproduction in most fungal species proceeds by the development of a cup- or saucer-shaped fruiting body called an apothecium, which releases fungal spores to its surrounding. Procreation in lichens is more problematic, in that the fungal offspring must also receive the right algal symbiont if they are to survive. The most common form of dispersion in lichen is by the accidental breaking off of small pieces of the thallus called isidia, which are then spread by wind to new substrates. In some species, small outgrowths of the thallus known as soralia arise, composed of both fungi and algae and surrounded by hyphae, to form soredia, which after dispersion give rise to a new thallus.

Ecological and Economic Importance
Lichens not only demonstrate some basic ecological concepts, such as mutualism and symbiosis, but also are excellent bioindicators of air pollution, as many species are particularly sensitive to certain contaminants in their surroundings, such as sulfur dioxide. They also play an important role in tundra biomes, functioning as a major source of food for reindeer and cattle in Lapland. One species of lichen (*Umbilicaria esculenta*) is considered a delicacy in Japan. Historically, lichens have been used as pigments for the dying of wool. The medical properties of some species of lichens for lung disease and rabies have led to a renewed interest in them.

David W. Rudge

See also: Animal-plant interactions; Coevolution; Communities: ecosystem interactions; Communities: structure; Convergence and divergence; Food chains and webs; Mycorrhizae; Symbiosis; Trophic levels and ecological niches.

Sources for Further Study
Brodo, Irwin M., Sylvia Duran Sharnoff, and Stephen Sharnoff. *Lichens of North America.* New Haven, Conn.: Yale University Press, 2001.

Lichens

Dobson, Frank. *Lichens: An Illustrated Guide.* Richmond, Surrey, England: Richmond, 1981.

Hale, Mason E., Jr. *The Biology of Lichens.* 3d ed. London: Edward Arnold, 1983.

Hawksworth, D. L., and D. J. Hill. *The Lichen-Forming Fungi.* New York: Blackie & Son, 1984.

Purvis, William. *Lichens.* Washington, D.C.: Smithsonian Institution Press, 2000.